Springer Monographs in Mathematics

Paul-Hermann Zieschang

Theory of Association Schemes

Springer

Paul-Hermann Zieschang
Department of Mathematics
University of Texas at Brownsville
Brownsville, TX 78520
USA
E-mail: zieschang@utb.edu

Library of Congress Control Number: 2005930450

Mathematics Subject Classification (2000): 05E30, 20N99

ISSN 1439-7382
ISBN-10 3-540-26136-2 Springer Berlin Heidelberg New York
ISBN-13 978-3-540-26136-0 Springer Berlin Heidelberg New York

Springer is a part of Springer Science+Business Media
springeronline.com

Printed in The Netherlands

Typesetting: by the author and TechBooks using a Springer LaTeX macro package

Cover design: *design & production* GmbH, Heidelberg

Printed on acid-free paper SPIN: 11371977 41/TechBooks 5 4 3 2 1 0

Preface

The present text is an introduction to the theory of association schemes. We start with the definition of an association scheme (or a scheme as we shall say briefly), and in order to do so we fix a set and call it X.

We write 1_X to denote the set of all pairs (x, x) with $x \in X$. For each subset r of the cartesian product $X \times X$, we define r^* to be the set of all pairs (y, z) with $(z, y) \in r$. For x an element of X and r a subset of $X \times X$, we shall denote by xr the set of all elements y in X with $(x, y) \in r$.

Let us fix a partition S of $X \times X$ with $\emptyset \notin S$ and $1_X \in S$, and let us assume that $s^* \in S$ for each element s in S. The set S is called a *scheme on* X if, for any three elements p, q, and r in S, there exists a cardinal number a_{pqr} such that $|yp \cap zq^*| = a_{pqr}$ for any two elements y in X and z in yr.

The notion of a scheme generalizes naturally the notion of a group, and we shall base all our considerations on this observation. Let us, therefore, briefly look at the relationship between groups and schemes.

Let S be a scheme, and let P and Q be nonempty subsets of S. We define PQ to be the set of all elements s in S for which there exist elements p in P and q in Q satisfying $1 \leq a_{pqs}$. If P possesses an element p with $\{p\} = P$ and Q an element q satisfying $\{q\} = Q$, we write pq instead of PQ. The set PQ will be called the *complex product* of P and Q. The associated operation on the set of all nonempty subsets of S will be referred to as the *complex multiplication in* S.

It follows right from the definition of the complex multiplication that, for any two elements p and q of a scheme S, p^*q is not empty. An element s of S will be called *thin* if s^*s contains exactly one element (which then must be the identity element of S). A nonempty subset of S will be called *thin* if each of its elements is thin.

Given a thin scheme S, it is easy to see that the set S^γ of all sets $\{s\}$ with $s \in S$ is a group with respect to the restriction of the complex multiplication in S to S^γ. Conversely, let G be a group. For each element g in G, we define

g^τ to be the set of all pairs (e, f) of elements of G satisfying $eg = f$. Then the set G^τ of all sets g^τ with $g \in G$ is a thin scheme on G.[1]

It turns out that, for each thin scheme S, the scheme $S^{\gamma\tau}$ is isomorphic to S. Conversely, for each group G, the group $G^{\tau\gamma}$ is isomorphic to G.[2] This establishes a one-to-one correspondence between groups and thin schemes. We call this one-to-one correspondence the *group correspondence*. The group correspondence allows us to view the class of groups as a distinguished class of schemes, namely as the class of thin schemes.

The group correspondence suggests the development of a general structure theory of schemes based on concepts similar to those which have been so successful in group theory. In fact, we shall follow this idea right from the beginning of our text. In particular, we subdivide our theory according to the shape of group theory as it presents itself nowadays. This means that, after three introductory chapters, we focus separately on the decomposition theory of schemes (including their local theory), their representation theory, and their theory of generators.

In order to get a first impression about the contents of the twelve individual chapters, we shall now briefly preview each one separately.[3]

For any three elements p, q, and r of a scheme S, the cardinal number a_{pqr} will be called the *structure constant* of p, q, and r in S. Structure constants, complex products, and the relationship between these two notions form the subject of the first chapter. Most of these results, in particular those of Section 1.4, are fundamental for the further development of the theory.

The complex product gives rise to the notion of a so-called closed subset, and it is this concept on which we focus in the second chapter of this monograph.

A nonempty subset R of a scheme is called *closed* if $p^*q \subseteq R$ for any two elements p and q in R.

For each thin scheme S, the group correspondence establishes a one-to-one correspondence between the closed subsets of S and the subgroups of S^γ. This means, in particular, that the notion of a closed subset generalizes that of a subgroup. More importantly, closed subsets retain some of the interesting properties of subgroups and transfer them to scheme theory. For instance, the set of all left cosets of a closed subset of a scheme S in S (defined with the help of the complex multiplication in S) is a partition of S.[4] Furthermore,

[1] The first of these two observations will be proved as Theorem 5.5.1, the second one as Theorem 5.5.2.

[2] The first of these two observations will be proved as Theorem 5.5.3, the second one as Theorem 5.5.4.

[3] We shall say more about the various sections at the beginning of the individual chapters.

[4] This observation enables us to generalize the notion of a 'coset geometry' from group theory to scheme theory. In fact, the relationship between 'buildings' (in the sense of Jacques Tits) and those schemes to which we shall refer later as

Richard Dedekind's 'modularity law' holds for closed subsets. The treatment of so-called length functions, which one obtains from generating sets of closed subsets, is another subject which allows one to mimic group theoretic techniques in scheme theory. Finally, the notion of a closed subset leads us to the notion of an involution in scheme theory.

A non-identity element of a scheme is called an *involution* if it is contained in a closed subset of cardinality 2.

In the third chapter of this monograph, we look at closed subsets generated by specific subsets. Our investigation leads us to the definitions of commutator subsets and thin residue of a closed subset, as well as to other characteristic closed subsets of schemes.

Closed subsets generated by sets of involutions turn out to be an interesting subject, especially if one imposes appropriate extra conditions on the set of the generating involutions. As an example, we introduce constrained sets of involutions; as another example, we look at sets of involutions satisfying the exchange condition. A constrained set of involutions which satisfies the exchange condition will be called a Coxeter set.

At the end of the third chapter, we give some basic information about Coxeter sets. A more rigorous treatment of Coxeter sets will be given in the last two chapters of this monograph.

The aforementioned exchange condition is defined in such a way that, via the group correspondence, its thin version is equivalent to the well-known group theoretic exchange condition which distinguishes the Coxeter groups within the class of all groups generated by involutions. In addition to the definition of closed subsets and the notion of an involution, the exchange condition is a further example of our general philosophy that, via the group correspondence, each scheme theoretic definition or statement applied to thin schemes has a natural and well-known group theoretic analogue.

There is one significant difference between subgroups of groups and closed subsets of schemes. In order for a quotient structure of a group to be a group one has to 'factor out' a normal subgroup. In scheme theory (at least as it applies to finite sets) one may factor out *any* closed subset without leaving the class of schemes. In this manner one obtains a quotient scheme for each closed subset of a scheme of finite valency, irrespective of whether the closed subset is normal or not. As a consequence, there is much more space for inductive reasoning.

Quotient schemes are introduced in the fourth chapter of this monograph. Factorization over non-normal closed subsets provides us with a particularly smooth approach to a generalization of Ludwig Sylow's theorems on finite

---

Coxeter schemes is based on the idea of looking at coset geometries of schemes; cf. [43; Theorem E]. However, in this monograph, we shall ignore this geometric aspect completely.

groups to scheme theory; cf. Theorem 4.5.3, Theorem 4.5.5, and Theorem 4.5.7. Generalizing Sylow's theorems on finite groups to scheme theory requires, of course, an appropriate generalization of Sylow subgroups to closed Sylow subsets in scheme theory. It is clear that our definition of closed Sylow subsets is yet another example of our general philosophy that each scheme theoretic definition or result applied to thin schemes has a corresponding group theoretic definition or result.

In the fifth chapter, we introduce morphisms and, more specifically, homomorphisms. We establish the natural relationship between quotient schemes and homomorphisms, thereby generalizing Emmy Noether's group theoretic Homomorphism Theorem and her two Isomorphism Theorems to scheme theory. The Second Isomorphism Theorem enables us to show that, up to isomorphism, schemes of finite valency admit only one so-called composition series; cf. Theorem 5.4.2. This generalizes a well-known theorem on finite groups due to Camille Jordan and Otto Hölder. Similar to group theory, this theorem gives rise to a scheme theoretic notion of composition factors, as well as emphasizes a notion of simple schemes. We finish the fifth chapter with an investigation of schemes which contain thin composition factors.

Morphisms are related to faithful maps, which lead naturally to the notion of a faithfully embedded closed subset. Such subsets provide an appropriate language for an attempt to establish so-called recognition theorems. These theorems deal with the question of which schemes are quotient schemes of thin schemes. We shall come back to recognition theorems and their role in scheme theory later in this preface.

In the sixth chapter, we introduce faithful maps and faithfully embedded closed subsets. In particular, we define a closed subset to be schurian if it is faithfully embedded in itself. This chapter is also the place where we prove the first recognition theorems.

The seventh chapter is devoted to products. We define direct products of closed subsets of schemes, quasi-direct products of schemes, and, following an approach due to Sejeong Bang, Mitsugu Hirasaka, and Sung-Yell Song, semidirect products of schemes. One of the main results of this chapter is a theorem of Pamela Ferguson and Alexandre Turull on indecomposable schemes. This theorem is similar to the group theoretic theorem of Wolfgang Krull, Erhard Schmidt, and Robert Remak for commutative groups. Semidirect products will play a role in our investigation of spherical Coxeter sets of cardinality 2 (see the last chapter).

The eighth chapter of this monograph deals with thin schemes. In this chapter, we collect basic and well-known ring theoretic concepts and facts which are needed for the representation theory of schemes of finite valency.

The ninth chapter is an introduction to the basic ideas of representation theory of schemes of finite valency. It is based on the analysis of the previous chapter. Generalizing the notion of a group ring we define scheme rings. We give sufficient criteria for a scheme ring to be semisimple (thereby generalizing Heinrich Maschke's important result on the semisimplicity of the group ring of a finite group). At the end, we apply a few of our representation theoretic results to finite schemes with three elements.

In the tenth chapter, we study schemes generated by a set of two involutions. The main result of this chapter is a characterization of schemes which have finite valency and are generated by a Coxeter set of cardinality 2. Referring to results obtained in the previous chapter we obtain, as a consequence, a representation theoretic characterization of such schemes.

The flow of the tenth chapter shows in a particularly convincing way how smooth the exchange condition emerges from the general theory of generators in scheme theory.

In the final two chapters of this monograph, we focus on constrained sets of involutions which satisfy the exchange condition. In other words, we look at Coxeter sets. We mentioned earlier that the exchange condition generalizes the well-known group theoretic exchange condition which distinguishes the Coxeter groups among the groups generated by involutions. In a strong sense, this justifies our use of the term 'Coxeter scheme over L' when referring to any scheme generated by a constrained set L of involutions satisfying the scheme theoretic exchange condition.

The eleventh chapter is an outline of a general theory of Coxeter schemes. It turns out that most of the basic observations on Coxeter schemes are natural generalizations of the corresponding ones in the theory of Coxeter groups. In fact, the treatment of so-called parabolic subgroups can be taken over almost word-by-word to parabolic subsets. This close relationship between Coxeter schemes and Coxeter groups has a firm mathematical basis, and it leads us to a general question in scheme theory which we had postponed earlier and would like to briefly address here.

The way in which the group correspondence produces thin schemes from groups can be generalized. In fact, each subgroup H of a group G gives rise to a scheme on the set of all left cosets of H in G. For each element g in G, we define g^H to be the set of all pairs (eH, fH) where e and f are elements in G satisfying $eg = f$. It is easy to see (and well known) that the set $G//H$ of all sets g^H with $g \in G$ is a scheme on the set G/H of all left cosets of H in G. The case $\{1\} = H$ is one with which we are already familiar from the group correspondence. (In this case, we obtain thin schemes.) The general case leads to quotient schemes of thin schemes in the same way the case $\{1\} = H$ leads to thin schemes via the group correspondence.

It is well known (and easy to see) that schemes are not necessarily isomorphic to quotient schemes of thin schemes. However, it seems that a general scheme theoretic characterization of such quotient schemes is out of reach. It is for this reason that one might instead ask for specific sufficient conditions for a scheme to be isomorphic to a quotient scheme of a thin scheme.

Recall that in the sixth chapter, we defined a closed subset to be schurian if it is faithfully embedded in itself. From this definition we easily obtain that a scheme is schurian if and only if it is isomorphic to a quotient scheme of a thin scheme.

Let us now return, as earlier promised, to a discussion on recognition theorems.

Suppose we are given a condition (or a collection of conditions) $\mathcal{S}$ under which a scheme S turns out to be schurian. It is then natural to ask for a group theoretic condition $\mathcal{G}$ which is satisfied by any two groups H and G such that H is a subgroup of G and $G//H$ satisfies $\mathcal{S}$. If, conversely, $G//H$ satisfies $\mathcal{S}$ for any two groups H and G, H a subgroup of G, satisfying $\mathcal{G}$, then we call the theorem which says that $\mathcal{S}$ implies $\mathcal{G}$ a *recognition theorem.*

As beautifully and completely developed as group theory is nowadays, one of the major goals of scheme theory must be to establish recognition theorems in which large classes of schemes (different from the class of all thin schemes) are recognized as quotients of specific classes of groups. In this way, schurian schemes may well serve as classifying categories for specific classes of groups, so that characterization theorems or classification theorems in group theory would appear as part of a general scheme theory.

In the first part of the final chapter, we show that the exchange condition is almost sufficient for a scheme to be schurian. We prove that Coxeter schemes S over a set L of involutions are always schurian if S is finite, if L has at least three elements, and if L does not contain thin elements. More precisely, we show in Theorem 12.3.4 that, for each Coxeter scheme S satisfying the above three conditions, there exists a group G and a so-called Borel subgroup B such that, via the group correspondence, S is isomorphic to $G//B$. Since we shall conversely show that, for each group G possessing a Borel subgroup B, the quotient scheme $G//B$ is a Coxeter scheme satisfying the above three conditions, Theorem 12.3.4 may be regarded as one of our recognition theorems.[5]

[5] Developing a theory of coset geometries (as mentioned in the previous footnote), one obtains from our recognition theorem of Coxeter schemes Jacques Tits' main result on buildings of 'spherical type'. This theorem asserts that each such building is associated with a group if it is 'thick' and of 'rank' at least 3.

In fact, our interest in Coxeter schemes is partially motivated by the fact that such schemes provide an alternate language for Tits' theory of buildings, a language in which buildings can be treated as algebraic objects simultaneously with other algebraic objects such as groups. The emphasis on Coxeter schemes themselves, rather than on the elements of the underlying set ('chambers' in the language of geometers), shows that many of the traditional concepts in the theory of buildings are actually not needed to investigate buildings. In particular, the 'free monoid', traditionally associated with each building as a substantial tool in each introduction to buildings (cf., e.g., [38] or [40]), is not part of the theory of Coxeter schemes the way it is developed in the last two chapters of this monograph.

Theorem 6.4.5 and Theorem 6.5.3 are two further recognition theorems which deal with involutions. Theorem 6.4.5 relates to George Glauberman's Z^*-Theorem a specific class of schemes of finite valency generated by elements of valency 2.

After having established the recognition theorem for Coxeter schemes, we present a well-known result of Walter Feit and Graham Higman on finite generalized polygons. (We provide a scheme theoretic version of a proof given by Robert Kilmoyer and Louis Solomon.) At the end of the final chapter, we prove the theorems of Stanley Payne and Udo Ott on polarities of finite generalized quadrangles and finite generalized hexagons. They appear here as results on semidirect products.

At this point, we conclude our treatment of schemes, although we acknowledge that there are quite a few more aspects of scheme theory which should have their place in a comprehensive textbook but which are not included in our monograph. Among the more theoretical sources of information about schemes (of finite valency), I would like to mention, in particular, the following two research areas.

First, there are numerous results on characters of schemes of finite valency which have not found their way into this text, although they provide an impressive insight into the rich and prosperous theory of schemes. In particular, modular representation theory of schemes of finite valency, as it has been developed mainly by Akihide Hanaki, has been excluded completely.[6]

The other large source of information about schemes (of finite valency) comes from the numerous interesting results on table algebras which have not been considered in this monograph. (Table algebras generalize the notion of schemes of finite valency.) Most of these results are due to Zvi Arad, Harvey Blau, and Mikhail Muzychuk.

It is mainly for these two reasons that this monograph should not be considered to be a comprehensive text on scheme theory. It is much more an attempt to present those concepts and results in scheme theory which seem to have the potential to convince the reader that scheme theory is a young theory the florescence of which is still to come.

For the remainder of this text, we fix a set and denote it by X. The letter S will always stand for a scheme on X. We shall always write 1 instead of 1_X.

The condition which gives us the structure constants for a scheme, will be called the *regularity condition.*

Except for a few isolated cases, the letters v, w, x, y, and z are reserved to denote elements in X. The letters p, q, r, s, t, and u will mostly stand for elements of S. (Clearly, the letter p will also often be used to denote prime

---

[6] This excludes one of the most beautiful results on schemes of finite valency, the theorem of Akihide Hanaki and Katsuhiro Uno which says that schemes are commutative if their valency is a prime number; cf. [19; Theorem 3.3].

numbers.) Involutions of S will be denoted by h, k, or l, a set of involutions mostly by H, K, or L. The letters O, P, Q, and R will usually denote subsets of S, the letters T, U, V, and W closed subsets of S.

I would like to close my introductory remarks by thanking two colleagues to whom I owe research stays which allowed me to concentrate on scheme theory over longer periods of times without any academic duties.

Ernest Shult gave me the opportunity to stay at Kansas State University, Manhattan (Kansas), during the academic year 1991/92 as a fellow of the *Max-Kade-Foundation*, New York (U. S. A.).

Eiichi Bannai invited me to visit Kyushu University, Fukuoka (Japan), during the academic year 1996/97 as a fellow of the *Long-Term Invitation Fellowship Program for Research in Japan* rewarded by the Japan Society for the Promotion of Science (JSPS), Tokyo (Japan).

Several colleagues have influenced this text directly or indirectly through stimulating correspondence and discussion. I would like to mention Harvey Blau from Northern Illinois University, DeKalb (Illinois), Akihide Hanaki from Shinshu University, Matsumoto (Japan), Mitsugu Hirasaka from Pusan National University, Pusan (Republic of Korea), Mikhail Muzychuk from Netanya Academic College, Netanya (Israel), Sung-Yell Song from Iowa State University, Ames (Iowa), Paul Terwilliger from University of Wisconsin, Madison (Wisconsin), and Andrew Woldar from Villanova University, Villanova (Pennsylvania). It is, in particular, Mitsugu Hirasaka and Mikhail Muzychuk to whom I owe many general insights, as well as a number of small observations of which I was not immediately aware.

Waterneverstorf, May 2005 *Paul-Hermann Zieschang*

Contents

1

Basic Facts

This introductory chapter deals with the arithmetic of structure constants and its consequences for complex products.

1.1 Structure Constants

In this section, we establish a collection of fundamental equations about structure constants. We shall also look at structural consequences derived from these equations.

Lemma 1.1.1 *For any two elements p and q of S, the following hold.*

(i) *We have $a_{1pq} = \delta_{pq}$ and $a_{p1q} = \delta_{pq}$.*

(ii) *For each element s in S, $a_{pqs} = a_{q^*p^*s^*}$.*

(iii) *For any two elements t and u in S,*

$$\sum_{s\in S} a_{pqs}a_{stu} = \sum_{s\in S} a_{psu}a_{qts}.$$

PROOF. (i) This follows immediately from the definition of the structure constants of S.

(ii) Let y be an element in X, and let z be an element in ys. Then, by definition, $|yp \cap zq^*| = a_{pqs}$ and $y \in zs^*$. From $y \in zs^*$ we obtain $|zq^* \cap yp^{**}| = a_{q^*p^*s^*}$. Since $p^{**} = p$, the two equations yield $a_{pqs} = a_{q^*p^*s^*}$.

(iii) Let y be an element in X, and let z be an element in yu. We count in two different ways the pairs (v, w) in $(yp \times zt^*)$ such that $u \in vq$. Then the desired equation follows from the definition of the structure constants.

Let s be an element in S. We write n_s instead of a_{ss^*1} and call n_s the *valency* of s. Note that, as s is assumed to be nonempty, $1 \leq n_s$.

For each nonempty subset R of S, we define n_R to be the sum of the cardinal numbers n_r with $r \in R$. We call n_R the *valency* of R.

Note that $|X| = n_S$.

Let P and Q be nonempty subsets of S such that $P \subseteq Q$. Then $n_P \leq n_Q$. If Q has finite valency, we have $n_P = n_Q$ if and only if $P = Q$.

Lemma 1.1.2 *For each element s in S, the following hold.*

(i) *For each element x in X, $|xs| = n_s$.*

(ii) *We have $|s| = n_S n_s$.*

(iii) *If S has finite valency, $n_{s^*} = n_s$.*

PROOF. (i) Let x be an element in X. Then, as $s^{**} = s$, $|xs| = a_{ss^*1}$, so that the claim follows from the definition of n_s.

(ii) This follows from (i) together with the observation that $|X| = n_S$.

(iii) From (ii) we know that $|s^*| = n_{s^*} n_S$ and that $|s| = n_s n_S$. Thus, as $|s^*| = |s|$, $n_{s^*} n_S = n_s n_S$. Thus, assuming n_S to be finite we obtain $n_{s^*} = n_s$.

Lemma 1.1.3 *For any two elements p and q of S, the following hold.*

(i) *If $1 \leq a_{pq^*1}$, then $p = q$.*

(ii) *For each element s in S, $a_{psq} n_q = a_{qs^*p} n_p$.*

(iii) *We have*

$$\sum_{s \in S} a_{psq} = n_p = \sum_{s \in S} a_{sp^*q}.$$

(iv) *We have*

$$\sum_{s \in S} a_{pqs} n_s = n_p n_q.$$

PROOF. (i) This follows from the fact that, by definition, the intersection of any two different elements of S is empty.

(ii) Applying Lemma 1.1.1(iii) to s, q^*, and 1 in the role of q, t, and u we obtain $a_{psq} n_q = n_p a_{sq^*p^*}$; cf. (i). From Lemma 1.1.1(ii) we know that $a_{sq^*p^*} = a_{qs^*p}$. Thus, $a_{psq} n_q = a_{qs^*p} n_p$.

(iii) Let y be an element in X, and let z be an element in yq. Then, as yp is the disjoint union of the sets $yp \cap zs^*$ with $s \in S$,

$$\sum_{s \in S} a_{psq} = \sum_{s \in S} |yp \cap zs^*| = |yp| = n_p;$$

cf. Lemma 1.1.2(i).

The second equation follows from the first one together with Lemma 1.1.1(ii).

(iv) From (ii) we obtain

$$\sum_{s\in S} a_{pqs}n_s = \sum_{s\in S} a_{sq^*p}n_p.$$

Thus, the claim follows from the second equation of (iii).

Lemma 1.1.4 *For any two elements p and q of S, the following hold.*

(i) *For each element s in S with $n_{s^*} = n_s$, we have $a_{spq}n_q = a_{s^*qp}n_p$.*

(ii) *Let t and u be elements in S, and let us assume that $n_{q^*} = n_q$, that $n_{t^*} = n_t$, and that $n_{u^*} = n_u$. Then we have*

$$\sum_{s\in S} a_{pqs}a_{tus}n_s = \sum_{s\in S} a_{p^*ts}a_{qu^*s}n_s.$$

(iii) *If q has finite valency, $a_{qp^*q} = a_{qpq}$.*

PROOF. (i) From Lemma 1.1.3(ii) we know that $a_{spq}n_q = a_{qp^*s}n_s$ and that $a_{s^*qp}n_p = a_{pq^*s^*}n_{s^*}$. From Lemma 1.1.1(ii) we know that $a_{qp^*s} = a_{pq^*s^*}$. Thus, the desired equation follows from our hypothesis that $n_{s^*} = n_s$.

(ii) From $n_{t^*} = n_t$, (i), and Lemma 1.1.1(ii) we obtain

$$\sum_{s\in S} a_{pqs}a_{tus}n_s = \sum_{s\in S} a_{pqs}a_{t^*su}n_u = \sum_{s\in S} a_{q^*p^*s^*}a_{s^*tu^*}n_u.$$

From $n_{q^*} = n_q$ and (i) we obtain

$$\sum_{s\in S} a_{p^*ts}a_{qu^*s}n_s = \sum_{s\in S} a_{p^*ts}a_{q^*su^*}n_{u^*} = \sum_{s\in S} a_{q^*su^*}a_{p^*ts}n_{u^*}.$$

Thus, the claim follows from Lemma 1.1.1(iii).

(iii) From Lemma 1.1.3(ii) we know that $a_{qp^*q}n_q = a_{qpq}n_q$, so that our claim follows from the hypothesis that q has finite valency.

There is a natural way to generalize the notion of a structure constant, which in turn allows generalization of some of our previous results. We shall not make use of such generalizations very often. However, there are a few places where it is convenient to work with these numbers.

Let s be an element in S, let n be an integer with $3 \leq n$, and let $r_1, \ldots, r_n$ be elements in S. We inductively define integers $a_{r_1\ldots r_n s}$ by

$$a_{r_1\ldots r_n s} := \sum_{q\in S} a_{r_1\ldots r_{n-1}q}a_{qr_n s}.$$

Our next lemma generalizes Lemma 1.1.1(iii), (ii) and Lemma 1.1.3(iv).

Lemma 1.1.5 *Let n be an integer with $2 \leq n$, and let $r_1, \ldots, r_n$ be elements in S.*

(i) *Let s be an element in S. Then, if $3 \leq n$,*

$$a_{r_1 \ldots r_n s} = \sum_{q \in S} a_{r_1 q s} a_{r_2 \ldots r_n q}.$$

(ii) *For each element s in S, we have $a_{r_1 \ldots r_n s} = a_{r_n^* \ldots r_1^* s^*}$.*

(iii) *We have*

$$\sum_{s \in S} a_{r_1 \ldots r_n s} n_s = n_{r_1} \cdots n_{r_n}.$$

PROOF. (i) If $n = 3$, the claim is just a restatement of Lemma 1.1.1(iii) (with r_1, r_2, r_3, and s in the role of p, q, t, and u). Therefore, we assume that $4 \leq n$. Assuming that the claim holds for $n - 1$, we obtain

$$\begin{aligned} a_{r_1 \ldots r_n s} &= \sum_{q \in S} a_{r_1 \ldots r_{n-1} q} a_{q r_n s} = \sum_{q \in S} \Big(\sum_{p \in S} a_{r_1 p q} a_{r_2 \ldots r_{n-1} p} \Big) a_{q r_n s} \\ &= \sum_{p \in S} a_{r_2 \ldots r_{n-1} p} \sum_{q \in S} a_{r_1 p q} a_{q r_n s} = \sum_{p \in S} a_{r_2 \ldots r_{n-1} p} \sum_{q \in S} a_{r_1 q s} a_{p r_n q} \\ &= \sum_{q \in S} a_{r_1 q s} \Big(\sum_{p \in S} a_{r_2 \ldots r_{n-1} p} a_{p r_n q} \Big) = \sum_{q \in S} a_{r_1 q s} a_{r_2 \ldots r_n q}. \end{aligned}$$

(The fourth equation follows from Lemma 1.1.1(iii).)

(ii) If $n = 2$, the claim is just a restatement of Lemma 1.1.1(ii) (with r_1 and r_2 in the role of p and q). Therefore, we assume that $3 \leq n$.

Assuming that the claim holds for $n - 1$, we obtain

$$a_{r_1 \ldots r_n s} = \sum_{q \in S} a_{r_1 \ldots r_{n-1} q} a_{q r_n s} = \sum_{q \in S} a_{r_n^* q^* s^*} a_{r_{n-1}^* \ldots r_1^* q^*} = a_{r_n^* \ldots r_1^* s^*}.$$

(The second equation follows from Lemma 1.1.1(ii), the last one from (i).)

(iii) If $n = 2$, the claim is just a restatement of Lemma 1.1.3(iv) (with r_1 and r_2 in the role of p and q). Therefore, we assume that $3 \leq n$.

We have

$$\begin{aligned} \sum_{s \in S} a_{r_1 \ldots r_n s} n_s &= \sum_{s \in S} \Big(\sum_{q \in S} a_{r_1 \ldots r_{n-1} q} a_{q r_n s} \Big) n_s = \sum_{q \in S} a_{r_1 \ldots r_{n-1} q} \sum_{s \in S} a_{q r_n s} n_s \\ &= \sum_{q \in S} a_{r_1 \ldots r_{n-1} q} (n_q n_{r_n}) = \Big(\sum_{q \in S} a_{r_1 \ldots r_{n-1} q} n_q \Big) n_{r_n} = n_{r_1} \cdots n_{r_n}. \end{aligned}$$

(The third equation follows from Lemma 1.1.3(iv).)

Lemma 1.1.6 *Assume that S has finite valency. Then we have*

$$\sum_{p\in S}\sum_{q\in S}\frac{a_{q^*pqs}}{n_{q^*}} = n_S.$$

for each element s in S.

PROOF. For any two elements q and s in S, we have

$$\sum_{p\in S}\frac{a_{q^*pqs}}{n_{q^*}} = \sum_{p\in S}\sum_{r\in S}\frac{a_{q^*pr}a_{rqs}}{n_{q^*}} = \sum_{r\in S}(\sum_{p\in S}a_{q^*pr})\frac{a_{rqs}}{n_{q^*}} = \sum_{r\in S}a_{rqs} = n_{q^*}.$$

(The third equation follows from the first equation of Lemma 1.1.3(iii), the last equation follows from the second equation of Lemma 1.1.3(iii).)

1.2 Symmetric Elements

An element s in S is called *symmetric* if $s^* = s$. Let us see how information about the structure constants of S relates to this notion of symmetry.

Lemma 1.2.1 *If S has odd valency, all symmetric elements in $S\setminus\{1\}$ have even valency.*

PROOF. Let s be a symmetric element in $S\setminus\{1\}$. Then $y \in zs$ for any two elements y and z in X with $z \in ys$. Thus, as $1 \neq s$, $|s|$ is even.

On the other hand, we have $|s| = n_s n_S$; cf. Lemma 1.1.2(ii). Thus, as n_S is assumed to be odd, n_s must be even.

For each subset R of S, we define R^* to be the set of all elements s in S with $s^* \in R$.

Lemma 1.2.2 *Assume that each element of S has finite valency and that no element in $S\setminus\{1\}$ is symmetric. Then each element in S has odd valency.*

PROOF. By hypothesis, there exists a subset R of S such that $\{R^*, R\}$ is a partition of $S\setminus\{1\}$. Thus, for each element s in S,

$$n_s = \sum_{r\in S}a_{srs} = a_{s1s} + 2\sum_{r\in R}a_{srs} = 1 + 2\sum_{r\in R}a_{srs}.$$

(The first equation follows from the first equation of Lemma 1.1.3(iii), the second one from Lemma 1.1.4(iii), and the last one from the second equation of Lemma 1.1.1(i).)

The following lemma is a special case of [1; Theorem 2.2(i)].

Lemma 1.2.3 *Assume that S has finite valency, and let k denote the highest 2-power dividing each n_s with $s \in S \setminus \{1\}$. Assume that $1 \neq k$, and let s be an element in $S \setminus \{1\}$ such that $n_s k^{-1}$ is odd. Then s is symmetric.*

PROOF. Let p and q be elements in $S \setminus \{1\}$. Then, by Lemma 1.1.4(ii),

$$\sum_{s \in S} a_{pp^*s} a_{qq^*s} n_s = \sum_{s \in S} a_{p^*qs} a_{p^*qs} n_s.$$

Thus,

$$n_p n_q + \sum_{s \in S \setminus \{1\}} a_{pp^*s} a_{qq^*s} n_s = \delta_{pq} n_{p^*}^2 + \sum_{s \in S \setminus \{1\}} (a_{p^*qs})^2 n_s.$$

We are assuming that $1 \neq k$. Thus, $n_p n_q k^{-1}$ as well as $\delta_{pq} n_{p^*}^2 k^{-1}$ are even. Thus, modulo 2,

$$\sum_{s \in S \setminus \{1\}} a_{pp^*s} a_{qq^*s} n_s k^{-1} \equiv \sum_{s \in S \setminus \{1\}} (a_{p^*qs})^2 n_s k^{-1}.$$

Let us denote by R the set of all elements s in $S \setminus \{1\}$ such that $n_s k^{-1}$ is odd. Then, for each element s in $S \setminus R \setminus \{1\}$, $n_s k^{-1}$ is even. Thus, modulo 2,

$$\begin{aligned} \sum_{r \in R} a_{pp^*r} a_{qq^*r} n_r k^{-1} &\equiv \sum_{s \in S \setminus \{1\}} a_{pp^*s} a_{qq^*s} n_s k^{-1} \equiv \sum_{s \in S \setminus \{1\}} (a_{p^*qs})^2 n_s k^{-1} \\ &\equiv \sum_{s \in S \setminus \{1\}} a_{p^*qs} n_s k^{-1} = \sum_{s \in S} a_{p^*qs} n_s k^{-1} - \delta_{pq} n_{p^*} k^{-1}. \end{aligned}$$

According to Lemma 1.1.3(iv), the last difference is equal to

$$n_{p^*} n_q k^{-1} - \delta_{pq} n_{p^*} k^{-1}.$$

Let us now assume that $p \in R$. From $p \in R$ we obtain that $n_{p^*} k^{-1}$ is odd. From $1 \neq q$ we obtain that n_q is even. Thus, modulo 2,

$$n_{p^*} n_q k^{-1} - \delta_{pq} n_{p^*} k^{-1} = n_{p^*} k^{-1} (n_q - \delta_{pq}) \equiv n_q - \delta_{pq} \equiv \delta_{pq}.$$

Note, finally, that $n_{r^*} k^{-1}$ is odd for each element r in R. Thus, modulo 2,

$$\sum_{r \in R} a_{pp^*r} a_{qq^*r} \equiv \sum_{r \in R} a_{pp^*r} a_{qq^*r} n_r k^{-1} \equiv \delta_{pq}.$$

Let us now look at the matrix

$$A := (a_{pp^*q})_{pq}$$

where p and q are elements in R. What we just saw means that A is regular.

For any two elements p and q in R, we have $a_{pp^*q^*} = a_{pp^*q}$; cf. Lemma 1.1.1(ii). Moreover, from Lemma 1.1.2(iii), we obtain $R^* = R$. Thus, as A is regular, R contains no non-symmetric element. (Otherwise, A would have two equal columns.)

A subset R of S is called *symmetric* if all of its elements are symmetric.

Corollary 1.2.4 *Assume that S has finite valency. Assume that all elements in $S \setminus \{1\}$ have the same valency and that this valency is even. Then S is symmetric.*

PROOF. This is an immediate consequence of Lemma 1.2.3.

A closed subset T of S is called *commutative* if $a_{pqr} = a_{qpr}$ for any three elements p, q, and r in T.

Lemma 1.2.5 *If S is symmetric, S is commutative.*

PROOF. For any three elements p, q, and r in S, we have $a_{pqr} = a_{q^*p^*r^*} = a_{qpr}$; cf. Lemma 1.1.1(ii).

1.3 The Complex Product

Recall that, for any two nonempty subsets P and Q of S, the complex product PQ of P and Q is defined to be the set of all elements s in S such that there exist elements p in P and q in Q satisfying $1 \leq a_{pqs}$.

Whenever s is an element in S and T a nonempty subset of S, we write sT instead of $\{s\}T$ and Ts instead of $T\{s\}$.

There are a few basic facts about the complex product. Occasionally, we shall quote these results without further reference.

Firstly, for each nonempty subset R of S, we have $R1 = R = 1R$.

Secondly, for any three nonempty subsets P, Q, and R of S with $P \subseteq Q$, we have $RP \subseteq RQ$ and $PR \subseteq QR$.

From the second observation we obtain $R(P \cup Q) = RP \cup RQ$ and $(P \cup Q)R = PR \cup QR$ for any three nonempty subsets P, Q, and R of S.

A similar observation is that, for any two nonempty subsets P and Q of S, PQ is equal to the union of the sets Pq with $q \in Q$ as well as to the union of the sets pQ with $p \in P$.

From the second observation we also obtain $R(P \cap Q) \subseteq RP \cap RQ$ and $(P \cap Q)R \subseteq PR \cap QR$ for any three nonempty subsets P, Q, and R of S. However, in general we do not have $RP \cap RQ \subseteq R(P \cap Q)$ or $PR \cap QR \subseteq (P \cap Q)R$. In Lemma 2.2.1, we shall give sufficient conditions for these equations to hold.

The following lemma is basic for the further development of the theory.

Lemma 1.3.1 *For any three nonempty subsets P, Q, and R of S, we have $(PQ)R = P(QR)$.*

PROOF. Let us first convince ourselves that $(PQ)R \subseteq P(QR)$. In order to do so, we pick an element s in $(PQ)R$, and we shall prove that $s \in P(QR)$.

From $s \in (PQ)R$ we obtain elements t in PQ and r in R such that $1 \leq a_{trs}$. Since $t \in PQ$, we find elements p in P and q in Q such that $1 \leq a_{pqt}$. It follows that $1 \leq a_{pqt}a_{trs}$.

Since $1 \leq a_{pqt}a_{trs}$, there exists an element u in S such that $1 \leq a_{pus}a_{qru}$; cf. Lemma 1.1.1(iii). From $1 \leq a_{pus}a_{qru}$ we obtain $1 \leq a_{pus}$ and $1 \leq a_{qru}$. From $1 \leq a_{qru}$, $q \in Q$, and $r \in R$ we obtain $u \in QR$. Thus, as $1 \leq a_{pus}$ and $p \in P$, $s \in P(QR)$.

The inclusion $P(QR) \subseteq (PQ)R$ is obtained similarly.

Lemma 1.3.2 *For any two nonempty subsets P and Q of S, the following hold.*

(i) *We have $1 \in P^*Q$ if and only if $P \cap Q$ is not empty.*
(ii) *If $1 \in PQ$, $1 \in QP$.*
(iii) *We have $(PQ)^* = Q^*P^*$.*

PROOF. (i) By definition, $1 \in P^*Q$ means that there exist elements p in P and q in Q such that $1 \leq a_{p^*q1}$. From Lemma 1.1.3(i), we know that $1 \leq a_{p^*q1}$ if and only if $p = q$.

(ii) Assume that $1 \in PQ$. Then, by (i), $P^* \cap Q$ is not empty. It follows that $Q^* \cap P$ is not empty. Thus, by (i), $1 \in QP$.

(iii) Let s be an element in $(PQ)^*$. Then, by definition, $s^* \in PQ$. Thus, there exist elements p in P and q in Q such that $1 \leq a_{pqs^*}$. Thus, by Lemma 1.1.1(ii), $1 \leq a_{q^*p^*s}$. Thus, as $q^* \in Q^*$ and $p^* \in P^*$, $s \in Q^*P^*$.

Since s has been chosen arbitrarily in $(PQ)^*$, we have shown that $(PQ)^* \subseteq Q^*P^*$. The inclusion $Q^*P^* \subseteq (PQ)^*$ is obtained similarly.

Lemma 1.3.3 *For any three nonempty subsets P, Q, and R of S, we have the following.*

(i) *The set $P \cap QR$ is empty if and only if $Q \cap PR^*$ is empty.*
(ii) *The set $P \cap RQ$ is empty if and only if $Q \cap R^*P$ is empty.*
(iii) *The set $P^* \cap RQ$ is empty if and only if $Q^* \cap PR$ is empty.*

PROOF. (i) Assume that $P \cap QR$ is not empty, and let us pick an element p in P such that $p \in QR$. Since $p \in QR$, there exist elements q in Q and r in R

such that $1 \leq a_{qrp}$. Thus, by Lemma 1.1.3(ii), $1 \leq a_{pr^*q}$. Thus, as $p \in P$ and $r \in R$, $q \in PR^*$. It follows that $q \in Q \cap PR^*$.

(ii) Assume that $P \cap RQ$ is not empty. Then, by Lemma 1.3.2(iii), $P^* \cap Q^*R^*$ is not empty. Thus, by (i), $Q^* \cap P^*R$ is not empty. Thus, by Lemma 1.3.2(iii), $Q \cap R^*P$ is not empty.

(iii) By (i), $P^* \cap RQ$ is empty if and only if $R \cap P^*Q^*$ is empty. By (ii), $R \cap P^*Q^*$ is empty if and only if $Q^* \cap PR$ is empty.

We shall often apply Lemma 1.3.3 to three sets each of which contains only one element. In this case (for instance), the first part of the lemma says that, for any three elements p, q, and r in S, $p \in qr$ if and only if $q \in pr^*$.[1]

Let n be an integer with $3 \leq n$, and let $R_1, \ldots, R_n$ be nonempty subsets of S. We inductively define $R_1 \cdots R_n$ to be the complex product $(R_1 \cdots R_{n-1})R_n$.

From Lemma 1.3.1 we easily obtain $R_1 \cdots R_n = R_1(R_2 \cdots R_n)$. With the help of Lemma 1.3.2(iii) we similarly obtain $(R_1 \cdots R_n)^* = R_n^* \cdots R_1^*$.

Lemma 1.3.4 *Let O, P, Q, and R be nonempty subsets of S. Then the set $OP \cap QR$ is empty if and only if $O^*Q \cap PR^*$ is empty.*

PROOF. From Lemma 1.3.3(i) we know that $OP \cap QR$ is empty if and only if $Q \cap OPR^*$ is empty. However, by Lemma 1.3.3(ii), $Q \cap OPR^*$ is empty if and only if $PR^* \cap O^*Q$ is empty.

Lemma 1.3.5 *Let n be an integer with $2 \leq n$, let s be an element in S, and let $R_1, \ldots, R_n$ be nonempty subsets of S. Then the following statements are equivalent.*

(a) *We have $s \in R_1 \cdots R_n$.*

(b) *There exist elements $s_0, \ldots, s_n$ in S such that $s_0 = 1$, $s_n = s$, and, for each element i in $\{1, \ldots, n\}$, $s_i \in s_{i-1}R_i$.*

(c) *For each element i in $\{1, \ldots, n\}$, there exists an element r_i in R_i such that $1 \leq a_{r_1 \ldots r_n s}$.*

PROOF. There is nothing to show if $2 = n$. Thus, we assume that $3 \leq n$.

(a) $\Rightarrow$ (b) Let us assume that $s \in R_1 \cdots R_n$. Then, by definition, there exists an element q in $R_1 \cdots R_{n-1}$ such that $s \in qR_n$.

From $q \in R_1 \cdots R_{n-1}$ we obtain, by induction, elements $s_0, \ldots, s_{n-1}$ in S such that $s_0 = 1$, $s_{n-1} = q$, and, for each element i in $\{1, \ldots, n-1\}$, $s_i \in s_{i-1}R_i$.

Setting $s_n := s$ we obtain from $s_{n-1} = q$ and $s \in qR_n$ that $s_n \in s_{n-1}R_n$.

(b) $\Rightarrow$ (c) By induction, there exists, for each element i in $\{1, \ldots, n-1\}$, an element r_i in R_i such that $1 \leq a_{r_1 \ldots r_{n-1} s_{n-1}}$. From $s \in s_{n-1}R_n$ we also obtain

[1] Recall that we write pq instead of $\{p\}\{q\}$ whenever p and q are elements in S.

an element r_n in R_n such that $1 \leq a_{s_{n-1}r_n s}$. Thus,

$$1 \leq a_{r_1 \ldots r_{n-1}s_{n-1}} a_{s_{n-1}r_n s} \leq a_{r_1 \ldots r_n s}.$$

(c) $\Rightarrow$ (a) Let us assume that $1 \leq a_{r_1 \ldots r_n s}$. Then, by definition, there exists an element q in S such that $1 \leq a_{r_1 \ldots r_{n-1} q}$ and $1 \leq a_{q r_n s}$.

From $1 \leq a_{r_1 \ldots r_{n-1} q}$ we obtain, by induction, that $q \in R_1 \cdots R_{n-1}$. From $1 \leq a_{q r_n s}$ and $r_n \in R_n$ we obtain $s \in qR_n$. Thus, as $q \in R_1 \cdots R_{n-1}$, $s \in R_1 \cdots R_n$.

Let s be an element of S, and let R be a nonempty subset of S.

Instead of $R\{s\}$ we write Rs. Similarly, we write sR instead of $\{s\}R$. We define

$$R^s := \{r \in S \mid sr \subseteq Rs\}.$$

Note that $sR^s \subseteq Rs$. The following lemma gives more details about R^s.

Lemma 1.3.6 *Let s be an element in S, and let R be a nonempty subset of S. Then the following hold.*

(i) *We have $R^s \subseteq s^* Rs$.*

(ii) *If $1 \in R$, $1 \in R^s$.*

(iii) *Let P and Q be nonempty subsets of S such that $PQ \subseteq R$. Then we have $P^s Q^s \subseteq R^s$.*

PROOF. (i) From $sR^s \subseteq Rs$ we obtain

$$R^s \subseteq s^* s R^s \subseteq s^* Rs.$$

(ii) Let us assume that $1 \in R$. Then we have $s \in Rs$. Thus, by definition, $1 \in R^s$.

(iii) Let r be an element in $P^s Q^s$. Then there exist elements t in P^s and u in Q^s such that $r \in tu$. Since $t \in P^s$, $st \subseteq Ps$. Since $u \in Q^s$, $su \subseteq Qs$. Thus, as $r \in tu$ and $PQ \subseteq R$,

$$sr \subseteq stu \subseteq Psu \subseteq PQs \subseteq Rs,$$

so that $sr \in Rs$. Thus, by definition, $r \in R^s$.

Lemma 1.3.7 *Let p and q be elements in S, and let R be a nonempty subset of S such that $1 \in R$ and $RR \subseteq R$. Then, if $Rp = Rq$, $R^p = R^q$.*

PROOF. Let s be an element in R^p. Then, by definition, $ps \subseteq Rp$.

Let us assume that $Rp = Rq$. Then, as we are assuming that $1 \in R$, we obtain $q \in Rp$, and this implies $qs \subseteq Rps$.

From $ps \subseteq Rp$ and $qs \subseteq Rps$ we obtain $qs \subseteq RRp$. Thus, as we are assuming that $RR \subseteq R$, we have that $qs \subseteq Rp$. Since we are assuming that $Rp = Rq$, this implies $qs \subseteq Rq$. Thus, $s \in R^q$.

So far, we have seen that $R^p \subseteq R^q$. The proof for $R^q \subseteq R^p$ is similar.

Let us now, at the end of this section, look at the relationship between subsets of X and complex products. In order to do this we fix a nonempty subset of S and call it R.

For each element x in X, we define xR to be the union of the sets xr with $r \in R$.

Let x be an element in X, and let Q be a subset of S such that $Q \cap R$ is not empty. Since elements of S are pairwise disjoint, $x(Q \cap R) = xQ \cap xR$.

For each nonempty subset Y of X, we define YR to be the union of the sets yR with $y \in Y$.

Let Z be a nonempty subset of X. It is clear that, for each nonempty subset Y of Z, $YR \subseteq ZR$. Moreover, we have $ZQ \subseteq ZR$ for each nonempty subset Q of R.

Lemma 1.3.8 *Let Y be a nonempty subset of X, and let P and Q be nonempty subsets of S. Then $(YP)Q = Y(PQ)$.*

PROOF. Let x be an element in $(YP)Q$. Then, by definition, there exist elements z in YP and q in Q such that $x \in zq$. Since $z \in YP$, there exist elements y in Y and p in P such that $z \in yp$.

Let us denote by s the uniquely determined element in S which satisfies $x \in ys$. Then, as $z \in yp \cap xq^*$, $1 \leq a_{pqs}$. Thus, by definition, $s \in pq$, so that

$$x \in ys \subseteq y(pq) \subseteq y(PQ) \subseteq Y(PQ).$$

Since x has been chosen arbitrarily in $(YP)Q$, we have shown that $(YP)Q \subseteq Y(PQ)$.

Conversely, let x be an element in $Y(PQ)$. Then, by definition, there exist elements y in Y and s in PQ such that $x \in ys$. Since $s \in PQ$, there exist elements p in P and q in Q such that $1 \leq a_{pqs}$. Thus, as $x \in ys$, $yp \cap xq^*$ is not empty. Let z be an element in $yp \cap xq^*$. Then

$$x \in zq \subseteq zQ \subseteq (yp)Q \subseteq (YP)Q.$$

Since x has been chosen arbitrarily in $Y(PQ)$, we have shown that $Y(PQ) \subseteq (YP)Q$.

Lemma 1.3.9 *Let n be an integer with $2 \leq n$, let s be an element in S, and let $R_1, \ldots, R_n$ be nonempty subsets of S. Then the following statements are equivalent.*

(a) *We have* $s \in R_1 \cdots R_n$.

(b) *Let* y *be an element in* X, *and let* z *be an element in* ys. *Then there exist elements* x_0, $\ldots$, x_n *in* X *such that* $x_0 = y$, $x_n = z$, *and, for each element* i *in* $\{1, \ldots, n\}$, $x_i \in x_{i-1}R_i$.

(c) *There exist elements* x_0, $\ldots$, x_n *in* X *such that* $x_n \in x_0 s$ *and, for each element* i *in* $\{1, \ldots, n\}$, $x_i \in x_{i-1}R_i$.

PROOF. There is nothing to show if $1 = n$. Thus, we assume that $2 \leq n$.

(a) $\Rightarrow$ (b) Let us assume that $s \in R_1 \cdots R_n$. Then, by definition, there exists an element r in $R_1 \cdots R_{n-1}$ such that $s \in rR_n$.

From $z \in ys$ and $s \in rR_n$ we obtain $z \in yrR_n$. Thus, referring to Lemma 1.3.8 we obtain an element w in yr such that $z \in wR_n$.

From $r \in R_1 \cdots R_{n-1}$ and $z \in wR_n$ we obtain, by induction, elements x_0, $\ldots$, x_{n-1} in X such that $x_0 = y$, $x_{n-1} = w$, and, for each element i in $\{1, \ldots, n-1\}$, $x_i \in x_{i-1}R_i$.

Setting $x_n := z$ we obtain from $x_{n-1} = w$ and $z \in wR_n$ that $x_n \in x_{n-1}R_n$.

(b) $\Rightarrow$ (c) This follows from the fact that s is not empty.

(c) $\Rightarrow$ (a) Let us assume that there exist elements x_0, $\ldots$, x_n in X such that $x_n \in x_0 s$ and, for each element i in $\{1, \ldots, n\}$, $x_i \in x_{i-1}R_i$. Let us further denote by r the element in S which satisfies $x_{n-1} \in x_0 r$.

From $x_n \in x_{n-1}R_n$ and $x_{n-1} \in x_0 r$ we obtain $x_n \in x_0 r R_n$. By induction, we have that $r \in R_1 \cdots R_{n-1}$. Thus, we conclude that $x_n \in x_0 R_1 \cdots R_n$. Thus, as $x_n \in x_0 s$, $s \in R_1 \cdots R_n$.

1.4 Complex Products and Valencies

In this section, we establish a few results relating complex products to valencies. The letters P and Q will stand for nonempty subsets of S, each having finite valency.

We start with a generalization of Lemma 1.1.3(iv).

Lemma 1.4.1 *We have*

$$\sum_{s \in PQ} \Big(\sum_{p \in P} \sum_{q \in Q} a_{pqs}\Big) n_s = n_P n_Q.$$

PROOF. From Lemma 1.1.3(iv) we know that

$$\sum_{s \in PQ} a_{pqs} n_s = n_p n_q$$

for any two elements p in P and q in Q. Thus, we have

$$\sum_{p\in P}\sum_{q\in Q}(\sum_{s\in PQ} a_{pqs}n_s) = \sum_{p\in P}\sum_{q\in Q} n_p n_q,$$

and that proves the lemma.

Lemma 1.4.2 *We have $n_{PQ} \leq n_P n_Q$.*

PROOF. For each element s in PQ, we have

$$1 \leq \sum_{p\in P}\sum_{q\in Q} a_{pqs}.$$

Thus, the claim follows from Lemma 1.4.1.

Lemma 1.4.3 *For each element s in S, we have*

$$\sum_{p\in P}\sum_{q\in Q} a_{pqs} = n_P$$

if and only if $P^*s \subseteq Q$.

PROOF. Let s be an element in S. From the first equation of Lemma 1.1.3(iii) we obtain

$$\sum_{p\in P}\sum_{r\in S} a_{prs} = n_P.$$

Thus, the equation in question holds if and only if, for any two elements p in P and r in $S \setminus Q$, $a_{prs} = 0$.

For any two elements p in P and r in S, we have $a_{prs} = 0$ if and only if $s \notin pr$. Moreover, by Lemma 1.3.3(ii), $s \notin pr$ is equivalent to $r \notin p^*s$. Thus, the equation in question holds if and only if, for any two elements p in P and r in $S \setminus Q$, $r \notin p^*s$. This means that $P^*s \subseteq Q$.

The following two consequences of Lemma 1.4.3 are among the most frequently quoted results in this monograph.

Lemma 1.4.4 *The following statements hold.*

(i) *We have $n_Q \leq n_{PQ}$.*
(ii) *We have $n_Q = n_{PQ}$ if and only if $Q = P^*PQ$.*

PROOF. From Lemma 1.4.1, together with the first equation of Lemma 1.1.3(iii), we obtain

$$n_P n_Q = \sum_{s\in PQ} n_s(\sum_{p\in P}\sum_{q\in Q} a_{pqs}) \leq \sum_{s\in PQ} n_s n_P = n_{PQ}n_P.$$

However, we are assuming that n_P and n_Q are finite. Thus, by Lemma 1.4.2, n_{PQ} is finite, so that we have (i).

On the other hand, we know from Lemma 1.4.3 that, for each element s in PQ,

$$\sum_{p \in P} \sum_{q \in Q} a_{pqs} = n_P$$

if and only if $P^*s \subseteq Q$. Thus, $n_Q = n_{PQ}$ if and only if $P^*PQ \subseteq Q$.

The following lemma generalizes Lemma 1.4.4. One obtains Lemma 1.4.4 by setting $\{1\} = R$ in Lemma 1.4.5.

Lemma 1.4.5 *Let R be a nonempty subset of S, and assume that R has finite valency. Then the following hold.*

(i) *We have $n_Q \leq n_{PQR}$.*

(ii) *We have $n_Q = n_{PQR}$ if and only if $Q = P^*PQRR^*$.*

PROOF. (i) From Lemma 1.4.4(i) we know that $n_Q \leq n_{PQ}$ and that $n_{(PQ)^*} \leq n_{R^*(PQ)^*}$. From Lemma 1.3.2(iii) we know that $(PQR)^* = R^*(PQ)^*$. Finally, as n_P, n_Q, and n_R are assumed to be finite, we have $n_{(PQ)^*} = n_{PQ}$ and $n_{(PQR)^*} = n_{PQR}$; cf. Lemma 1.1.2(iii). Thus, $n_Q \leq n_{PQR}$.

(ii) From (i) we know that $n_Q \leq n_{QR} \leq n_{PQR}$. Thus, we have $n_Q = n_{PQR}$ if and only if $n_Q = n_{QR}$ and $n_{QR} = n_{PQR}$.

The first of these two conditions is equivalent to $n_{Q^*} = n_{R^*Q^*}$; cf. Lemma 1.3.2(iii). By Lemma 1.4.4(ii), $n_{Q^*} = n_{R^*Q^*}$ is equivalent to $Q^* = RR^*Q^*$. However, according to Lemma 1.3.2(iii), this is equivalent to $Q = QRR^*$.

From Lemma 1.4.4(ii) we also know that $n_{QR} = n_{PQR}$ is equivalent to $QR = P^*PQR$.

Note finally that the two equations $Q = QRR^*$ and $QR = P^*PQR$ hold if and only if $Q = P^*PQRR^*$.

We shall often apply Lemma 1.4.5 to the case where $\{1\} = P$.

Corollary 1.4.6 *If $n_Q = n_{P^*Q}$, then $n_{PP^*} \leq n_Q$.*

PROOF. Let us assume that $n_Q = n_{P^*Q}$. Then, by Lemma 1.4.4(ii), $Q = PP^*Q$. In particular, $n_Q = n_{PP^*Q}$. On the other hand, by Lemma 1.4.5(i), $n_{PP^*} \leq n_{PP^*Q}$. Thus, $n_{PP^*} \leq n_Q$.

1.5 Complex Products of Subsets of Cardinality 1

Let p and q be elements in S. Recall that we write pq instead of $\{p\}\{q\}$.

Lemma 1.5.1 *An element s in S is thin if and only if $1 = n_s$.*

PROOF. By definition, an element s of S is thin if and only if $\{1\} = s^*s$. The equation $\{1\} = s^*s$ says that $r \notin s^*s$ for each element r in $S \setminus \{1\}$.

Let r be an element in S. Then, by Lemma 1.3.3(ii), $r \notin s^*s$ is equivalent to $s \notin sr$. Moreover, $s \notin sr$ means that $a_{srs} = 0$.

Thus, s is thin if and only if $a_{s1s} = n_s$; cf. the first equation of Lemma 1.1.3(iii). Thus, the claim follows from the second equation of Lemma 1.1.1(i).

Lemma 1.5.2 *Let p and q be elements in S such that n_p and n_q are finite. Then $|p^*q|$ is less than or equal to the greatest common divisor of n_p and n_q.*

PROOF. Let us denote by m the greatest common divisor of n_p and n_q. Then there exists an integer n coprime to n_q such that $n_p = mn$.

Let s be an element in S. Then, by Lemma 1.1.3(ii), $a_{psq}n_q = a_{qs^*p}n_p$. Thus, as $n_p = mn$, n divides $a_{psq}n_q$. Thus, as n_q and n are coprime, n divides a_{psq}.

Let s be an element in p^*q. Then, by Lemma 1.3.3(ii), $q \in ps$, and this means that $1 \leq a_{psq}$. Thus, as n divides a_{psq}, $n \leq a_{psq}$. Thus, by the first equation of Lemma 1.1.3(iii),

$$|p^*q|n \leq \sum_{s \in p^*q} a_{psq} \leq n_p = mn.$$

This finishes the proof of the lemma.

The following lemma is related to Lemma 1.3.3(ii).

Lemma 1.5.3 *Let p, q and r be elements in S such that $\{p\} = rq$. Assume that n_q is finite and that $n_p \leq n_q$. Then $\{q\} = r^*p$.*

PROOF. We are assuming that $\{p\} = rq$. Thus, $n_p = n_{rq}$. Thus, assuming that $n_p \leq n_q$, we obtain from Lemma 1.4.4(i) that $n_q = n_{rq}$.

We are assuming that n_q is finite. According to Lemma 1.4.5(i), we also have that $n_r \leq n_{rq} = n_p \leq n_q$, so that n_r is finite, too. Thus, as $n_q = n_{rq}$, $\{q\} = r^*rq$; cf. Lemma 1.4.4(ii). Thus, our claim follows from $\{p\} = rq$.

Lemma 1.5.4 *Let s be an element of S, and assume that s has finite valency. Assume that there exist elements p and q in S with $n_p \leq n_q$ and $n_p \leq n_s \leq a_{spq}$. Then $n_s = n_{s^*s}$.*

PROOF. From the first equation of Lemma 1.1.3(iii) we know that $a_{spq} \leq n_s$. Thus, as we are assuming that $n_s \leq a_{spq}$ we obtain $a_{spq} = n_s$. Thus, by Lemma 1.4.3, $\{p\} = s^*q$.

From $p \in s^*q$ we obtain $q \in sp$; cf. Lemma 1.3.3(ii). From $q \in sp$ we obtain $n_q \leq n_s n_p$; cf. Lemma 1.1.3(iv). Thus, as $n_p \leq n_s$ and n_s is assumed to be finite, n_q must be finite. Thus, as we are assuming that $n_p \leq n_q$, we obtain from $\{p\} = s^*q$ that $\{q\} = sp$; cf. Lemma 1.5.3. It follows that $\{p\} = s^*sp$. Thus, by Lemma 1.4.5(i),

$$n_s \leq n_{s^*s} \leq n_{s^*sp} = n_p,$$

so that our claim follows from our hypothesis that $n_p \leq n_s$.

Lemma 1.5.5 *Let p and q be elements in S with $|p^*q| = 1$. Assume that p and q have finite valency and that $n_{p^*} = n_{p^*q}$ and that $n_q = n_{p^*q}$. Then we have $pp^* = qq^*$, $n_{p^*} = n_{pp^*}$, and $n_q = n_{qq^*}$.*

PROOF. We are assuming that $n_{p^*} = n_{p^*q}$. Thus, by Lemma 1.4.5(ii), $\{p^*\} = p^*qq^*$. It follows that $pp^* = pp^*qq^*$. Thus, as $qq^* \subseteq pp^*qq^*$, $qq^* \subseteq pp^*$.

Similarly, one obtains from $n_q = n_{p^*q}$ that $pp^* \subseteq qq^*$, so that $pp^* = qq^*$.

From $n_q = n_{p^*q}$ we also obtain $n_{pp^*} \leq n_q$; cf. Corollary 1.4.6. On the other hand, we know from Lemma 1.4.5(i) that $n_q \leq n_{qq^*}$. Thus, as $pp^* = qq^*$, $n_q = n_{qq^*}$.

From $n_{p^*} = n_{p^*q} = n_q$, $n_q = n_{qq^*}$, and $pp^* = qq^*$ one obtains $n_{p^*} = n_{pp^*}$.

Let s be an element in S. According to Lemma 1.5.1, we have $\{1\} = s^*s$ if and only if $1 = n_s$. Let us now see what happens to s^*s if $n_s = 2$.

Lemma 1.5.6 *Let s be an element in S with $n_s = 2$ and $n_{s^*} = 2$. Then the following hold.*

(i) *There exists a symmetric element r in $s^*s \setminus \{1\}$ such that $n_r \leq 2$ and $\{1, r\} = s^*s$.*

(ii) *We have $n_{s^*s} = 2$ or $n_{s^*s} = 3$.*

PROOF. (i) We are assuming that $n_s = 2$. Thus, by Lemma 1.5.1, s is not thin, and that means that $2 \leq |s^*s|$. On the other hand, as $n_s = 2$, we obtain from Lemma 1.5.2 that $|s^*s| \leq 2$, so that we have $|s^*s| = 2$.

From Lemma 1.3.2(i) we know that $1 \in s^*s$. Thus, as $|s^*s| = 2$, there exists an element r in $s^*s \setminus \{1\}$ such that $\{1, r\} = s^*s$.

From $\{1, r\} = s^*s$ we obtain $r^* = r$; cf. Lemma 1.3.2(iii).

Applying Lemma 1.1.3(iv) to s^* and s instead of p and q, we obtain

$$a_{s^*s1} + a_{s^*sr}n_r \leq n_{s^*}n_s = 4.$$

Thus, as $a_{s^*s1} = n_{s^*} = 2$, $n_r \leq 2$.

(ii) This follows immediately from (i).

Let s be an element in S with $n_s = 2$ and $n_{s^*} = 2$. Then, by Lemma 1.5.6(ii), $n_{s^*s} = 2$ or $n_{s^*s} = 3$. Lemma 1.5.4 gives a sufficient condition for $n_{s^*s} = 2$.

2

Closed Subsets

As mentioned in the preface of this monograph closed subsets play an important role in scheme theory. Via the group correspondence, closed subsets generalize the notion of a subgroup. They also generalize some of the important properties of subgroups from group theory to scheme theory.

The first section of this chapter is a collection of general observations on closed subsets most of which are straightforward generalizations of facts on subgroups. For instance, we show that the set of all 'double cosets' of two closed subset of S is a partition of S; cf. Lemma 2.1.3. We also introduce transversals (as they have been introduced in [45]), and, at the end if this section, we show that closed subsets give rise to subschemes.

The second section starts with Dedekind's modularity law for schemes and includes a selection of consequences of this law. The results will be useful in Section 2.4.

In the third section of this chapter, we investigate the relationship between closed subsets of S and structure constants of S.

In Section 2.4, we apply some of the previously obtained results on closed subsets in order to derive a sufficient condition for a closed subset to be maximal.

In Section 2.5, we define the normalizer and the strong normalizer of closed subsets. In the last of the six sections of this chapter, we introduce conjugates of closed subsets. Conjugates are related to normalizers and strong normalizers and will play a role in Section 4.4 when we investigate Sylow subsets.

2.1 Basic Facts

Recall that a nonempty subset R of S is called closed if $R^*R \subseteq R$.

Note that a nonempty subset R of S is closed if and only if $p^*q \subseteq R$ for any two elements p and q in R.

There are a few elementary facts about closed subsets which we occasionally shall quote without reference. Let us look at these facts first.

Firstly, it is obvious that $\{1\}$ and S are closed.

Secondly, let T be a closed subset of S. Then, by definition, T is not empty. Thus, by Lemma 1.3.2(i), $1 \in T^*T \subseteq T$. It follows that $1 \in T$.

Since $1 \in T$,

$$T^* = T^*1 \subseteq T^*T \subseteq T.$$

Thus, $T^* \subseteq T$. Thus, as $T^{**} = T$, $T \subseteq T^*$, whence $T^* = T$.

From $T^* = T$ and $T^*T \subseteq T$ we obtain $TT \subseteq T$.

Note finally that, as each closed subset of S contains 1 as an element, intersections of closed subsets of S are closed.

Lemma 2.1.1 *Let T and U be closed subsets of S. Then TU is closed if and only if $TU = UT$.*

PROOF. Let us first assume that TU is closed. Then, we have $(TU)^* = TU$. Now recall that, by Lemma 1.3.2(iii), $(TU)^* = U^*T^*$. Thus, $TU = U^*T^*$. However, as T and U are assumed to be closed, we have $T^* = T$ and $U^* = U$. Thus, $TU = UT$.

Conversely, if $TU = UT$,

$$(TU)^*TU = U^*T^*TU \subseteq U^*TU = U^*UT \subseteq UT = TU.$$

(The first equation follows from Lemma 1.3.2(iii).) Therefore, TU is closed.

Lemma 2.1.2 *Let T and U be closed subsets of S. Then we have $\{1\} = T \cap U$ if and only if, for each element s in TU, there exist uniquely determined elements t in T and u in U such that $s \in tu$.*

PROOF. Let us first assume that $\{1\} = T \cap U$, and let us fix an element s in TU. Then, by definition, there exist elements t in T and u in U such that $s \in tu$.

Let us now pick elements t' in T and u' in U such that $s \in t'u'$. We have to show that $t' = t$ and $u' = u$.

From $s \in tu \cap t'u'$ we obtain that $t^*t' \cap uu'^*$ is not empty; cf. Lemma 1.3.4. Since T is assumed to be closed, $t^*t' \subseteq T$. Similarly, as U is assumed to be closed, $uu'^* \subseteq U$. It follows that $t^*t' \cap uu'^* \subseteq T \cap U = \{1\}$. Therefore, $1 \in t^*t'$ and $1 \in uu'^*$. Thus, by Lemma 1.3.2(i), $t' = t$ and $u' = u$.

Conversely, let s be an element in $T \cap U$. Then $s \in TU$ and $1s = s = s1$. Thus, as $1 \in T \cap U$, $s = 1$.

Lemma 2.1.3 *For any two closed subsets T and U of S, $\{TsU \mid s \in S\}$ is a partition of S.*

Proof. Let p and q be elements in S such that $p \in TqU$. From $p \in TqU$ we obtain $TpU \subseteq TqU$. (Recall that $1 \in T$ and $1 \in U$.) Thus, it is enough to show that $q \in TpU$.

Since $p \in TqU$, there exists an element s in qU such that $p \in Ts$. From $s \in qU$ we obtain $q \in sU$; cf. Lemma 1.3.3(i). From $p \in Ts$ we obtain $s \in Tp$; cf. Lemma 1.3.3(ii). From $q \in sU$ and $s \in Tp$ we obtain $q \in TpU$.

For each nonempty subset R of S, we define

$$X/R := \{xR \mid x \in X\}.$$

For any two nonempty subsets P and Q of S with $P \subseteq Q$, we define

$$Q/P := \{qP \mid q \in Q\}.$$

Let x be an element in X, and let R be a nonempty subset of S. Recall that xR is our notation for the union of the sets xr with $r \in R$.

Lemma 2.1.4 *Let R be a subset of S with $1 \in R$. Then the following statements are equivalent.*

(a) *The set R is closed.*

(b) *The set X/R is a partition of X.*

(c) *The set S/R is a partition of X.*

Proof. (a) $\Rightarrow$ (b) We pick elements y and z in X such that $y \in zR$. We shall be done if we succeed in showing that $yR = zR$.

Since $y \in zR$, $yR \subseteq zR$. (This follows from Lemma 1.3.8 together with the hypothesis that R is closed.) From $y \in zR$ we also obtain $z \in yR^* = yR$. Thus, we conclude, as before, that $zR \subseteq yR$.

(b) $\Rightarrow$ (c) We fix elements p and q in S with $p \in qR$. We shall be done if we succeed in showing that $pR = qR$.

From $p \in qR$ we obtain an element r in R such that $1 \leq a_{qrp}$. Thus, there exist elements y in X and z in yp such that $yq \cap zr^*$ is not empty. Let us pick an element x in $yq \cap zr^*$. Since $z \in xr$, our hypothesis that X/R is a partition yields $xR = zR$.

In order to show that $pR \subseteq qR$ we pick an element s in pR. From $s \in pR$ we obtain $p \in sR^*$; cf. Lemma 1.3.3(i). Thus, as $z \in yp$, $z \in ysR^*$. Thus, there exists an element w in ys such that $z \in wR^*$.

From $z \in wR^*$ we obtain $w \in zR$. Thus, as $xR = zR$, $w \in xR$. Thus, as $x \in yq$, $w \in yqR$. Thus, as $w \in ys$, $s \in qR$.

Now, as s has been chosen arbitrarily in pR, we have shown that $pR \subseteq qR$. The inclusion $qR \subseteq pR$ is obtained similarly.

(c) ⇒ (a) Let r be an element in R. Then $1 \in r^*r \subseteq r^*R$. However, we are assuming that $1 \in R$. Thus, as S/R is assumed to be a partition of S, $r^*R = R$. Thus, as r has been chosen arbitrarily in R, we have shown that $R^*R \subseteq R$.

Let T be a closed subset of S.

We call the elements in X/T *left cosets* of T *in* X. For each closed subset U of S with $T \subseteq U$, the elements of U/T will be referred to as *left cosets* of T *in* U.

A subset R of S is called a *transversal* of T *in* X if $|yR \cap zT| = 1$ for any two elements y and z in X.

Let U and V be closed subsets of S, and let us assume that $T \subseteq V$ and $U \subseteq V$. A subset R of V is called a *transversal* of T and U *in* V if $|R \cap TvU| = 1$ for each element v in V.

Let U be a closed subset of S such that $T \subseteq U$. A transversal of $\{1\}$ and T in U is called a *left transversal* of T *in* U, a transversal of T and $\{1\}$ in U is called a *right transversal* of T *in* U.

Theorem 2.1.5 *For each closed subset T of S, we have the following.*

(i) *Let R be a transversal of T in X. Then R is a left transversal of T in S and, for each element r in R, $\{1\} = r^*r \cap T$.*

(ii) *Let R be a left transversal of T in S. Assume that, for each element r in R, $\{1\} = r^*r \cap T$. Then, R is a transversal of T in X.*

Proof. (i) Let R be a transversal of T in X, and let s be an element in S. We shall see that $|R \cap sT| = 1$

Let y be an element in X, and let z be an element in ys. Since R is assumed to be a transversal of T in X, $yR \cap zT$ is not empty. Thus, as $z \in ys$, there exists an element r in R such that $s \in rT$. From $s \in rT$ we obtain $r \in sT$; cf. Lemma 1.3.3(i). Thus, as $r \in R$, $1 \leq |R \cap sT|$.

Let us now show that $|R \cap sT| \leq 1$. In order to do so we fix elements p and q in $R \cap sT$. We shall see that $p = q$.

Since $p, q \in sT$, we have $s \in pT \cap qT$; cf. Lemma 1.3.3(i). Since $s \in pT$ and $z \in ys$, $z \in ypT$. Thus, there exists an element v in yp such that $z \in vT$. Since $v \in yp$ and $p \in R$, $v \in yR$. Thus, as $z \in vT$, $v \in yR \cap zT$.

Similarly, as $s \in qT$ and $z \in ys$, there exists an element w in yq such that $z \in wT$. Since $w \in yq$ and $q \in R$, $w \in yR$. Thus, as $z \in wT$, $w \in yR \cap zT$.

From $v \in yR \cap zT$ and $w \in yR \cap zT$ we obtain $v = w$. (Recall that R is assumed to be a transversal of T in X.) Thus, as $v \in yp$ and $w \in yq$, $p = q$.

So far, we have seen that R is a left transversal of T in S.

It is clear that $1 \in r^*r \cap T$. In order to show that $r^*r \cap T \subseteq \{1\}$, we pick an element t in $r^*r \cap T$. We shall see that $t = 1$.

Let y be an element in X, and let z be an element in yt. Then, as $t \in r^*r$, $z \in yr^*r$. Thus, there exists an element x in yr^* such that $z \in xr$. It follows that $y \in xr \cap zt^* \subseteq xr \cap zT$. Thus, as $z \in xr \cap zT$, $y = z$. (Recall that R is assumed to be a transversal of T in X.) Thus, as $z \in yt$, we conclude that $t = 1$.

(ii) Let y and z be elements in X. We have to show that $|yR \cap zT| = 1$.

Let s be the uniquely determined element in S which satisfies $z \in ys$. Since R is assumed to be a left transversal of T in S, there exists an element r in $R \cap sT$. From $r \in sT$ we obtain $s \in rT$; cf. Lemma 1.3.3(i). Thus, as $z \in ys$, $z \in yrT$. Thus, there exists an element x in yr with $z \in xT$. It follows that $x \in yr \cap zT$. In particular, as $r \in R$, $1 \leq |yR \cap zT|$.

We still have to show that $|yR \cap zT| \leq 1$. In order to do this we pick two elements v and w in $yR \cap zT$. We shall see that $v = w$.

From $v \in zT$ and $w \in zT$ we obtain $w \in vT$; cf. Lemma 2.1.4. Thus, there exists an element t in T such that $w \in vt$.

From $v \in yR$ we obtain an element q in R with $v \in yq$. From $w \in vt$ and $v \in yq$ we obtain $w \in yqt$. Thus, there exists an element p in qt such that $w \in yp$.

From $w \in yR$ and $w \in yp$ we obtain $p \in R$. From $p \in qt$ and $t \in T$ we obtain $p \in qT$. Thus, as $|R \cap qT| \leq 1$ and $q \in R \cap qT$, $q = p$. Thus, as $p \in qt$, $q \in qt$. Thus, by Lemma 1.3.3(ii), $t \in q^*q$. Thus, as $t \in T$, our hypothesis leads to $t = 1$. Thus, as $w \in vt$, $v = w$.

Lemma 2.1.6 *Let R be a subset of S such that $1 \in R$ and $RR \subseteq R$. Then, if R has finite valency, R is closed.*

PROOF. We are assuming that $RR \subseteq R$. Thus, $n_{RR} \subseteq n_R$. On the other hand, R is assumed to have finite valency. Thus, by Lemma 1.4.4(i), $n_R \subseteq n_{RR}$. It follows that $n_R = n_{RR}$. Thus, by Lemma 1.4.4(ii), $R = R^*RR$. Thus, as we are assuming that $1 \in R$, $R^*R \subseteq R$.

Corollary 2.1.7 *Let T be a closed subset of S, and assume that T^s has finite valency. Then T^s is closed.*

PROOF. From $1 \in T$ we obtain $1 \in T^s$; cf. Lemma 1.3.6(ii). From $TT \subseteq T$ we obtain $T^sT^s \subseteq T^s$; cf. Lemma 1.3.6(iii). Thus, our claim is a consequence of Lemma 2.1.6.

We shall now see that closed subsets of S give rise to new schemes, the so-called subschemes.

Let Y be a subset of X. For each element s in S, we set

$$s_Y := s \cap (Y \times Y).$$

For each subset R of S, we define R_Y to be the set of all sets r_Y with $r \in R$. The following theorem is obvious.

Theorem 2.1.8 *Let x be an element in X, and let T be a closed subset of S. Then we have the following.*

(i) *For each element t in T, $(t_{xT})^* = (t^*)_{xT}$.*

(ii) *The set T_{xT} is a scheme on xT.*

(iii) *For any three elements p, q, and r in T, we have $a_{p_{xT}q_{xT}r_{xT}} = a_{pqr}$.*

Let x be an element in X, and let T be a closed subset of S. We call T_{xT} the *subscheme* of S *defined by* xT.

The first part of the following lemma describes the (obvious) relationship between the complex multiplication in S and the one in T_{xT}.

Lemma 2.1.9 *Let x be an element in X, let T be a closed subset of S, and let R be a nonempty subset of T. Then we have the following.*

(i) *For each nonempty subset Q of T, $Q_{xT}R_{xT} = (QR)_{xT}$.*

(ii) *The set R_{xT} is closed if and only if R is closed.*

Proof. (i) Let t be an element in T. We have $t_{xT} \in Q_{xT}R_{xT}$ if and only if there exist elements q in Q and r in R such that $1 \leq a_{q_{xT}r_{xT}t_{xT}}$.

By Theorem 2.1.8(iii), $1 \leq a_{q_{xT}r_{xT}t_{xT}}$ is equivalent to $1 \leq a_{qrt}$. Thus, by definition, $t_{xT} \in Q_{xT}R_{xT}$ is equivalent to $t \in QR$. However, we have $t \in QR$ if and only if $t_{xT} \in (QR)_{xT}$.

(ii) From Theorem 2.1.8(i) and (i) we know that

$$(R_{xT})^* R_{xT} = (R^*)_{xT} R_{xT} = (R^* R)_{xT}.$$

Thus, R_{xT} is closed if and only if $(R^*R)_{xT} \subseteq R_{xT}$. However, this is equivalent to $R^*R \subseteq R$, and that means that R is closed.

2.2 Dedekind Identities

The following lemma is a special case of a general observation made by Richard Dedekind in 1900; cf. [8; Theorem VIII].

Lemma 2.2.1 *Let P and Q be nonempty subsets of S, and let T be a closed subset of S. Then we have the following.*

(i) *If $P \subseteq T$, $T \cap PQ = P(T \cap Q)$.*

(ii) *If $Q \subseteq T$, $T \cap PQ = (T \cap P)Q$.*

PROOF. (i) In order to show that $T \cap PQ \subseteq P(T \cap Q)$, we pick an element t in $T \cap PQ$. Since $t \in PQ$, there exists an element q in Q such that $t \in Pq$. Thus, by Lemma 1.3.3(iii), $q^* \in t^*P$. Thus, as T is assumed to be closed, $P \subseteq T$ yields $q \in T$. Thus, as $t \in Pq$, $t \in P(T \cap Q)$.

Conversely, as T is assumed to be closed, $P \subseteq T$ yields $P(T \cap Q) \subseteq T$. Thus, as $P(T \cap Q) \subseteq PQ$, $P(T \cap Q) \subseteq T \cap PQ$.

(ii) Applying (i) to Q^* and P^* in the role of P and Q, this follows from Lemma 1.3.2(iii).

Corollary 2.2.2 *Let P and Q be nonempty subsets of S, and let T be a closed subset of S. Assume that PT and QT are closed. Then we have $(P \cap QT)(Q \cap PT) = PQ \cap QT \cap PT$.*

PROOF. Since T is assumed to be a closed subset of S, we have $1 \in T$. Thus, $P \subseteq PT$. Thus, as PT is assumed to be closed, Lemma 2.2.1(i) yields

$$PT \cap (P \cap QT)Q = (P \cap QT)(PT \cap Q).$$

From $1 \in T$ we also obtain $Q \subseteq QT$. Thus, as QT is assumed to be closed, Lemma 2.2.1(ii) yields

$$QT \cap PQ = (QT \cap P)Q.$$

The desired equation follows easily from the last two equations.

Corollary 2.2.3 *Let P and Q be nonempty subsets of S, and let T be a closed subset of S such that $Q \subseteq T$. Then we have the following.*

(i) *If $T \subseteq QPQ$, $Q(P \cap T)Q = T$.*
(ii) *If $P \cup Q = PQ \cap QP$, $(P \cap T) \cup Q = (P \cap T)Q \cap Q(P \cap T)$.*

PROOF. (i) Since we are assuming that $Q \subseteq T$, we obtain from Lemma 2.2.1 that

$$Q(P \cap T)Q = (QP \cap T)Q = QPQ \cap T.$$

Thus, if $T \subseteq QPQ$, $Q(P \cap T)Q = T$.

(ii) We are assuming that $Q \subseteq T$. Thus, $(P \cap T) \cup Q = (P \cup Q) \cap T$. On the other hand, we know from Lemma 2.2.1 that

$$PQ \cap T \cap QP = (P \cap T)Q \cap Q(P \cap T).$$

Thus, if $P \cup Q = PQ \cap QP$, $(P \cap T) \cup Q = (P \cap T)Q \cap Q(P \cap T)$.

2.3 Structure Constants

Let s be an element in S, let n be an integer with $2 \leq n$, and let $R_1, \ldots, R_n$ be nonempty subsets of S. We define

$$a_{R_1 \dots R_n s} := \sum_{r_1 \in R_1} \cdots \sum_{r_n \in R_n} a_{r_1 \dots r_n s}.$$

If, for some element i in $\{1, \dots, n\}$, the set R_i contains an element r_i with $\{r_i\} = R_i$, we replace the index R_i in $a_{R_1 \dots R_n s}$ with r_i.

Lemma 2.3.1 *Let p and q be elements in S, let T be a closed subset of S, and assume that T has finite valency. Then the following hold.*

(i) *If $q \in pT$, $a_{pTp} = a_{pTq}$.*

(ii) *We have $a_{pTTq} = a_{pTq} n_T$.*

PROOF. (i) Let x be an element in X, and let z be an element in xq. Then, as $q \in pT$, $z \in xpT$. Thus, there exists an element y in xp such that $z \in yT$. Since $z \in yT$, $yT = zT$; cf. Lemma 2.1.4. In particular, $xp \cap yT = xp \cap zT$. Thus, taking into account that $\{yt^* \mid t \in T\}$ is a partition of yT and that $\{zt^* \mid t \in T\}$ is a partition of zT the equation follows from the fact that $y \in xp$ and $z \in xq$.

(ii) Referring to Lemma 1.1.1(iii) we obtain

$$a_{pTs} a_{sTq} = \sum_{t \in T} \sum_{u \in T} a_{pts} a_{suq} = \sum_{t \in T} \sum_{u \in T} a_{psq} a_{tus} = a_{psq} \sum_{t \in T} \sum_{u \in T} a_{tus}.$$

Thus, referring to Lemma 1.1.3(iii) (in order to obtain the third equation) we obtain

$$\begin{aligned} a_{pTTq} &= \sum_{s \in S} a_{pTs} a_{sTq} = \sum_{s \in S} a_{psq} \sum_{t \in T} \sum_{u \in T} a_{tus} \\ &= \sum_{t \in S} a_{psq} \sum_{t \in T} n_t = \sum_{t \in S} a_{psq} n_T = a_{pTq} n_T. \end{aligned}$$

This finishes the proof of (ii).

The following lemma generalizes Lemma 2.3.1(i). Lemma 2.3.1(i) is obtained from Lemma 2.3.2 by setting $T = \{1\}$ and $U = T$.

Lemma 2.3.2 *Let p be an element in S, let T and U be closed subsets of S, and let q be an element in TpU. Then, if T and U have finite valency, $a_{TpUp} = a_{TpUq}$.*

PROOF. Let us set $R := Tp \cap qU$. Then

$$a_{TpUq} = \sum_{t \in T} \sum_{u \in U} a_{tpuq} = \sum_{t \in T} \sum_{u \in U} \sum_{r \in R} a_{tpr} a_{ruq} = \sum_{r \in R} a_{Tpr} a_{rUq}.$$

From Lemma 2.3.1(i) together with Lemma 1.1.1(ii) we obtain $a_{Tpr} = a_{Tpp}$ for each element r in Tp. From Lemma 2.3.1(i) we obtain $a_{rUq} = a_{rUr}$ for each element r in qU. Thus,

$$a_{TpUq} = \sum_{r\in R} a_{Tpp}a_{rUr} = a_{Tpp}\sum_{r\in R} a_{rUr}.$$

This proves that $a_{TpUq} = a_{TpUq'}$ for any two elements q and q' in TpU with $q' \in qU$. Similarly, one shows that $a_{TpUq} = a_{TpUq'}$ for any two elements q and q' in TpU such that $q' \in Tq$. This finishes the proof of the lemma.

Lemma 2.3.3 *Let s be an element in S, and let T and U be closed subsets of S. Assume that s, T, and U have finite valency. Then we have $a_{TsUs}n_{TsU} = n_T n_s n_U$.*

PROOF. From Lemma 2.3.2 we obtain

$$a_{TsUs}n_{TsU} = a_{TsUs}\sum_{r\in TsU} n_r = \sum_{r\in TsU} a_{TsUs}n_r = \sum_{r\in TsU} a_{TsUr}n_r.$$

From Lemma 1.1.5(iii) we know that

$$\sum_{r\in S} a_{tsur}n_r = n_t n_s n_u.$$

Note also that, for each element $r \in S\setminus tsu$, $a_{tsur} = 0$; cf. Lemma 1.3.5. Thus,

$$n_T n_s n_U = \sum_{t\in T}\sum_{u\in U}\sum_{r\in TsU} a_{tsur}n_r = \sum_{r\in TsU}(\sum_{t\in T}\sum_{u\in U} a_{tsur})n_r.$$

Thus, the lemma follows from the definition of a_{TsUr}.

Lemma 2.3.4 *Let s be an element in S, and let T be a closed subset of S. Assume that s and T have finite valency. Then the following hold.*

(i) *The integer n_T divides n_{sT}.*

(ii) *We have $a_{sTs}n_{sT} = n_s n_T$.*

(iii) *The integer $(n_T)^{-1}n_{sT}$ divides n_s.*

PROOF. (i) Let y be an element in X. Then we have $|ysT| = n_{sT}$. On the other hand, we have $zT \subseteq ysT$ and $|zT| = n_T$ for each element z in ysT. Thus, the claim follows from Lemma 2.1.4.

(ii) This is Lemma 2.3.3 in the case where $T = \{1\}$ and $U = T$.

(iii) From (ii) we know that $a_{sTs}n_{sT} = n_s n_T$. Since n_s and n_T are assumed to be finite, we obtain from Lemma 1.4.2 that n_{sT} is finite. Thus, our claim follows from (i).

Lemma 2.3.5 *Let p, q, and s be elements of S, and let T be a closed subset of S. Assume that p, q, and T have finite valency. Then we have*

$$\sum_{u\in pT}\sum_{v\in qT} a_{pTp}a_{qTq}a_{uvs} = a_{pTqTs}.$$

PROOF. We have

$$\sum_{v\in qT} a_{qTq}a_{uvs} = \sum_{v\in qT} a_{qTv}a_{uvs} = \sum_{v\in S} a_{qTv}a_{uvs}$$
$$= \sum_{v\in S} a_{Tq^*v^*}a_{v^*u^*s^*} = a_{Tq^*u^*s^*} = a_{uqTs}.$$

(The first equation follows from Lemma 2.3.1(i), the third and the fifth equation follow from Lemma 1.1.5(ii).) Thus,

$$\sum_{u\in pT}\sum_{v\in qT} a_{pTp}a_{qTq}a_{uvs} = \sum_{u\in pT} a_{pTp}\sum_{v\in qT} a_{qTq}a_{uvs}$$
$$= \sum_{u\in pT} a_{pTu}a_{uqTs} = \sum_{u\in S} a_{pTu}a_{uqTs} = a_{pTqTs},$$

and this proves the lemma.

For thin schemes, the second part of the following result is associated with the name of Joseph-Louis Lagrange.

Lemma 2.3.6 *Let T and U be closed subsets of S, and assume that n_T and n_U are finite. Then the following hold.*

(i) *We have $n_T n_U = n_{TU} n_{T\cap U}$.*

(ii) *If $T \subseteq U$, n_T divides n_U.*

(iii) *Assume that n_S is finite and that $(n_T)^{-1}n_S$ and $(n_U)^{-1}n_S$ are coprime. Then $TU = S$.*

PROOF. (i) For any two elements t in T and u in U, we have

$$a_{t1u1} = a_{tu1} = \delta_{tu^*}n_t;$$

cf. Lemma 1.1.3(i). Thus, the claim follows from Lemma 2.3.3.

(ii) Let R be a left transversal of T in U. Then, for each element r in R, n_T divides n_{rT}; cf. Lemma 2.3.4(i).

On the other hand, we know from Lemma 2.1.4 that $\{rT \mid r \in R\}$ is a partition of U. Thus, n_U is the sum of the integers n_{rT} with $r \in R$. Therefore n_T divides n_U.

(iii) From (ii) we know that $n_{T\cap U}$ divides n_T. Thus, $(n_T)^{-1}n_S$ divides $(n_{T\cap U})^{-1}n_S$. Similarly, we obtain that $(n_U)^{-1}n_S$ divides $(n_{T\cap U})^{-1}n_S$.

On the other hand, we are assuming that $(n_T)^{-1}n_S$ and $(n_U)^{-1}n_S$ are coprime. Thus,

$$(n_T)^{-1}n_S(n_U)^{-1}n_S \leq (n_{T\cap U})^{-1}n_S.$$

From this we obtain $n_{T\cap U}n_S \leq n_T n_U$. Thus, by (i), $n_{T\cap U}n_S \leq n_{T\cap U}n_{TU}$. It follows that $n_S \leq n_{TU}$, and that means that $TU = S$.

Assume S to have finite valency, and let $\mathcal{T}$ be a finite set of closed subsets of S, and let us denote by U the intersection of the elements in $\mathcal{T}$. From Lemma 2.3.6(i) we inductively deduce that

$$(n_U)^{-1}n_S \leq \prod_{T\in\mathcal{T}} (n_T)^{-1}n_S.$$

Let p and q be elements in S. Recall that $\{p\}^q$ stands for the set of all elements s in S which satisfy $qs \subseteq pq$. In particular, for each element s in S, $\{1\}^s$ is the set of all elements r in S which satisfy $\{s\} = sr$.

Lemma 2.3.7 *Let s be an element in S such that n_{s^*} and n_s are finite. Then we have the following.*

(i) *The set $\{1\}^s$ is closed.*

(ii) *The integer $n_{\{1\}^s}$ divides n_s.*

(iii) *Let T be a closed subset of S such that n_T is finite and n_s and n_T are coprime. Then $\{1\} = \{1\}^s \cap T$.*

Proof. (i) We are assuming that n_{s^*} and n_s are finite. Thus, by Lemma 1.4.2, n_{s^*s} is finite. From Lemma 1.3.6(i) we know that $\{1\}^s \subseteq s^*s$. Thus, $\{1\}^s$ has finite valency. Thus, by Corollary 2.1.7, $\{1\}^s$ is closed.

(ii) From (i) we know that $\{1\}^s$ is closed, and from the definition of $\{1\}^s$ we obtain $\{s\} = s\{1\}^s$. Thus, by Lemma 2.3.4(i), $n_{\{1\}^s}$ divides n_s.

(iii) In (i) we saw that $\{1\}^s$ is closed. Thus, by Lemma 2.3.6(ii), $n_{\{1\}^s\cap T}$ divides $n_{\{1\}^s}$ and n_T.

From (ii) we know that $n_{\{1\}^s}$ divides n_s. Thus, as we are assuming n_s and n_T to be coprime, we conclude that $1 = n_{\{1\}^s\cap T}$. It follows that $\{1\} = \{1\}^s \cap T$.

Recall that a non-identity element s of S is called an involution if $\{1, s\}$ is closed.

Note that involutions are symmetric.

Lemma 2.3.8 *For each involution l of S, the following hold.*

(i) *We have $a_{lll} = n_l - 1$.*

(ii) *If l is not thin, $l \in ll$.*

(iii) *Let p and q be elements in S such that $p \neq q$ and $\{q\} = pl$. Then $a_{plq} = 1$.*

Proof. (i) Since l is assumed to be an involution, we have $ll \subseteq \{1, l\}$.

Let s be an element in $S\setminus\{1, l\}$. Then $s \notin ll$. Thus, by Lemma 1.3.3(i), $l \notin sl^*$. Thus, by definition, $a_{sl^*l} = 0$.

Since s has been chosen arbitrarily in $S \setminus \{1, l\}$, we now obtain from the second equation of Lemma 1.1.3(iii) that $a_{1l^*l} + a_{ll^*l} = n_l$. Since $l^* = l$, this implies $a_{1ll} + a_{lll} = n_l$. But, by the first equation of Lemma 1.1.1(i), $a_{1ll} = 1$. Therefore, $a_{lll} = n_l - 1$.

(ii) Since l is assumed to be an involution, we have $ll \subseteq \{1, l\}$. Assuming l not to be thin means that $\{1\} \neq ll$. Thus, we must have that $l \in ll$.

(iii) Let y be an element in X, let z be an element in yq, and let us assume that $2 \leq a_{plq}$. Then there exist elements v and w in $yp \cap zl$ such that $v \neq w$.

From $v \in yp$ we obtain $y \in vp^*$. Thus, as $w \in yp$, $w \in vp^*p$. From $v \in zl$ we obtain $z \in vl$. Thus, as $w \in zl$, $w \in vll$. Thus, as $v \neq w$, $w \in vl$.

From $w \in vp^*p$ and $w \in vl$ we obtain $l \in p^*p$. Thus, by Lemma 1.3.3(ii), $p \in pl$, contrary to $p \neq q$ and $\{q\} = pl$.

Let p be a prime number. An element s in S is called *p-valenced* if n_s is a power of p. A nonempty subset of S is called *p-valenced* if all of its elements are p-valenced. A nonempty p-valenced subset R of S is called a *p-subset* of S if n_R is a power of p.

Lemma 2.3.9 *Let p be a prime number, and let T be a closed p-subset of S. Then there exists an element t in $T \setminus \{1\}$ such that $1 = n_t$.*

PROOF. Since T is assumed to be p-valenced, n_t is a power of p whenever t is an element in T. By definition, n_T is the sum of the integers n_t with $t \in T$.

Since T is assumed to be a p-subset of S, n_T is a power of p. On the other hand, as we are assuming T to be closed, $1 \in T$. Thus, as $n_1 = 1$, we obtain from the previous paragraph that $T \setminus \{t\}$ contains an element t with $1 = n_t$.

2.4 Maximal Closed Subsets

Let T and U be closed subsets of S such that $T \subseteq U$ and $T \neq U$. We call T a *maximal* closed subset of U if T and U are the only closed subsets of U containing T as a subset.

In this section, we give a sufficient condition for a closed subset of S to be maximal.

Lemma 2.4.1 *Let y and z be elements in X, and let T and U be closed subsets of S such that $z \in yUTU$. Then there exists an element x in X such that $xT \cap yU$ and $xT \cap zU$ are not empty.*

PROOF. Since $z \in yUTU$, there exists an element x in yU such that $z \in xTU$. From $x \in yU$ (and $x \in xT$) we obtain $x \in xT \cap yU$.

Since $z \in xTU$, there exists an element w in xT such that $z \in wU$. From $z \in wU$ we obtain $w \in zU$; cf. Lemma 2.1.4. Thus, $w \in xT \cap zU$.

Lemma 2.4.2 *Let v, w, y, and z be elements in X. Let T and U be closed subsets of S satisfying $T \cup U = TU \cap UT$. Assume that none of the sets $vT \cap yU$, $vT \cap zU$, $wT \cap yU$, and $wT \cap zU$ is empty. Then, if $yU \neq zU$, $vT = wT$.*

PROOF. We are assuming that $vT \cap yU$ is not empty. Thus, there exists an element y' in $vT \cap yU$. Similarly, we find elements v' in $vT \cap zU$, w' in $wT \cap yU$, and z' in $wT \cap zU$.

Since $v' \in vT$ and $y' \in vT$, $v' \in y'T$; cf. Lemma 2.1.4. Similarly, as $v' \in zU$, $z \in v'U$, so that $z \in y'TU$. Thus, as $z' \in zU$, $z' \in y'TU$.

Similarly, we obtain from $w' \in yU$, $y' \in yU$, $w' \in wT$, and $z' \in wT$ that $z' \in y'UT$.

From $z' \in y'TU$ and $z' \in y'UT$ we obtain $z' \in y'(TU \cap UT)$. Thus, as we are assuming that $T \cup U = TU \cap UT$, we conclude that $z' \in y'(T \cup U)$.

Let us now assume that $yU \neq zU$. Then, as $z' \in zU$ and $y' \in yU$, $z' \notin y'U$; cf. Lemma 2.1.4. Thus, as $z' \in y'(T \cup U)$, we must have $z' \in y'T$. Thus, referring to Lemma 2.1.4 once more we obtain from $y' \in vT$ and $z' \in wT$ that $vT = wT$.

For the remainder of this section, we assume S to have finite valency.

Lemma 2.4.3 *Let T, U, and V be closed subsets of S such that $V = UTU$ and $T \cup U = TU \cap UT$. Then*

$$((n_U)^{-1}n_V - 1)n_{T\cap U} = ((n_{T\cap U})^{-1}n_T - 1)n_U.$$

PROOF. We fix an element x in X, we set $\tau := (n_{T\cap U})^{-1}n_T$, and we set $\upsilon := (n_{T\cap U})^{-1}n_U$.

Referring to Lemma 2.1.4 and to Lemma 2.3.6(ii) we find exactly τ elements wU in X/U with $\emptyset \neq xT \cap wU$. Similarly, there exist exactly υ elements wT in X/T with $\emptyset \neq wT \cap xU$.

With the help of Lemma 2.4.1 and Lemma 2.4.2 we count in two different ways the pairs (yT, zU) satisfying $z \in xV$ and

$$yT \cap xU \neq \emptyset \neq yT \cap zU.$$

Then we get $(n_U)^{-1}n_V - 1 = (\tau - 1)\upsilon$. From this we obtain the desired equation.

Lemma 2.4.4 *Let T and U be closed subsets of S such that $T \neq S$, $UTU = S$, and $T \cup U = TU \cap UT$. Then, we have $n_T \leq n_U$.*

Proof. There is nothing to show if $U = S$. Therefore, we assume that $U \neq S$. In this case, we obtain elements y and z in X such that $yU \neq zU$. Since we are assuming that $UTU = S$, we obtain from Lemma 2.4.1 an element w in X such that

$$wT \cap yU \neq \emptyset \neq wT \cap zU.$$

Since we are assuming that $T \neq S$, there exists an element x in X such that $wT \cap xU$ is empty; cf. Lemma 2.4.2.

We set $\tau := (n_{T\cap U})^{-1} n_T$ and $\upsilon := (n_{T\cap U})^{-1} n_U$.

Then, there exist τ elements vU in X/U such that $\emptyset \neq wT \cap vU$. For each of these elements, there exists an element $v'T$ in X/T such that

$$v'T \cap vU \neq \emptyset \neq v'T \cap xU;$$

cf. Lemma 2.4.1. Moreover, for any two different elements uU and vU in X/U with

$$wT \cap uU \neq \emptyset \neq wT \cap vU,$$

we have $u'T \neq v'T$; cf. Lemma 2.4.2. Thus, $\tau \leq \upsilon$, and from this we conclude that $n_T \leq n_U$.

The following theorem is the main result of this section. It is [44; Theorem].

Theorem 2.4.5 *Let U be a closed subset of S such that $U \neq S$. Assume that there exists a closed subset T of S which satisfies $T \neq S$, $UTU = S$, and $T \cup U = TU \cap UT$. Then U is a maximal closed subset of S.*

Proof. Let V be a closed subset of S such that $U \subseteq V$. We shall be done if we succeed in showing that $V \in \{U, S\}$.

We set $\tau := (n_{T\cap U})^{-1} n_T$ and $\upsilon := (n_{T\cap U})^{-1} n_U$. Then, by Lemma 2.4.3,

$$n_S = n_U(1 + (\tau - 1)\upsilon).$$

Next, we set $\delta := (n_{T\cap U})^{-1} n_{T\cap V}$. Then, by Corollary 2.2.3 and Lemma 2.4.3,

$$n_V = n_U(1 + (\delta - 1)\upsilon).$$

Finally, we set $\sigma := (n_V)^{-1} n_S$. Then, the two above equations yield

$$1 + (\tau - 1)\upsilon = (1 + (\delta - 1)\upsilon)\sigma.$$

It follows that

$$(\tau - 1 - (\delta - 1)\sigma)\upsilon = \sigma - 1.$$

Let us now first assume that $2 \leq \delta$ and that $2 \leq \sigma$. Since $2 \leq \sigma$, the last equation yields $(\delta - 1)\sigma \leq \tau - 1$ and $\upsilon \leq \sigma - 1$. Since $2 \leq \delta$ and $(\delta - 1)\sigma \leq \tau - 1$,

$\sigma \leq \tau - 1$. Thus, as $\upsilon \leq \sigma - 1$, $\upsilon \leq \tau - 1$. Thus, $n_U \leq n_T - 1$, contrary to Lemma 2.4.4.

This contradiction shows that we have $1 = \delta$ or $1 = \sigma$.

Let us now assume that $1 = \delta$. Then we have $n_{T \cap U} = n_{T \cap V}$. Since $U \subseteq V$, this yields $T \cap U = T \cap V$, so that we have $T \cap V \subseteq U$. It follows that $U = U(T \cap V)U$.

On the other hand, as we are assuming that $UTU = S$, we have that $U(T \cap V)U = V$; cf. Corollary 2.2.3(i). Thus, as $U = U(T \cap V)U$, $U = V$.

Note, finally, that $1 = \sigma$ is equivalent to $V = S$.

2.5 Normalizer and Strong Normalizer

Let P and Q be subsets of S, with Q nonempty. We set

$$N_P(Q) := \{p \in P \mid Qp \subseteq pQ\}.$$

The set $N_P(Q)$ is called the *normalizer* of Q *in* P.

Let R be a nonempty subset of S. We say that an element s in S *normalizes* the set R if $s \in N_S(R)$.

Let T and U be closed subsets of S such that $T \subseteq U$. The closed subset T is said to be a *normal* closed subset *of* U if $U \subseteq N_S(T)$. In this case, we also say that T is *normal in* U.

For each nonempty subset R of S, we obviously have $1 \in N_S(R)$.

Note that we have $T \subseteq N_S(T)$ for each closed subset T of S. However, in general, $N_S(T)$ is not closed.

Lemma 2.5.1 *For each closed subset T of S, the following hold.*

(i) *Let R be a subset of S, and assume that $R \cap T$ is not empty. Then $N_T(R) \subseteq N_T(R \cap T)$.*

(ii) *We have $TN_S(T) \subseteq N_S(T)$.*

PROOF. (i) Let t be an element in $N_T(R)$. Then

$$(R \cap T)t \subseteq Rt \cap T \subseteq tR \cap T = t(R \cap T);$$

cf. Lemma 2.2.1(i). Thus, $t \in N_T(R \cap T)$.

(ii) Let s be an element in $TN_S(T)$. Then there exists an element r in $N_S(T)$ such that $s \in Tr$.

Since $r \in N_S(T)$, $Tr \subseteq rT$. Thus, as $s \in Tr$, $s \in rT$. Thus, by Lemma 2.1.3, $Ts = Tr$ and $rT = sT$. Thus, as $Tr \subseteq rT$, $Ts \subseteq sT$. Thus, by definition, $s \in N_S(T)$.

Lemma 2.5.2 *For each closed subset T of S, we have the following.*

(i) *For each element s in S, $s^* \in N_S(T)$ is equivalent to $T \subseteq T^s$.*

(ii) *Let s be an element in S such that s^*, $s \in N_S(T)$. Then $Ts = sT$.*

(iii) *For each closed subset U of S with $T \subseteq N_S(U)$, TU is closed.*

(iv) *Assume T to be normal in S, and let U be a normal closed subset of S. Then TU is a normal closed subset of S.*

PROOF. (i) Let s be an element in S. We have $s^* \in N_S(T)$ if and only if $Ts^* \subseteq s^*T$. According to Lemma 1.3.2(iii), $Ts^* \subseteq s^*T$ is equivalent to $sT \subseteq Ts$, and this means that $T \subseteq T^s$.

(ii) We are assuming that $s^* \in N_S(T)$, and that means that $Ts^* \subseteq s^*T$. Thus, by Lemma 1.3.2(iii), $sT \subseteq Ts$. Since we are also assuming that $s \in N_S(T)$, we have $Ts \subseteq sT$, too.

(iii) Considering Lemma 2.1.1 this is a consequence of (ii).

(iv) This follows from (iii).

Let P be a subset of S.

For each element s in S, we set

$$C_P(s) := \{p \in P \mid sp = ps\}.$$

The set $C_P(s)$ is called the *centralizer* of s *in* P.

For each nonempty subset Q of S, we define $C_P(Q)$ to be the intersection of the sets $C_P(q)$ with $q \in Q$. The set $C_P(Q)$ is called the *centralizer* of Q *in* P.

Note that we have $C_P(Q) \subseteq N_P(Q)$ for each nonempty subset Q of S.

Lemma 2.5.3 *Let T and U be closed subsets of S such that $T \subseteq N_S(U)$, $U \subseteq N_S(T)$, and $\{1\} = T \cap U$. Then $T \subseteq C_S(U)$.*

PROOF. Let t be an element in T, let u an element in U, and let s an element in tu. We shall be done if we succeed in showing that $s \in ut$.

From $s \in tu$ (together with the hypothesis that $U \subseteq N_S(T)$) we obtain $s \in uT$. Thus, there exists an element p in T such that $s \in up$.

From $s \in up$ (together with the hypothesis that $T \subseteq N_S(U)$) we obtain $s \in pU$. Thus, there exists an element q in U such that $s \in pq$. Thus, as $s \in tu$, Lemma 2.1.2 yields $t = p$. Thus, as $s \in up$, $s \in ut$.

Lemma 2.5.4 *Let p and q be elements of S, let T be a normal closed subset of S, and assume that p, q, and T have finite valency. Then the following hold.*

(i) *We have $a_{Tpq} = a_{pTq}$.*

(ii) *For each element s in S, we have $a_{pTqs} = a_{pqTs}$.*

(iii) *For each element s in S, we have*

$$\sum_{u\in pT}\sum_{v\in qT} a_{pTp}a_{qTq}a_{uvs} = n_T \sum_{w\in sT} a_{pqw}a_{wTw}.$$

Proof. (i) From Lemma 2.5.2(ii) we know that $Ts = sT$. Thus, our claim follows from Lemma 2.3.3.

(ii) From Lemma 1.1.1(iii) we obtain

$$a_{pTqs} = \sum_{r\in S} a_{pTr}a_{rqs} = \sum_{r\in S} a_{prs}a_{Tqr}$$

and

$$a_{pqTs} = \sum_{r\in S} a_{pqr}a_{rTs} = \sum_{r\in S} a_{prs}a_{qTr}.$$

Thus, the claim follows from (i).

(iii) From Lemma 2.3.5 we know that

$$\sum_{u\in pT}\sum_{v\in qT} a_{pTp}a_{qTq}a_{uvs} = a_{pTqTs}.$$

Moreover, from (ii) and Lemma 2.3.1(ii) we easily deduce that

$$a_{pTqTs} = a_{pqTTs} = n_T a_{pqTs}.$$

Thus,

$$\sum_{u\in pT}\sum_{v\in qT} a_{pTp}a_{qTq}a_{uvs} = n_T a_{pqTs}.$$

On the other hand, we obtain from Lemma 2.3.1(i) that

$$\sum_{w\in sT} a_{pqw}a_{wTw} = \sum_{w\in sT} a_{pqw}a_{wTs} = a_{pqTs}.$$

This finishes the proof of (iii).

Let us now introduce a third operator on the set of all nonempty subsets of S. Again, let P and Q be subsets of S, and let us assume Q to be not empty. We set

$$K_P(Q) := \{p \in P \mid p^*Qp \subseteq Q\}.$$

The set $K_P(Q)$ is called the *strong normalizer* of Q *in* P.

For each nonempty subset R of S, we obviously have $1 \in K_S(R)$.

Note also that we have $T \subseteq K_S(T)$ for each closed subset T of S.

Lemma 2.5.5 *For each nonempty subset R of S, we have $K_S(R) \subseteq N_S(R)$.*

PROOF. Let s be an element in $K_S(R)$. Then, by definition, $s^*Rs \subseteq R$. Thus, as $1 \in s^*s$, $Rs \subseteq ss^*Rs \subseteq sR$. It follows that $Rs \subseteq sR$, and that means that $s \in N_S(R)$.

There are cases where strong normalizer and normalizer coincide. Let P be a thin subset of S, and let Q be a nonempty subset of S. Then we obviously have $K_P(Q) = N_P(Q)$.

Lemma 2.5.6 *For any two closed subsets T and U of S, the following hold.*

(i) *We have $K_S(T) \cap K_S(U) \subseteq K_S(T \cap U)$.*

(ii) *If $T \subseteq N_S(U)$, $K_S(T) \cap N_S(U) \subseteq K_S(TU)$.*

(iii) *If $T \subseteq U$, $K_S(T) \cap N_S(U) \subseteq K_S(U)$.*

PROOF. (i) Let s be an element in $K_S(T) \cap K_S(U)$. Then

$$s^*(T \cap U)s \subseteq s^*Ts \cap s^*Us \subseteq T \cap U.$$

Thus, $s \in K_S(T \cap U)$.

(ii) Let s be an element in $K_S(T) \cap N_S(U)$. Then

$$s^*TUs \subseteq s^*TsU \subseteq TU.$$

Thus, $s \in K_S(TU)$.

(iii) Since $U \subseteq N_S(U)$, this follows from (i).

Let T and U be closed subsets of S such that $T \subseteq U$. The closed subset T is said to be a *strongly normal* closed subset *of U* if $U \subseteq K_S(T)$. In this case, we also say that T is *strongly normal in U*.

Lemma 2.5.5 says that, if a closed subset T of S is strongly normal in a closed subset U of S, then T is normal in U.

Lemma 2.5.7 *Let T and U be closed subsets of S such that T is strongly normal in U. Let V be a closed subset of S such that $U \subseteq N_S(V)$. Then TV is strongly normal in UV.*

PROOF. We are assuming that $U \subseteq N_S(V)$. Thus, by Lemma 2.5.2(iii), UV is closed. Similarly, we obtain from $T \subseteq U$ and $U \subseteq N_S(V)$ that TV is closed.

In order to show that TV is strongly normal in UV, we pick an element s in UV. We shall see that $s^*TVs \subseteq TV$.

Since $s \in UV$, there exist elements u in U and v in V such that $s \in uv$. From $u \in U$ and $U \subseteq N_S(V)$ we obtain $Vu \subseteq uV$. From $u \in U$ and $U \subseteq K_S(T)$ we obtain $u^*Tu \subseteq T$. Thus,

$$s^*TVs \subseteq v^*u^*TVuv \subseteq v^*u^*TuVv \subseteq v^*TVv = TV.$$

(Recall that, by Lemma 1.3.2(iii), $s^* \in v^*u^*$.)

Lemma 2.5.8 *Let R be a nonempty subset of S, and let T be a closed subset of S. Then the following hold.*

(i) *For any two elements p and q in $K_T(R)$, we have $pq \subseteq K_T(R)$.*

(ii) *If T has finite valency, $K_T(R)$ is closed.*

PROOF. (i) Let p and q be elements in $K_T(R)$. Then we have

$$s^*Rs \subseteq q^*p^*Rpq \subseteq q^*Rq \subseteq R$$

for each element s in pq, and this says that $s \in K_T(R)$.

(ii) Considering that $1 \in K_T(R)$ this follows from (i) and Lemma 2.1.6.

Note that $K_S(\{1\})$ is the set of the thin elements of S.

For each subset R of S, we set

$$O_\vartheta(R) := \{r \in R \mid 1 = n_r\},$$

and call it the *thin radical* of R.

Lemma 2.5.9 *For each closed subset T of S, the following hold.*

(i) *We have $O_\vartheta(T) = K_T(\{1\})$.*

(ii) *For any two elements p and q in $O_\vartheta(T)$, we have $pq \subseteq O_\vartheta(T)$.*

(iii) *If T has finite valency, $O_\vartheta(T)$ is closed.*

PROOF. (i) Since $K_T(\{1\})$ is the set of all thin elements of T, our claim is an immediate consequence of Lemma 1.5.1.

(ii) From (i) we know that $O_\vartheta(S) = K_S(\{1\})$. Thus, our claim is a consequence of Lemma 2.5.8(i).

(iii) This follows from (i) together with Lemma 2.5.8(ii).

2.6 Conjugates of Closed Subsets

Let P and Q be nonempty subsets of S. We call Q a *conjugate* of P if there exists an element s in S such that $Q = s^*Ps$.

Lemma 2.6.1 *Let s be an element in S, and let T be a closed subset of S such that $ss^* \subseteq T$. Then the following hold.*

(i) *We have $T^s = s^*Ts$.*

(ii) *The set* s^*Ts *is closed.*

(iii) *We have* $K_S(T^s) = K_S(T)^s$.

PROOF. (i) From Lemma 1.3.6(i) we know that $T^s \subseteq s^*Ts$. However, assuming that $ss^* \subseteq T$ we obtain $ss^*Ts \subseteq Ts$, so that, by definition, $s^*Ts \subseteq T^s$.

(ii) Clearly, we have $1 \in s^*Ts$ and $(s^*Ts)^* = s^*Ts$. However, assuming that $ss^* \subseteq T$ we also obtain $s^*Tss^*Ts \subseteq s^*Ts$.

(iii) According to (i), it is enough to show that $K_S(s^*Ts) = s^*K_S(T)s$. (Recall that $T \subseteq K_S(T)$.)

Let q be an element in $s^*K_S(T)s$. Then there exists an element p in $K_S(T)$ such that $q \in s^*ps$. Thus, we obtain from $ss^* \subseteq T$ and $p \in K_S(T)$ that

$$q^*s^*Tsq \subseteq s^*p^*ss^*Tss^*ps \subseteq s^*p^*Tps \subseteq s^*Ts.$$

Thus, $q \in K_S(s^*Ts)$.

Since q has been chosen arbitrarily in $s^*K_S(T)s$, we thus have shown that $s^*K_S(T)s \subseteq K_S(s^*Ts)$.

What we have shown so far also applies to s^* and s^*Ts instead of s and T. (Note that, by (ii), s^*Ts is closed.) Thus, we have

$$sK_S(s^*Ts)s^* \subseteq K_S(ss^*Tss^*) = K_S(T).$$

Thus, $K_S(s^*Ts) \subseteq s^*sK_S(s^*Ts)s^*s \subseteq s^*K_S(T)s$.

Lemma 2.6.2 *Let* s *be an element in* S, *and let* T *be a closed subset of* S. *Assume that* s^*, s, *and* T *have finite valency. Then the following hold.*

(i) *We have* $ss^* \subseteq T$ *if and only if* $n_T = n_{s^*Ts}$.

(ii) *Assume that* $n_T = n_{s^*Ts}$, *and let* U *be a closed subset of* S *with* $T \subseteq U$. *Then* $n_U = n_{s^*Us}$.

PROOF. (i) This is an application of Lemma 1.4.5(ii) to $\{s^*\}$, T, and $\{s\}$ in the role of P, Q, and R.

(ii) Assuming $n_T = n_{s^*Ts}$ we obtain from (i) that $ss^* \subseteq T$. Thus, as we are assuming that $T \subseteq U$, $ss^* \subseteq U$, so that, again by (i), $n_U = n_{s^*Us}$.

Lemma 2.6.3 *Let* T *and* V *be closed subsets of* S *such that* $T \subseteq V$. *Assume* V *to have finite valency and set* $U := K_V(T)$. *Then we have the following.*

(i) *Let* p *and* q *be elements in* V *such that* $Up = Uq$. *Then* $pp^* \subseteq T$ *if and only if* $qq^* \subseteq T$.

(ii) *The number of sets* s^*Ts *with* $s \in V$ *and* $ss^* \subseteq T$ *is equal to the number of sets* Us *with* $s \in V$ *and* $ss^* \subseteq T$.

PROOF. We are assuming that V has finite valency. Thus, by Lemma 2.5.8(ii), U is closed.

(i) Since we are assuming that $Up = Uq$, there exists an element u in U such that $p \in uq$. Thus, by Lemma 1.3.3(ii), $q \in u^*p$. Thus, assuming that $pp^* \subseteq T$, we obtain

$$qq^* \subseteq u^*pp^*u \subseteq u^*Tu \subseteq T.$$

Similarly, one obtains $pp^* \subseteq T$ from $qq^* \subseteq T$.

(ii) Let p and q be elements in V, and let us first assume that $p^*Tp \subseteq q^*Tq$ and that $qq^* \subseteq T$. Then, $qp^*Tpq^* \subseteq qq^*Tqq^* \subseteq T$. It follows that $pq^* \subseteq U$. Thus, by Lemma 1.3.3(i), $p \in Uq$. Thus, by Lemma 2.1.4, $Up = Uq$.

Let us now assume that $Up = Uq$. Then there exists an element u in U such that $p \in uq$. It follows that

$$p^*Tp \subseteq q^*u^*Tuq \subseteq q^*Tq.$$

Similarly, one shows that $q^*Tq \subseteq p^*Tp$, so that $p^*Tp = q^*Tq$.

Lemma 2.6.4 *Let T and U be closed subsets of S such that $\{1\} = T \cap U$. Let t be an element in T, and let u be an element in U such that $1 = |tu|$. Then the following hold.*

(i) *The set $u^*tu \cap T$ contains at most one element.*

(ii) *If $U \subseteq N_S(T)$, $u^*tu \cap T$ contains exactly one element.*

PROOF. (i) We are assuming that tu contains exactly one element. Let us call this element s, and let us pick two elements p and q in $u^*tu \cap T$.

From $p \in u^*tu$ and $\{s\} = tu$ we obtain $p \in u^*s$. Thus, by Lemma 1.3.3(ii), $s \in up$. Similarly, we obtain $s \in uq$, so that $s \in up \cap uq$.

From $s \in up \cap uq$ we obtain that $u^*u \cap pq^*$ is not empty; cf. Lemma 1.3.4. Thus, as $u^*u \cap pq^* \subseteq T \cap U = \{1\}$, $\{1\} = u^*u \cap pq^*$. It follows that $1 \in pq^*$, so that, according to Lemma 1.3.2(i), $p = q$.

(ii) Let us assume that $U \subseteq N_S(T)$. Then $tu \subseteq uT$. Thus, as tu is assumed to contain only one element, there exists an element r in T such that $tu \subseteq ur$. Thus, by Lemma 1.3.3(ii), $r \in u^*tu$. Now the claim follows from (i).

Lemma 2.6.5 *Let p and q be elements in S such that n_{p^*} and n_q are finite and coprime, and let T be a closed subset of S such that $q \in p^*Tp$. Then $q \in T^p$.*

PROOF. We are assuming that $q \in p^*Tp$. Thus, by Lemma 1.3.3(ii), $p \in Tpq$. However, we are assuming that n_{p^*} and n_q are coprime. Thus, by Lemma 1.5.2, $|pq| = 1$.

From $p \in Tpq$ and $|pq| = 1$ we obtain $pq \subseteq Tp$; cf. Lemma 1.3.3(ii). Thus, by definition, $q \in T^p$.

Corollary 2.6.6 *Let s be an element in S such that n_{s^*} is finite, and let T and U be closed subsets of S. Assume that, for each element u in $s^*Ts \cap U$, n_u is finite and n_{s^*} and n_u are coprime. Then $T^s \cap U = s^*Ts \cap U$.*

PROOF. We are assuming that, for each element u in $s^*Ts \cap U$, n_{s^*} and n_u are coprime. Thus, by Lemma 2.6.5, $s^*Ts \cap U \subseteq T^s$, so that the desired equation follows from Lemma 1.3.6(i).

Let us see what happens if we apply Corollary 2.6.6 to $\{1\}$ and S in place of T and U.

Corollary 2.6.7 *Let s be an element in S such that n_{s^*} is finite. Assume that, for each element r in s^*s, n_r is finite and n_{s^*} and n_r are coprime. Then the following hold.*

(i) *We have $\{1\}^s = s^*s$.*

(ii) *The set s^*s is closed.*

(iii) *We have $n_s = n_{s^*s}$.*

PROOF. (i) We are assuming that, for each element r in s^*s, n_{s^*} and n_r are coprime. Thus, by Corollary 2.6.6, $\{1\}^s = s^*s$.

(ii) From (i) we obtain $\{s\} = ss^*s$. Thus, by Lemma 1.3.2(iii), $\{s^*\} = s^*ss^*$. Thus, by Lemma 1.4.5(i), $n_s \leq n_{s^*}$. Thus, as n_{s^*} is assumed to be finite, n_s must be finite. Thus, by Lemma 1.4.2, n_{s^*s} is finite, so that, by (i), $\{1\}^s$ has finite valency. Thus, referring to Corollary 2.1.7 we obtain (ii) from (i).

(iii) From (i) we know that $\{s\} = ss^*s$. Thus, by Lemma 1.4.4(ii), $n_s = n_{s^*s}$.

Let s be an element in S. For each subset R of S, we define $D_R(s)$ to be the set of all elements r in R such that $r^*r \subseteq s^*s$.

Lemma 2.6.8 *Assume S to have finite valency. Let s be an element in S such that, for each element r in s^*s, n_{s^*} and n_r are coprime.*

Let T be a closed subset of $D_S(s^*)$. Assume that $T \subseteq T^s$ and that, for each element t in T, n_{s^*} and n_t are coprime. Then $T \subseteq K_S(s^*s)$.

PROOF. Let t be an element in T. We have to show that $t \in K_S(s^*s)$.

Since $t \in T$, n_{s^*} and n_t are coprime. Thus, by Lemma 1.5.2, $|st| = 1$. On the other hand, we are assuming that $T \subseteq T^s$. Thus, as $t \in T$, $t \in T^s$, and that means that $st \subseteq Ts$.

Since $|st| = 1$ and $st \subseteq Ts$, there exists an element r in T such that $st \subseteq rs$. From $st \subseteq rs$ we obtain $t^*s^*st \subseteq s^*r^*rs$.

From $r \in T$ and $T \subseteq D_S(s^*)$ we obtain $r \in D_S(s^*)$. Thus, $r^*r \subseteq ss^*$. Thus, $s^*r^*rs \subseteq s^*ss^*s$. But, by Corollary 2.6.7(ii), s^*s is closed. Thus, $s^*r^*rs \subseteq s^*s$. Thus, as $t^*s^*st \subseteq s^*r^*rs$, $t^*s^*st \subseteq s^*s$.

3

Generating Subsets

Let R be a subset of S. We define $\langle R \rangle$ to be the intersection of all closed subsets of S which contain R. We say that R *generates* $\langle R \rangle$ or that $\langle R \rangle$ is *generated* by R.

Recall that the intersection of closed subsets is closed. Therefore $\langle R \rangle$ is closed for each subset R of S.

Note also that $\langle P \rangle \subseteq \langle Q \rangle$ for any two subsets P and Q of S with $P \subseteq Q$.

The first section of this chapter is a collection of basic facts about generating sets of closed subsets. We introduce the length function which one obtains from generating subsets, and we establish a connection between generating subsets and commutator subsets.

In the second section, we introduce the thin residue of a closed subset. The thin residue is a specific commutator subset. We look at the thin residue of a complex product of two closed subsets, at the complex product of the thin residue and the thin radical, and at multiple thin residues of closed subsets.

In the third section of this chapter, we investigate closed subsets of S generated by elements of valency 2. We start with closed subsets of S which are generated by a single symmetric element of valency 2. After that, we shall look at more general cases.

Generating sets of elements of S are particularly interesting if they consist of involutions. In fact, a major part of this monograph deals with closed subsets generated by distinguished sets of involutions. Section 3.4 deals with general aspects of closed subsets generated by involutions.

In the last two sections of this chapter, we impose specific conditions on sets of involutions. In Section 3.5, we look at constrained sets of involutions, and in the last of the six sections of this chapter, we look at constrained sets of involutions which satisfy the exchange condition. According to what we said in the preface of this monograph, such sets of involutions are called Coxeter sets.

It might be worth mentioning that, via the group correspondence, most of the results about Coxeter sets which we compile in the last section are natural generalizations of well-known facts on Coxeter groups.

Some of the results in the last two sections of this chapter, in particular Theorem 3.6.4 and Theorem 3.6.6, foreshadow the importance of closed subsets generated by involutions in scheme theory.

3.1 Basic Facts

Let R be a nonempty subset of S. We set $R^0 := \{1\}$. For each positive integer n, we inductively define $R^n := R^{n-1}R$.

Lemma 3.1.1 *For each nonempty subset R of S, we have the following.*

(i) *The set $\langle R\rangle$ is the union of the sets $(R^* \cup R)^n$ with n a non-negative integer.*

(ii) *If $\langle R\rangle$ has finite valency, $\langle R\rangle$ is the union of the sets R^n with n a non-negative integer.*

PROOF. (i) We set $P := R^* \cup R$, and we define Q to be the union of all sets P^n such that n is a non-negative integer. We have to show that $\langle R\rangle = Q$.

From Lemma 1.3.2(iii) we obtain $(P^n)^* = (P^*)^n$ for each non-negative integer n. Thus, as $P^* = P$, we obtain $(P^n)^* = P^n$ for each non-negative integer n. It follows that, for any two non-negative integers l and m,

$$(P^l)^*P^m = P^lP^m = P^{l+m} \subseteq Q.$$

Thus, Q is closed. Thus, as $R \subseteq Q$, $\langle R\rangle \subseteq Q$.

Conversely, for each non-negative integer n, we have $P^n \subseteq \langle P\rangle^n \subseteq \langle P\rangle = \langle R\rangle$. Therefore, $Q \subseteq \langle R\rangle$.

(ii) Let us denote by Q the union of all sets R^n such that n is a non-negative integer. Then $1 \in Q$, $R \subseteq Q$, $Q^2 \subseteq Q$, and, by (i), $Q \subseteq \langle R\rangle$.

We are assuming $\langle R\rangle$ to have finite valency. Thus, as $Q \subseteq \langle R\rangle$, Q has finite valency. Thus, as $1 \in Q$ and $Q^2 \subseteq Q$, Q is closed; cf. Lemma 2.1.6. Thus, as $R \subseteq Q$, $\langle R\rangle \subseteq Q$.

From $Q \subseteq \langle R\rangle$ and $\langle R\rangle \subseteq Q$ we obtain $\langle R\rangle = Q$, and that finishes the proof of (ii).

Let R be a nonempty subset of S.

Let s be an element in $\langle R\rangle$. Then, by Lemma 3.1.1(i), there exists a non-negative integer n such that $s \in (R^* \cup R)^n$. In the following, we shall denote by $\ell_R(s)$ the smallest non-negative integer n satisfying $s \in (R^* \cup R)^n$. We call $\ell_R(s)$ the *length* of s *with respect to* R.

Since $1 \in R^0$, $\ell_R(1) = 0$. Note also that we have $\ell_R(s^*) = \ell_R(s)$ for each element s in $\langle R \rangle$.

Lemma 3.1.2 *Let R be a nonempty subset of S, and let s be an element in $\langle R \rangle \setminus \{1\}$. Then there exist elements q in $\langle R \rangle$ and r in $R^* \cup R$ such that $s \in qr$ and $\ell_R(s) = \ell_R(q) + 1$.*

PROOF. We set $n := \ell_R(s)$. Then, by definition, $s \in (R^* \cup R)^n$. On the other hand, we are assuming that $1 \neq s$, so that $1 \leq n$. From $s \in (R^* \cup R)^n$ and $1 \leq n$ we obtain elements q in $(R^* \cup R)^{n-1}$ and r in $R^* \cup R$ such that $s \in qr$. From $q \in (R^* \cup R)^{n-1}$ we obtain $\ell_R(q) \leq n - 1$. From $n = \ell_R(s)$ and $s \in qr$ we obtain $n \leq \ell_R(q) + 1$.

Lemma 3.1.3 *Let s be an element in S, and let R be a nonempty subset of S with $R^* = R$. Then the following hold.*

(i) *Assume that $Rs \subseteq s\langle R \rangle$. Then $s \in N_S(\langle R \rangle)$.*

(ii) *We have $\langle R^s \rangle \subseteq \langle R \rangle^s$.*

PROOF. (i) Let us denote by Q the set of all elements q in $\langle R \rangle$ with $qs \not\subseteq s\langle R \rangle$. By way of contradiction, we assume that Q is not empty. We fix an element q in Q such that $\ell_R(q)$ is as small as possible.

Since $s \in s\langle R \rangle$, $1 \neq q$. Thus, by Lemma 3.1.2, there exist elements p in $\langle R \rangle$ and r in R such that $q \in pr$ and $\ell_R(q) = \ell_R(p) + 1$. Since $\ell_R(p) = \ell_R(q) - 1$, $p \notin Q$. Thus, as $p \in \langle R \rangle$, $ps \subseteq s\langle R \rangle$. Thus, as $rs \subseteq s\langle R \rangle$,

$$qs \subseteq prs \subseteq ps\langle R \rangle \subseteq s\langle R \rangle,$$

contradiction.

(ii) Let us denote by Q the set of all elements q in $\langle R^s \rangle$ with $sq \not\subseteq \langle R \rangle s$. By way of contradiction, we assume that Q is not empty. We fix an element q in Q such that $\ell_R(q)$ is as small as possible.

Since $s \in \langle R \rangle s$, $1 \neq q$. Thus, by Lemma 3.1.2, there exist elements p in $\langle R^s \rangle$ and r in R^s such that $q \in pr$ and $\ell_R(q) = \ell_R(p) + 1$. Since $\ell_R(p) = \ell_R(q) - 1$, $p \notin Q$. Thus, as $p \in \langle R^s \rangle$, $sp \subseteq \langle R \rangle s$.

On the other hand, as $r \in R^s$, $sr \subseteq Rs$. Thus,

$$sq \subseteq spr \subseteq \langle R \rangle sr \subseteq \langle R \rangle Rs = \langle R \rangle s,$$

contradiction.

Lemma 3.1.4 *Let s be an element in S, and let R be a nonempty subset of S such that $s^* Rs \subseteq \langle R \rangle$. Then $s \in K_S(\langle R \rangle)$.*

PROOF. Let us denote by Q the set of all elements q in $\langle R\rangle$ with $s^*qs \not\subseteq \langle R\rangle$. By way of contradiction, we assume that Q is not empty. We fix an element q in Q such that $\ell_R(q)$ is as small as possible.

We are assuming that $s^*Rs \subseteq \langle R\rangle$. Thus,

$$s^*s \subseteq s^*Rss^*R^*s \subseteq \langle R\rangle.$$

This shows that $1 \notin Q$. In particular, as $q \in Q$, $1 \neq q$. Thus, by Lemma 3.1.2, there exist elements p in $\langle R\rangle$ and r in $R^* \cup R$ such that $q \in pr$ and $\ell_R(q) = \ell_R(p) + 1$. Since $\ell_R(p) = \ell_R(q) - 1$, $p \notin Q$. Thus, as $p \in \langle R\rangle$, $s^*ps \subseteq \langle R\rangle$.

We are assuming that $s^*Rs \subseteq \langle R\rangle$. Thus, as $\langle R\rangle$ is closed, $(s^*Rs)^* \subseteq \langle R\rangle$. On the other hand, by Lemma 1.3.2(iii), $s^*R^*s = (s^*Rs)^*$. Thus, $s^*R^*s \subseteq \langle R\rangle$. It follows that $s^*(R^* \cup R)s \subseteq \langle R\rangle$. Thus, as $r \in R^* \cup R$,

$$s^*qs \subseteq s^*prs \subseteq s^*pss^*rs \subseteq \langle R\rangle,$$

contradiction.

Lemma 3.1.5 *Let R be a subset of S. Then we have the following.*

(i) *The set $R^* \cup R$ is thin if and only $\langle R\rangle$ is thin.*

(ii) *If R is symmetric, $\langle O_\vartheta(R)\rangle \subseteq O_\vartheta(\langle R\rangle)$.*

PROOF. (i) Since $R^* \cup R \subseteq \langle R\rangle$, $R^* \cup R$ is thin if $\langle R\rangle$ is thin.

Let us assume that $R^* \cup R$ is thin and that $\langle R\rangle$ is not thin. Assuming $\langle R\rangle$ not to be thin we find an element s in $\langle R\rangle$ such that s is not thin. Among the non-thin elements of $\langle R\rangle$ we fix s in such a way that $\ell_R(s)$ is as small as possible.

Since s is not thin, $1 \neq s$. Thus, by Lemma 3.1.2, there exist elements q in $\langle R\rangle$ and r in $R^* \cup R$ such that $s \in qr$ and $\ell_R(s) = \ell_R(q)+1$. Since $\ell_R(q) = \ell_R(s)-1$, the minimal choice of s forces q to be thin. Since $r \in R$ and R is assumed to be thin, r is thin. Thus, as $s \in qr$, s is thin; cf. Lemma 2.5.9(i), (ii).

(ii) We set $Q := O_\vartheta(R)$. Then $Q \subseteq O_\vartheta(\langle R\rangle)$. Thus, for each non-negative integer n, $Q^n \subseteq O_\vartheta(\langle R\rangle)$; cf. Lemma 2.5.9(ii). Thus, by Lemma 3.1.1(i), $\langle Q\rangle \subseteq O_\vartheta(\langle R\rangle)$.

Theorem 3.1.6 *Let p be a prime number, and let R be a subset of S. Assume that $n_r \leq p-1$ for each element r in $R^* \cup R$. Then, for each element s in $\langle R\rangle$, p does not divide n_s.*

PROOF. Let us denote by Q the set of all elements s in $\langle R\rangle$ such that p divides n_s. By way of contradiction, we assume that Q is not empty. We pick an element q in Q such that $\min \ell_R(Q) = \ell_R(q)$.

Since $q \in Q$, p divides n_q. Thus, $1 \neq q$. Thus, by Lemma 3.1.2, there exist elements t in $\langle R \rangle$ and r in $R^* \cup R$ such that $q \in tr$ and $\ell_R(q) = \ell_R(t) + 1$.

Since $r \in R^* \cup R$, $n_r \leq p-1$. Moreover, by the second equation of Lemma 1.1.3(iii), $a_{qr^*t} \leq n_r$. Thus, $a_{qr^*t} \leq p-1$. Now recall that, by Lemma 1.1.3(ii), $a_{trq}n_q = a_{qr^*t}n_t$. Moreover, as $q \in tr$, $1 \leq a_{trq}$. Thus, as p divides n_q and $a_{qr^*t} \leq p-1$, p must divide n_t.

On the other hand, as $\ell_R(t) = \ell_R(q) - 1$ and $\min \ell_R(Q) = \ell_R(q)$, $t \notin Q$. Thus, as $t \in \langle R \rangle$, p does not divide n_t. This contradiction finishes the proof.

Corollary 3.1.7 *Let R be a subset of S, and assume that each element in $R^* \cup R$ has valency 2. Then $\langle R \rangle$ is 2-valenced.*

PROOF. Let p be an odd prime number, and let s be an element in $\langle R \rangle$. Then, by Theorem 3.1.6, p does not divide n_s. Thus, n_s is a power of 2.

Let T be a closed subset of S such that $\{1\} \neq T$. We define $\Phi(T)$ to be the intersection of all maximal closed subsets of T. Note that $\Phi(T)$ is closed. We call $\Phi(T)$ the *Frattini subset* of T.

Let T and U be closed subsets of S satisfying $TU = S$ and $U \neq S$. Then U is called a *supplement* of T *in* S.

Lemma 3.1.8 *Let T be a closed subset of S such that $\{1\} \neq T$. Then the following hold.*

(i) *Let R be a subset of T which satisfies $\langle \Phi(T) \cup R \rangle = T$. Then, we have $\langle R \rangle = T$.*

(ii) *The closed subset $\Phi(T)$ does not have a supplement in T.*

PROOF. (i) Suppose, by way of contradiction, that $\langle R \rangle \neq T$. Then there exists a maximal closed subset U of T such that $\langle R \rangle \subseteq U$. However, by definition of the Frattini subset, we have $\Phi(T) \subseteq U$. Therefore, we have $\langle \Phi(T) \cup R \rangle \subseteq U$, contrary to $\langle \Phi(T) \cup R \rangle = T$.

(ii) This follows immediately from (i).

For any two elements p and q in S, we define

$$[p, q] := p^*q^*pq$$

and call this set the *commutator* of p and q.

It is obvious that, for any two elements p and q in S, $[p, q]^* = [q, p]$.

Let P and Q be nonempty subsets of S.

We define $[P, Q]$ to be the closed subset of S generated by the union of all sets $[p, q]$ with $p \in P$ and $q \in Q$. The set $[P, Q]$ is called the *commutator subset* of P and Q.

If there exists an element q in Q with $\{q\} = Q$, we write $[P, q]$ instead of $[P, Q]$. Similarly, if there exists an element p in P with $\{p\} = P$, we write $[p, Q]$ instead of $[P, Q]$.

Note that $[P, Q] = [Q, P]$.

Lemma 3.1.9 *Let R be a nonempty subset of S, and let T be a closed subset of S. Then the following hold.*

(i) *We have $[R, T]T = T[R, T]$.*

(ii) *Let Q be the union of the sets Tr^*TrT with $r \in R$. Then $\langle Q\rangle = [R, T]T$.*

PROOF. (i) Let r be an element in R, and let t be an element in T. We shall show first that $[r, t]T \subseteq T[R, T]$.

Let s be an element in $[r, t]T$. Then, as $[r, t] = r^*t^*rt$ and $t \in T$, $s \in r^*t^*rT$. Thus, there exists an element q in T such that $s \in r^*t^*rq$. It follows that

$$s \in r^*t^*qrr^*q^*rq = r^*t^*qr[r, q].$$

Thus, there exists an element p in t^*q such that $s \in r^*pr[r, q]$. From $p \in t^*q$ and $t^*q \subseteq T$ we obtain $p \in T$. Thus,

$$s \in pp^*r^*pr[r, q] = p[p, r][r, q] \subseteq T[r, p]^*[r, q] \subseteq T[R, T].$$

Since s has been chosen arbitrarily in $[r, t]T$, we have shown that, for any two elements r in R and t in T, $[r, t]T \subseteq T[R, T]$.

From what we have seen so far we also obtain

$$[r, t]^*T = t^*r^*trT = t^*r^*trt^*T \subseteq t^*[r, t^*]T \subseteq t^*T[R, T] = T[R, T]$$

for any two elements r in R and t in T. Thus, by Lemma 3.1.1(i), $[R, T]T \subseteq T[R, T]$.

With the help of Lemma 1.3.2(iii) one obtains from $[R, T]T \subseteq T[R, T]$ also that $T[R, T] \subseteq [R, T]T$. Thus, $[R, T]T = T[R, T]$.

(ii) Let s be an element in $\langle Q\rangle \setminus \{1\}$. Then, by Lemma 3.1.1(i), there exists a positive integer n such that $s \in Q^n$. Thus, there exist elements $r_1, \ldots, r_n$ in R such that $s \in Tr_1^*Tr_1T \cdots Tr_n^*Tr_nT$. Thus, there exist elements $t_1, \ldots, t_n$ in T such that

$$s \in Tr_1^*t_1^*r_1T \cdots Tr_n^*t_n^*r_nT = T[r_1, t_1]T \cdots T[r_n, t_n]T \subseteq [R, T]T;$$

cf. (i). Since s has been chosen arbitrarily in $\langle Q\rangle \setminus \{1\}$, we have shown that $\langle Q\rangle \subseteq [R, T]T$.

On the other hand, it is clear that, for any two elements r in R and t in T, $[r, t] = r^*t^*rt \subseteq \langle Q\rangle$. Thus, as $T \subseteq Q$, $[R, T]T \subseteq \langle Q\rangle$.

From $\langle Q\rangle \subseteq [R, T]T$ and $[R, T]T \subseteq \langle Q\rangle$ we obtain $\langle Q\rangle = [R, T]T$.

3.2 The Thin Residue

In this section, S is assumed to have finite valency. The letter T will always stand for a closed subset of S.

We define $O^{\vartheta}(T)$ to be the intersection of all strongly normal closed subsets of T and call it the *thin residue* of T.

Note that $O^{\vartheta}(T)$ is closed.

Theorem 3.2.1 *The following statements hold.*

(i) *The set $O^{\vartheta}(T)$ is strongly normal in T.*

(ii) *We have $O^{\vartheta}(T) = [T, 1]$.*

(iii) *For each closed subset U of S with $T \subseteq U$, we have $O^{\vartheta}(T) \subseteq O^{\vartheta}(U)$.*

PROOF. (i) This follows from Lemma 2.5.6(i).

(ii) Let us denote by R the union of all sets t^*t with $t \in T$. We have to show that $O^{\vartheta}(T) = \langle R \rangle$.

Let p and q be elements in T. First of all, we shall prove that $q^*p^*pq \subseteq \langle R \rangle$.

Let t be an element in pq. Then, by Lemma 1.3.3(i), $p \in tq^*$, whence $t^*pq \subseteq t^*tq^*q \subseteq \langle R \rangle$. Since t has been chosen arbitrarily in pq, this yields $q^*p^*pq \subseteq \langle R \rangle$; cf. Lemma 1.3.2(iii).

Since p and q have been chosen arbitrarily in T, we have shown $t^*Rt \subseteq \langle R \rangle$ for each element t in T. Thus, by Lemma 3.1.4, $\langle R \rangle$ is strongly normal in T. Thus, by definition, $O^{\vartheta}(T) \subseteq \langle R \rangle$.

In order to show that $\langle R \rangle \subseteq O^{\vartheta}(T)$ it suffices to show that $R \subseteq O^{\vartheta}(T)$. (This is because $O^{\vartheta}(T)$ is closed.)

Let t be an element in T. Then, as $1 \in O^{\vartheta}(T)$, $t^*t \subseteq t^*O^{\vartheta}(T)t$. On the other hand, we know from (i) that $O^{\vartheta}(T)$ is strongly normal in T, so that $t^*O^{\vartheta}(T)t \subseteq O^{\vartheta}(T)$. Thus, $t^*t \subseteq O^{\vartheta}(T)$. Since t has been chosen arbitrarily in T, we have shown that $R \subseteq O^{\vartheta}(T)$.

(iii) This is an immediate consequence of (ii).

Lemma 3.2.2 *Let U be a closed subset of S such that $T \subseteq N_S(U)$. Then we have $O^{\vartheta}(T)U = O^{\vartheta}(TU)U$.*

PROOF. From Theorem 3.2.1(iii) we know that $O^{\vartheta}(T) \subseteq O^{\vartheta}(TU)$, and from this we obtain $O^{\vartheta}(T)U \subseteq O^{\vartheta}(TU)U$.

By Theorem 3.2.1(i), $O^{\vartheta}(T)$ is strongly normal in T. Moreover, we are assuming that $T \subseteq N_S(U)$. Thus, by Lemma 2.5.7, $O^{\vartheta}(T)U$ is strongly normal in TU. Thus, the definition of $O^{\vartheta}(TU)$ gives us $O^{\vartheta}(TU) \subseteq O^{\vartheta}(T)U$, and from this we obtain $O^{\vartheta}(TU)U \subseteq O^{\vartheta}(T)U$.

Lemma 3.2.3 *Let U be a closed subset of S such that TU is closed and $O^\vartheta(U) \subseteq T$. Then $O^\vartheta(TU)$ is the intersection of all strongly normal closed subsets of T which contain $O^\vartheta(U)$.*

PROOF. Let us denote by $\mathcal{V}$ the set of all strongly normal closed subsets of T which contain $O^\vartheta(U)$, and let us define W to be the intersection of all elements of $\mathcal{V}$. We have to show that $O^\vartheta(TU) = W$.

Let us first show that $O^\vartheta(TU) \subseteq W$. In order to do so we pick an element s in TU and an element V in $\mathcal{V}$. Since TU is assumed to be closed, $TU = UT$; cf. Lemma 2.1.1. Thus, as $s \in TU$, $s \in UT$. Thus, there exist elements t in T and u in U such that $s \in ut$. Since $u \in U$, $u^*u \subseteq O^\vartheta(U)$; cf. Theorem 3.2.1(ii). Since $V \in \mathcal{V}$, $t^*Vt \subseteq V$ and $O^\vartheta(U) \subseteq V$. Thus,

$$s^*s \subseteq t^*u^*ut \subseteq t^*O^\vartheta(U)t \subseteq V.$$

Now, as s has been chosen arbitrarily in TU, we conclude that $O^\vartheta(TU) \subseteq V$; cf. Theorem 3.2.1(ii). But also V has been chosen arbitrarily in $\mathcal{V}$. Therefore, we have shown that $O^\vartheta(TU) \subseteq W$.

Let us now show that, conversely, $W \subseteq O^\vartheta(TU)$.

From $O^\vartheta(TU) \subseteq W$ and $W \subseteq T$ we obtain $O^\vartheta(TU) \subseteq T$. In particular, $O^\vartheta(TU)$ is strongly normal in T. On the other hand, we know from Theorem 3.2.1(iii) that $O^\vartheta(U) \subseteq O^\vartheta(TU)$. Thus, $O^\vartheta(TU) \in \mathcal{V}$, so that, by definition, $W \subseteq O^\vartheta(TU)$.

Corollary 3.2.4 *Let U be a thin closed subset of S such that TU is closed. Then we have $O^\vartheta(T) = O^\vartheta(TU)$.*

PROOF. Since U is assumed to be thin, $\{1\} = O^\vartheta(U)$. Thus, the claim follows from Lemma 3.2.3.

Recall that a closed subset U of S is called a supplement of T in S if $TU = S$ and $U \neq S$. Referring to this terminology we obtain from Corollary 3.2.4 that, if $O_\vartheta(S)$ has a supplement in S, $O^\vartheta(S) \neq S$.

Lemma 3.2.5 *Assume that there exists a prime number p such that T is p-valenced. Then, if $\{1\} = O_\vartheta(O^\vartheta(T))$, $O^\vartheta(T)O_\vartheta(T) = T$.*

PROOF. Let t be an element in T. Then, by Theorem 3.2.1(ii), $t^*t \subseteq O^\vartheta(T)$. Thus, by Lemma 1.4.4(ii), $n_{O^\vartheta(T)} = n_{tO^\vartheta(T)}$.

We are assuming that T is p-valenced. Thus, assuming that $\{1\} = O_\vartheta(O^\vartheta(T))$, we obtain that p divides $n_{O^\vartheta(T)} - 1$. Thus, as $n_{O^\vartheta(T)} = n_{tO^\vartheta(T)}$, the set $O_\vartheta(T) \cap tO^\vartheta(T)$ is not empty. Thus, by Lemma 1.3.3(i), $t \in O^\vartheta(T)O_\vartheta(T)$.

We set $(O^\vartheta)^0(T) := T$. For each positive integer n, we inductively define

$$(O^\vartheta)^n(T) := O^\vartheta((O^\vartheta)^{n-1}(T)).$$

Note that $(O^\vartheta)^n(T)$ is closed for each non-negative integer n. Note also that, for each positive integer n,

$$(O^\vartheta)^n(T) \subseteq (O^\vartheta)^{n-1}(T).$$

Here is a generalization of Theorem 3.2.1(iii).

Lemma 3.2.6 *Let n be a non-negative integer, and let U be a closed subset of S such that $T \subseteq U$. Then $(O^\vartheta)^n(T) \subseteq (O^\vartheta)^n(U)$.*

PROOF. Our lemma is certainly true if $n = 0$. By induction, we may assume that $(O^\vartheta)^{n-1}(T) \subseteq (O^\vartheta)^{n-1}(U)$. Then, by Theorem 3.2.1(iii),

$$(O^\vartheta)^n(T) = O^\vartheta((O^\vartheta)^{n-1}(T)) \subseteq O^\vartheta((O^\vartheta)^{n-1}(U)) = (O^\vartheta)^n(U),$$

and that proves the lemma.

The first part of the following lemma generalizes Lemma 3.2.2.

Lemma 3.2.7 *Let n be a non-negative integer, and let U be a closed subset of S such that $T \subseteq N_S(U)$. Then the following hold.*

(i) *We have $(O^\vartheta)^n(T)U = (O^\vartheta)^n(TU)U$.*

(ii) *If $O^\vartheta(TU)U = TU$, $(O^\vartheta)^n(TU)U = TU$.*

PROOF. (i) There is nothing to show if $n = 0$. Therefore, we assume that $1 \leq n$.

Assume, by way of contradiction, that there exists a non-negative integer violating the claim. Then $1 \leq n$ and

$$(O^\vartheta)^{n-1}(T)U = (O^\vartheta)^{n-1}(TU)U.$$

Thus,

$$O^\vartheta((O^\vartheta)^{n-1}(T)U)U = O^\vartheta((O^\vartheta)^{n-1}(TU)U)U.$$

On the other hand, applying Lemma 3.2.2 to $(O^\vartheta)^{n-1}(T)$ in place of T, we obtain

$$(O^\vartheta)^n(T)U = O^\vartheta((O^\vartheta)^{n-1}(T))U = O^\vartheta((O^\vartheta)^{n-1}(T)U)U.$$

Finally, applying Lemma 3.2.2 to $(O^\vartheta)^{n-1}(TU)$ in place of T, we obtain

$$(O^\vartheta)^n(TU)U = O^\vartheta((O^\vartheta)^{n-1}(TU))U = O^\vartheta((O^\vartheta)^{n-1}(TU)U)U.$$

Thus, we have $(O^\vartheta)^n(T)U = (O^\vartheta)^n(TU)U$.

(ii) We are assuming that $T \subseteq N_S(U)$. Thus, we obtain from Lemma 2.5.1(ii) (together with Lemma 2.5.2(iii)) that $TU \subseteq N_S(U)$. In particular, $(O^\vartheta)^{n-1}(TU) \subseteq N_S(U)$. Thus, by Lemma 3.2.2,

$$O^\vartheta((O^\vartheta)^{n-1}(TU))U = O^\vartheta((O^\vartheta)^{n-1}(TU)U)U.$$

On the other hand, we may assume that $(O^\vartheta)^{n-1}(TU)U = TU$ and we are assuming that $O^\vartheta(TU)U = TU$. Thus,

$$O^\vartheta((O^\vartheta)^{n-1}(TU)U)U = TU.$$

Thus, as $O^\vartheta((O^\vartheta)^{n-1}(TU)) = (O^\vartheta)^n(TU)$, $(O^\vartheta)^n(TU)U = TU$.

3.3 Elements of Valency 2

In this section, we assume S to have finite valency. We shall first look at closed subsets of S which are generated by a single elements of valency 2.

Lemma 3.3.1 *Let s be a symmetric element of S, and assume that s has valency 2. Then each element of $\langle s\rangle$ has valency at most 2.*

Proof. Let us write ℓ instead of $\ell_{\{s\}}$ and Q to denote the set of all elements q in $\langle s\rangle$ with $3 \leq n_q$. By way of contradiction, we assume that Q is not empty. We fix an element q in Q such that $\min \ell(Q) = \ell(q)$.

Since $q \in Q$, $3 \leq n_q$. Thus, $1 \neq q$. Thus, as $q \in \langle s\rangle$, there exists an element p in $\langle s\rangle$ such that $q \in ps$ and $\ell(q) = \ell(p) + 1$; cf. Lemma 3.1.2.

Since $\ell(p) = \ell(q) - 1$ and $\min \ell(Q) = \ell(q)$, $p \notin Q$. Thus, as $p \in \langle s\rangle$, $n_p \leq 2$. Now recall that, by Lemma 1.1.3(ii), $a_{psq}n_q = a_{qs^*p}n_p$. Moreover, as $q \in ps$, $1 \leq a_{psq}$. Thus, as $n_p \leq 2$ and $3 \leq n_q$, $2 \leq a_{qs^*p}$.

We are assuming that $n_s = 2$. Thus, as $3 \leq n_q$, $q \neq s$. Thus, as $q \in ps$, $1 \neq p$. Thus, as $p \in \langle s\rangle$, there exists an element t in $\langle s\rangle$ such that $p \in ts$ and $\ell(p) = \ell(t) + 1$; cf. Lemma 3.1.2.

Since $p \in ts$ and $s^* = s$, $p \in ts^*$. Thus, $1 \leq a_{ts^*p}$. Since $\ell(p) = \ell(t) + 1$ and $\ell(q) = \ell(p) + 1$, $t \neq q$. Thus, as $2 \leq a_{qs^*p}$, $3 \leq a_{ts^*p} + a_{qs^*p}$. On the other hand, by the second equation of Lemma 1.1.3(iii), $a_{qs^*p} + a_{ts^*p} \leq n_s$, so that $3 \leq n_s$. This contradiction finishes the proof of the lemma.

Lemma 3.3.2 *Let s be a symmetric element of S, assume that s has valency 2, and let p and q be elements in $\langle s\rangle$ with $\ell_{\{s\}}(p) = \ell_{\{s\}}(q)$. Then $p = q$.*

Proof. Let us write ℓ instead of $\ell_{\{s\}}$ and Q to denote the set of all elements q in $\langle s\rangle$ such that there exists an element p in $\langle s\rangle$ with $\ell(p) = \ell(q)$ and $p \neq q$. By way of contradiction, we assume that Q is not empty, and we fix an element q in Q such that $\min \ell(Q) = \ell(q)$.

Since $q \in Q$, there exists an element p in $\langle s \rangle$ such that $\ell(p) = \ell(q)$ and $p \neq q$. Clearly, $1 \neq p$. Thus, as $p \in \langle s \rangle$, there exists an element t in $\langle s \rangle$ such that $p \in ts$ and $\ell(p) = \ell(t) + 1$; cf. Lemma 3.1.2. Similarly, as $1 \neq q$ and $q \in \langle s \rangle$, there exists an element u in $\langle s \rangle$ such that $q \in us$ and $\ell(q) = \ell(u) + 1$.

From $q \in us$ we obtain $u \in qs^*$; cf. Lemma 1.3.3(i). Thus, by definition, $1 \leq a_{qs^*u}$.

From $\ell(p) = \ell(t) + 1$, $\ell(q) = \ell(u) + 1$, and $\ell(p) = \ell(q)$ we obtain $\ell(t) = \ell(u)$. Thus, by the (minimal) choice of q, $t = u$. Thus, as $p \in ts$, $p \in us$. Thus, by Lemma 1.3.3(i), $u \in ps^*$. It follows that $1 \leq a_{ps^*u}$.

Since $1 \neq u$ and $u \in \langle s \rangle$, there exists an element r in $\langle s \rangle$ such that $u \in rs$ and $\ell(u) = \ell(r) + 1$; cf. Lemma 3.1.2. From $u \in rs$ and $s^* = s$ we obtain $u \in rs^*$. Thus, by definition, $1 \leq a_{rs^*u}$.

Note finally that, by the second equation of Lemma 1.1.3(iii),

$$a_{rs^*u} + a_{ps^*u} + a_{qs^*u} \leq n_s.$$

Thus, as $p \neq q$ and $n_s = 2$, we must have that $r = p$ or $r = q$. However, this contradicts $\ell(r) \leq \ell(p) - 1 = \ell(q) - 1$.

Recall that, for each element r in $\langle s \rangle$, $\ell_{\{s\}}(r^*) = \ell_{\{s\}}(r)$. Thus, Lemma 3.3.2 says, in particular, that all elements in $\langle s \rangle$ are symmetric.

Let us now look at the structure of closed subsets of S generated by an arbitrary (not necessarily symmetric) element of valency 2.

Lemma 3.3.3 *Let s be an element of S, assume that s has valency 2 and that $\{1\} = O_\vartheta(\langle s \rangle)$. Then s is symmetric and $n_{\langle s \rangle}$ is odd.*

PROOF. From $n_s = 2$ we obtain that $\langle s \rangle$ is 2-valenced; cf. Corollary 3.1.7. Thus, assuming that $\{1\} = O_\vartheta(\langle s \rangle)$, we obtain that $n_{\langle s \rangle}$ is odd.

From Lemma 1.5.6(i) we obtain an element r in $s^*s \backslash \{1\}$ such that $\{1, r\} = s^*s$. From $r \in s^*s$ we obtain $\langle r \rangle \subseteq \langle s \rangle$. Thus, as $n_{\langle s \rangle}$ is odd, $n_{\langle r \rangle}$ is odd; cf. Lemma 2.3.6(ii).

Since $s^*s = \{1, r\} \subseteq \langle r \rangle$, $n_{s\langle r \rangle} = n_{\langle r \rangle}$; cf. Lemma 1.4.4(ii). Thus, as $n_{\langle r \rangle}$ is odd, $n_{s\langle r \rangle}$ is odd. Thus, as $\langle s \rangle$ is 2-valenced, we obtain from $s\langle r \rangle \subseteq \langle s \rangle$ and $\{1\} = O_\vartheta(\langle s \rangle)$ that $1 \in s\langle r \rangle$. Thus, by Lemma 2.1.4, $s \in \langle r \rangle$. Thus, by Lemma 1.2.3, s is symmetric.

Lemma 3.3.4 *Let s be an element of S, assume that s has valency 2 and that $O^\vartheta(\langle s \rangle)$ has odd valency. Then $O^\vartheta(\langle s \rangle)O_\vartheta(\langle s \rangle) = \langle s \rangle$.*

PROOF. We are assuming that $n_s = 2$. Thus, by Lemma 1.5.6(i), there exists an element r in $\langle s \rangle \setminus \{1\}$ such that $n_r \leq 2$, $r^* = r$, and $\{1, r\} = s^*s$.

Let v be an element in X, let w be an element in vr. Since $w \in vr$ and $r \in s^*s$, $w \in vs^*s$. Thus, there exists an element x in vs^* such that $w \in xs$.

We are assuming that $O^{\vartheta}(\langle s\rangle)$ has odd valency. Thus, as $r \in s^*s \subseteq O^{\vartheta}(\langle s\rangle)$, $\langle r\rangle$ has odd valency; cf. Lemma 2.3.6(ii). Thus, as $r^* = r$ and $n_r \leq 2$, there exists an element q in $\langle r\rangle$ such that $n_q = 2$ and $r \in q^*q$.

Since $w \in vr$ and $r \in q^*q$, $w \in vq^*q$. Thus, there exists an element y in vq^* such that $w \in yq$. Let us denote by t the uniquely determined element in S satisfying $y \in xt$. Then $a_{sq^*t} = 2$.

We wish to show that $t \in O_{\vartheta}(\langle s\rangle)$. In order to do so we pick an element y' in xt.

Since $a_{sq^*t} = 2$, $|xs \cap y'q| = 2$. On the other hand, $xs = \{v, w\}$. Thus, $y' \in vq^* \cap wq^*$. Thus, as $\langle r\rangle$ has odd valency, $y' = y$.

Since y' has been chosen arbitrarily in xt, $xt = \{y\}$. It follows that $t \in O_{\vartheta}(S)$. Thus, as $s \in tq$ and $q \in \langle r\rangle \subseteq O^{\vartheta}(\langle s\rangle)$, $s \in O^{\vartheta}(\langle s\rangle)O_{\vartheta}(\langle s\rangle)$.

On the other hand, we know from Theorem 3.2.1(i) and Lemma 2.1.1 that $O^{\vartheta}(\langle s\rangle)O_{\vartheta}(\langle s\rangle)$ is closed. Thus, $O^{\vartheta}(\langle s\rangle)O_{\vartheta}(\langle s\rangle) = \langle s\rangle$.

Let us now look at the structure of closed subsets of S generated by an arbitrary set of elements of valency 2.

Theorem 3.3.5 *Let T be a closed subset of S generated by a set of elements of valency 2. Then, we have the following.*

(i) *If $O^{\vartheta}(T)$ has odd valency, $O^{\vartheta}(T)O_{\vartheta}(T) = T$.*

(ii) *If $\{1\} = O_{\vartheta}(T)$, $O^{\vartheta}(T) = T$.*

PROOF. (i) We are assuming that T contains a subset R of elements of valency 2 such that $\langle R\rangle = T$.

Let r be an element in R. Then, as $R \subseteq T$, $O^{\vartheta}(\langle r\rangle) \subseteq O^{\vartheta}(T)$; cf. Theorem 3.2.1(iii). Thus, as we are assuming $O^{\vartheta}(T)$ to have odd valency, $O^{\vartheta}(\langle r\rangle)$ has odd valency; cf. Lemma 2.3.6(ii). Thus, by Lemma 3.3.4,

$$O^{\vartheta}(\langle r\rangle)O_{\vartheta}(\langle r\rangle) = \langle r\rangle.$$

From $O^{\vartheta}(\langle r\rangle) \subseteq O^{\vartheta}(T)$ and $O_{\vartheta}(\langle r\rangle) \subseteq O_{\vartheta}(T)$ we also obtain

$$O^{\vartheta}(\langle r\rangle)O_{\vartheta}(\langle r\rangle) \subseteq O^{\vartheta}(T)O_{\vartheta}(T).$$

Thus, $r \in O^{\vartheta}(T)O_{\vartheta}(T)$.

Since r has been chosen arbitrarily in R, we have shown that

$$R \subseteq O^{\vartheta}(T)O_{\vartheta}(T).$$

Thus, as $O^{\vartheta}(T)O_{\vartheta}(T)$ is closed, the desired equation follows from $\langle R\rangle = T$.

(ii) We are assuming that T is generated by a set of elements all of which have valency 2. Thus, by Corollary 3.1.7, T is 2-valenced.

Since T is 2-valenced, our hypothesis that $\{1\} = O_\vartheta(T)$ forces T to have odd valency. Thus, by Lemma 2.3.6(ii), $O^\vartheta(T)$ has odd valency, so that our claim follows from (i).

3.4 Closed Subsets Generated by Involutions

In this section, the letter L stands for a set of involutions of S. We shall look at $\langle L\rangle$. Instead of ℓ_L we shall write ℓ.

Lemma 3.4.1 *Let n be a positive integer, and let $s_0, \ldots, s_n$ be elements in $\langle L\rangle$ such that, for each element i in $\{1,\ldots,n\}$, $s_i \in s_{i-1}L$.*
Assume that $\ell(s_0)+n = \ell(s_n)$. Then, for each element i in $\{0,\ldots,n\}$, $\ell(s_0)+i = \ell(s_i)$.

Proof. Our first claim is that, for each element i in $\{0,\ldots,n\}$, $s_i \in s_0L^i$.

The claim is obviously true if $i = 0$. Therefore, we assume that $i \in \{1,\ldots,n\}$. Then, assuming that $s_{i-1} \in s_0L^{i-1}$ we obtain from $s_i \in s_{i-1}L$ that $s_i \in s_0L^{i-1}L = s_0L^i$.

This shows that, for each element i in $\{0,\ldots,n\}$, $\ell(s_i) \leq \ell(s_0) + i$.

Let us now prove that, for each element i in $\{0,\ldots,n\}$, $s_n \in s_iL^{n-i}$.

The claim is obviously true if $i = n$. Therefore, we assume that $i \in \{0,\ldots,n-1\}$. Then, assuming that $s_n \in s_{i+1}L^{n-(i+1)}$ we obtain from $s_{i+1} \in s_iL$ that $s_n \in s_iLL^{n-(i+1)} = s_iL^{n-i}$.

This shows that $\ell(s_n) \leq \ell(s_i) + n - i$ for each element i in $\{0,\ldots,n\}$. Thus, as we are assuming that $\ell(s_0)+n = \ell(s_n)$, we obtain $\ell(s_0)+i \leq \ell(s_i)$ for each element i in $\{0,\ldots,n\}$.

From the definition of ℓ we obtain

$$\ell(s) \leq \ell(p) + \ell(q)$$

for any three elements p, q, and s in $\langle L\rangle$ with $s \in pq$, . The following lemma focuses on the case where equality holds.

Lemma 3.4.2 *Let p, q, and r be elements in $\langle L\rangle$ satisfying $r \in pq$ and $\ell(r) = \ell(p) + \ell(q)$. Let t and u be elements in $\langle L\rangle$ satisfying $q \in tu$ and $\ell(q) = \ell(t) + \ell(u)$. Then, there exists an element s in pt such that $r \in su$, $\ell(s) = \ell(p) + \ell(t)$, and $\ell(r) = \ell(s) + \ell(u)$.*

Proof. Since $r \in pq$ and $q \in tu$, $r \in ptu$. Thus, there exists an element s in pt such that $r \in su$.

Since $s \in pt$, $\ell(s) \leq \ell(p) + \ell(t)$. Since $r \in su$, $\ell(r) \leq \ell(s) + \ell(u)$. Thus, as we are assuming that $\ell(r) = \ell(p) + \ell(q)$ and that $\ell(q) = \ell(t) + \ell(u)$, we have

$$\ell(r) \leq \ell(s) + \ell(u) \leq \ell(p) + \ell(t) + \ell(u) = \ell(r).$$

It follows that $\ell(s) = \ell(p) + \ell(t)$ and $\ell(r) = \ell(s) + \ell(u)$.

Let q be an element in $\langle L \rangle$. We define $S_{-1}(q, L)$ to be the set of all elements r in $\langle L \rangle$ such that there exists an element p in $\langle L \rangle$ with $r \in pq$ and $\ell(r) = \ell(p) + \ell(q)$. By $S_1(q, L)$ we shall denote the set of all elements p in $\langle L \rangle$ such that there exists an element r in pq with $\ell(r) = \ell(p) + \ell(q)$.

As a consequence of the remark right before Lemma 3.4.2 we obtain that, in both of these definitions, the equation $\ell(r) = \ell(p) + \ell(q)$ can be replaced with the condition that $\ell(p) + \ell(q) \leq \ell(r)$.

Let s be an element in $\langle L \rangle$. For the remainder of this section, we shall write $S_{-1}(s)$ instead of $S_{-1}(s, L)$ and $S_1(s)$ instead of $S_1(s, L)$.

Note that, for each element s in $\langle L \rangle \setminus \{1\}$, $S_{-1}(s) \cap S_1(s)$ is empty. Note also that, for each element s in $\langle L \rangle \setminus \{1\}$, there exists an element l in L such that $s \in S_{-1}(l)$; cf. Lemma 3.1.2. Note, finally, that, for each element s in $\langle L \rangle$, $s \in S_{-1}(s)$ and $1 \in S_1(s)$.

Lemma 3.4.3 *For each element s in $\langle L \rangle$, the following hold.*

(i) *We have $S_{-1}(s) \subseteq S_1(s)s$.*

(ii) *We have $S_1(s) \subseteq S_{-1}(s)s^*$.*

PROOF. (i) Let q be an element in $S_{-1}(s)$. Then, by definition, there exists an element p in $\langle L \rangle$ such that $q \in ps$ and $\ell(q) = \ell(p) + \ell(s)$. Thus, by definition, $p \in S_1(s)$. Thus, as $q \in ps$, $q \in S_1(s)s$.

(ii) Let p be an element in $S_1(s)$. Then, by definition, there exists an element q in $\langle L \rangle$ such that $q \in ps$ and $\ell(q) = \ell(p) + \ell(s)$. Thus, by definition, $q \in S_{-1}(s)$. On the other hand, as $q \in ps$, $p \in qs^*$; cf. Lemma 1.3.3(i). Thus, $p \in S_{-1}(s)s^*$.

Lemma 3.4.4 *For any two elements p and q in $\langle L \rangle$, we have the following.*

(i) *If $p \in S_1(q)$, $q^* \in S_1(p^*)$.*

(ii) *If $S_{-1}(p) \cap S_1(q)$ is not empty, $p \in S_1(q)$.*

(iii) *If $q \in S_{-1}(p)$, $S_{-1}(q) \subseteq S_{-1}(p)$.*

(iv) *If $q \in S_{-1}(p)$, $S_1(q^*) \subseteq S_1(p^*)$.*

PROOF. (i) This follows from $\ell(p^*) = \ell(p)$ and $\ell(q^*) = \ell(q)$.

(ii) Let r be an element in $S_{-1}(p) \cap S_1(q)$. Since $r \in S_{-1}(p)$, there exists an element t in $\langle L \rangle$ such that $r \in tp$ and $\ell(r) = \ell(t) + \ell(p)$. Since $r \in S_1(q)$, there exists an element s in $\langle L \rangle$ such that $s \in rq$ and $\ell(s) = \ell(r) + \ell(q)$. Thus, by Lemma 3.4.2, there exists an element u in pq such that $s \in tu$, $\ell(u) = \ell(p) + \ell(q)$, and $\ell(s) = \ell(t) + \ell(u)$.

From $u \in pq$ and $\ell(u) = \ell(p) + \ell(q)$ we obtain $p \in S_1(q)$.

(iii) Let us assume that $q \in S_{-1}(p)$, and let us pick an element s in $S_{-1}(q)$. We shall show that $s \in S_{-1}(p)$.

Since $s \in S_{-1}(q)$, there exists an element u in $\langle L \rangle$ such that $s \in uq$ and $\ell(s) = \ell(u) + \ell(q)$. Since we are assuming that $q \in S_{-1}(p)$, there exists an element t in $\langle L \rangle$ such that $q \in tp$ and $\ell(q) = \ell(t) + \ell(p)$. Now, by Lemma 3.4.2, there exists an element r in ut such that $s \in rp$, $\ell(r) = \ell(u) + \ell(t)$, and $\ell(s) = \ell(r) + \ell(p)$.

From $s \in rp$ and $\ell(s) = \ell(r) + \ell(p)$ we obtain $s \in S_{-1}(p)$.

(iv) Let us assume that $q \in S_{-1}(p)$, and let us pick an element s in $S_1(q^*)$. We shall show that $s \in S_1(p^*)$.

Since $s \in S_1(q^*)$, $q \in S_1(s^*)$; cf. (i). Thus, as we are assuming that $q \in S_{-1}(p)$, $p \in S_1(s^*)$; cf. (ii). Thus, by (i), $s \in S_1(p^*)$.

For each subset R of $\langle L \rangle$, we define $S_{-1}(R)$ to be the intersection of the sets $S_{-1}(r)$ with $r \in \{1\} \cup R$. (Note that $S_{-1}(1) = \langle L \rangle$.)

Recall that, for each element s in $\langle L \rangle$, $s \in S_{-1}(s)$ and $1 \in S_1(s)$. The following lemma deals with the case where $s \in S_{-1}(L)$.

Lemma 3.4.5 *Let s be an element in $S_{-1}(L)$. Then the following hold.*

(i) *We have $\{1\} = S_1(s^*)$.*

(ii) *We have $\{s^*\} = S_{-1}(s^*)$.*

PROOF. (i) It is clear that $1 \in S_1(s^*)$. Let us assume, by way of contradiction, that $S_1(s^*)$ contains an element q with $1 \neq q$.

From $q \in S_1(s^*)$ we obtain an element t in qs^* such that $\ell(t) = \ell(q) + \ell(s^*)$.

From $1 \neq q$ we obtain elements p in $\langle L \rangle$ and l in L such that $q \in pl$ and $\ell(q) = \ell(p) + 1$; cf. Lemma 3.1.2.

From $s \in S_{-1}(L)$ we obtain an element r in $\langle L \rangle$ such that $s \in rl$ and $\ell(s) = \ell(r) + 1$.

From $t \in qs^*$, $q \in pl$, and $s \in rl$ we obtain

$$t \in pllr^* \subseteq pr^* \cup plr^*.$$

In particular, $\ell(t) \leq \ell(p) + 1 + \ell(r^*)$.

On the other hand, as $\ell(t) = \ell(q) + \ell(s^*)$, $\ell(q) = \ell(p) + 1$, and $\ell(s) = \ell(r) + 1$ we conclude that $\ell(t) = \ell(p) + 2 + \ell(r)$, contradiction.

(ii) This follows from Lemma 3.4.3(i) together with (i).

For each subset R of $\langle L \rangle$, we define $S_1(R)$ to be the intersection of the sets $S_1(r)$ with $r \in \{1\} \cup R$. (Note that $S_1(1) = \langle L \rangle$.)

Lemma 3.4.6 *Let p and q be elements in $\langle L\rangle$ such that $p \in S_1(q)$. Then the following hold.*

(i) *We have $S_1(pq) \subseteq S_1(p)$.*

(ii) *For each element r in $S_1(pq)$, there exists an element s in rp such that $s \in S_1(q)$.*

PROOF. We are assuming that $p \in S_1(q)$. Thus, by definition, there exists an element u in pq such that $\ell(u) = \ell(p) + \ell(q)$.

(i) Let s be an element in $S_1(pq)$. Then, as $u \in pq$, $s \in S_1(u)$. Thus, there exists an element r in su such that $\ell(r) = \ell(s) + \ell(u)$. Thus, by Lemma 3.4.2, there exists an element t in sp such that $\ell(t) = \ell(s) + \ell(p)$. It follows that $s \in S_1(p)$.

(ii) Let r be an element in $S_1(pq)$. Then, as $u \in pq$, $r \in S_1(u)$. Thus, there exists an element t in ru such that $\ell(t) = \ell(r) + \ell(u)$. Thus, by Lemma 3.4.2, there exists an element s in rp such that $t \in sq$ and $\ell(t) = \ell(s) + \ell(q)$. From $t \in sq$ and $\ell(t) = \ell(s) + \ell(q)$ we obtain $s \in S_1(q)$.

We say that L satisfies the *exchange condition* if, for any three elements h, k in L and s in $S_1(k)$, $h \in S_1(s)$ implies $hs \subseteq sk \cup S_1(k)$.

We shall investigate the exchange condition in Section 3.6 and, in more detail, in the last two chapters of this monograph. For the last two results of this section, we shall now look at a slightly different condition. We assume that, for any three elements h, k in L and s in $S_1(k)$, $h \in S_1(s)$ implies $hs \subseteq S_{-1}(k) \cup S_1(k)$.

Lemma 3.4.7 *For each element l in L, $S_{-1}(l) \cup S_1(l) = \langle L\rangle$.*

PROOF. Let us assume the claim to be false. Among the elements in $\langle L\rangle$ not contained in $S_{-1}(l) \cup S_1(l)$ we pick s in such a way that $\ell(s)$ is as small as possible.

Since $1 \in S_1(l)$ and $s \notin S_1(l)$, $1 \neq s$. Thus, by Lemma 3.1.2, there exist elements k in L and r in $\langle L\rangle$ such that $s \in kr$ and $\ell(s) = 1 + \ell(r)$. Since $\ell(s) = 1 + \ell(r)$, the (minimal) choice of s yields $r \in S_{-1}(l) \cup S_1(l)$.

Since $s \in kr$ and $\ell(s) = 1 + \ell(r)$, $s \in S_{-1}(r)$. Thus, as $s \notin S_{-1}(l)$, $r \notin S_{-1}(l)$; cf. Lemma 3.4.4(iii). Thus, as $r \in S_{-1}(l) \cup S_1(l)$, $r \in S_1(l)$.

Note also that, as $s \in kr$ and $\ell(s) = 1 + \ell(r)$, $k \in S_1(r)$.

From $k \in S_1(r)$ and $r \in S_1(l)$ (together with our hypothesis) we obtain $kr \subseteq S_{-1}(l) \cup S_1(l)$. Thus, as $s \in kr$, $s \in S_{-1}(l) \cup S_1(l)$. This contradiction concludes the proof of our lemma.

Lemma 3.4.8 *For each subset K of L, $S_1(K)\langle K\rangle = \langle L\rangle$.*

PROOF. Let us assume that $S_1(K)\langle K\rangle \neq \langle L\rangle$. Among the elements in $\langle L\rangle$ not contained in $S_1(K)\langle K\rangle$ we pick s such that $\ell(s)$ is as small as possible.

Since $s \notin S_1(K)\langle K\rangle$, $s \notin S_1(K)$. Thus, there exists an element k in K such that $s \notin S_1(k)$. Thus, by Lemma 3.4.7, $s \in S_{-1}(k)$. Thus, there exists an element r in $\langle L\rangle$ such that $s \in rk$ and $\ell(s) = \ell(r) + 1$.

Since $\ell(s) = \ell(r) + 1$, the (minimal) choice of s yields $r \in S_1(K)\langle K\rangle$. Thus, as $s \in rk$ and $k \in K$, $s \in S_1(K)\langle K\rangle$, contradiction.

3.5 Basic Results on Constrained Sets of Involutions

In this section, the letter L stands for a set of involutions of S. For each element s in $\langle L\rangle$, we write $S_1(s)$ instead of $S_1(s, L)$.

The set L is called *constrained* if, for any two elements q in $\langle L\rangle$ and p in $S_1(q)$, $1 = |pq|$.

For the remainder of this section, we assume L to be constrained. Instead of ℓ_L we shall write ℓ.

Lemma 3.5.1 *For any two elements q and r in $\langle L\rangle$, there exists at most one element p in $\langle L\rangle$ such that $r \in pq$ and $\ell(r) = \ell(p) + \ell(q)$.*

Proof. Let us fix an element r in $\langle L\rangle$. We shall denote by Q the set of all elements q in $\langle L\rangle$ such that there exist elements p and p' in $\langle L\rangle$ with $r \in pq$, $r \in p'q$, $\ell(r) = \ell(p) + \ell(q)$, $\ell(p') = \ell(p)$, and $p' \neq p$.

By way of contradiction, we assume that Q is not empty. We pick an element q in Q, and we do this in such a way that $\ell(q)$ is as small as possible.

Since $1 \notin Q$ and $q \in Q$, $1 \neq q$. Thus, by Lemma 3.1.2, there exist elements l in L and u in $\langle L\rangle$ such that $q \in lu$ and $\ell(q) = 1 + \ell(u)$. Thus, as $r \in pq$ and $\ell(r) = \ell(p) + \ell(q)$, there exists an element t in pl such that $r \in tu$, $\ell(t) = \ell(p) + 1$, and $\ell(r) = \ell(t) + \ell(u)$; cf. Lemma 3.4.2.

Similarly, we find an element t' in $p'l$ such that $r \in t'u$, $\ell(t') = \ell(p') + 1$, and $\ell(r) = \ell(t') + \ell(u)$.

Since $\ell(q) = 1 + \ell(u)$, the (minimal) choice of q yields $u \notin Q$. Thus, as $r \in tu$, $r \in t'u$, $\ell(r) = \ell(t) + \ell(u)$, and $\ell(r) = \ell(t') + \ell(u)$, $t' = t$.

We are assuming that L is constrained. Thus, as $t \in pl$ and $\ell(t) = \ell(p) + 1$, we conclude that $\{t\} = pl$. Similarly, $\{t'\} = p'l$. Thus, as $t' = t$, $p'l = pl$. It follows that $p' \in \{p, t\}$. Thus, as $t' = t$ and $\ell(t') = \ell(p') + 1$, $p' = p$. This contradiction finishes the proof of the lemma.

Lemma 3.5.2 *Let p, q, and r be elements in $\langle L\rangle$ such that $r \in pq$ and $\ell(r) = \ell(p) + \ell(q)$. Then $a_{pqr} = 1$.*

Proof. Let us denote by R the set of the elements r in $\langle L\rangle$ such that there exist elements p and q in $\langle L\rangle$ with $r \in pq$, $\ell(r) = \ell(p) + \ell(q)$, and $2 \le a_{pqr}$. By way of contradiction, we assume that R is not empty. We pick an element r in R, and we do this in such a way that $\ell(r)$ is as small as possible.

Since $r \in R$, there exist elements p and q in $\langle L\rangle$ such that $r \in pq$, $\ell(r) = \ell(p) + \ell(q)$, and $1 \neq a_{pqr}$. Since $r \in pq$ and $1 \neq a_{pqr}$, we have $2 \le a_{pqr}$. In particular, $1 \neq p$ and $1 \neq q$.

Since $1 \neq q$, there exist elements t in $\langle L\rangle$ and l in L such that $q \in tl$, and $\ell(q) = \ell(t) + 1$; cf. Lemma 3.1.2.

Since $r \in pq$, $\ell(r) = \ell(p) + \ell(q)$, $q \in tl$, and $\ell(q) = \ell(t) + 1$, there exists an element u in pt such that $r \in ul$, $\ell(u) = \ell(p) + \ell(t)$, and $\ell(r) = \ell(u) + 1$; cf. Lemma 3.4.2.

We are assuming that L is constrained. Thus, as $r \in ul$ and $\ell(r) = \ell(u) + 1$, $\{r\} = ul$.

From Lemma 1.1.1(iii) we know that

$$\sum_{s\in S} a_{pts}a_{slr} = \sum_{s\in S} a_{psr}a_{tls}.$$

Since $u \in pt$ and $\ell(u) = \ell(p) + \ell(t)$, $\{u\} = pt$. Thus, the left hand side of the above equation is equal to $a_{ptu}a_{ulr}$.

Since $q \in tl$ and $\ell(q) = \ell(t) + 1$, $\{q\} = tl$. Thus, the right hand side of the above equation is equal to $a_{pqr}a_{tlq}$.

The choice of r yields $a_{ptu} = 1$ and $a_{tlq} = 1$. (Recall that $1 \neq p$. Therefore, $\ell(q) \le \ell(r) - 1$.) Thus, $a_{ulr} = a_{pqr}$. Thus, as $2 \le a_{pqr}$, $2 \le a_{ulr}$. On the other hand, we have that $\{r\} = ul$. Thus, by Lemma 2.3.8(iii), $r = u$, contrary to $\ell(r) = \ell(u) + 1$.

Let us now look at the thin elements of $\langle L\rangle$.

Lemma 3.5.3 *Let R be a subset of $\langle L\rangle$ such that $O_\vartheta(L) \subseteq R$. Assume that $R^2 \subseteq R$ and that, for each element r in R, $r^*r \subseteq R$. Then $R \subseteq \langle L \cap R\rangle$.*

PROOF. Let us abbreviate $K := L \cap R$. We shall see that $R \subseteq \langle K\rangle$.

Suppose, by way of contradiction, that $R \not\subseteq \langle K\rangle$. Then, $R \setminus \langle K\rangle$ is not empty. We pick an element r in $R \setminus \langle K\rangle$, and we do this in such a way that $\ell(r)$ is as small as possible.

Since $r \notin \langle K\rangle$, $1 \neq r$. Thus, by Lemma 3.1.2, there exist elements q in $\langle L\rangle$ and l in L such that $r \in ql$ and $\ell(r) = \ell(q) + 1$. Since L is assumed to be constrained, we obtain from $r \in ql$ and $\ell(r) = \ell(q) + 1$ that $\{r\} = ql$.

Let us first prove that $l \in R$. If $1 = n_l$, this follows from our hypothesis that $O_\vartheta(L) \subseteq R$. If $2 \le n_l$, we obtain from Lemma 2.3.8(ii) that $l \in ll$. Thus, as $1 \in q^*q$ and $\{r\} = ql$, $l \in l^*q^*ql = r^*r$. However, we are assuming that $r^*r \subseteq R$. Thus, $l \in R$.

Since $r \in ql$, $q \in rl$; cf. Lemma 1.3.3(i). Thus, as $l \in R$, $q \in rR$. Thus, as we are assuming that $R^2 \subseteq R$, we obtain $q \in R$.

Recall that $\ell(r) = \ell(q) + 1$. Thus, the (minimal) choice of r yields $q \notin R \setminus \langle K \rangle$. Thus, as $q \in R$, $q \in \langle K \rangle$. Thus, as $r \in ql$ and $l \in K$, $r \in \langle K \rangle$. This contradiction proves the lemma.

Corollary 3.5.4 *For each closed subset T of $\langle L \rangle$, we have the following.*

(i) *If $O_\vartheta(L) \subseteq T$, $\langle L \cap T \rangle = T$.*

(ii) *If $O_\vartheta(L)$ is empty, $\langle L \cap T \rangle = T$.*

Proof. (i) Considering Lemma 2.5.9(ii) we obtain from Lemma 3.5.3 that $T \subseteq \langle L \cap T \rangle$. Since T is assumed to be closed we also have that $\langle L \cap T \rangle \subseteq T$.

(ii) This follows from (i).

Lemma 3.5.5 *The following statements hold.*

(i) *We have $\langle O_\vartheta(L) \rangle = O_\vartheta(\langle L \rangle)$.*

(ii) *The set $\langle L \rangle$ is thin if and only if L is thin.*

(iii) *We have $\{1\} = O_\vartheta(\langle L \rangle)$ if and only if $O_\vartheta(L)$ is empty.*

Proof. (i) From Lemma 1.5.2 we know that $O_\vartheta(\langle L \rangle)^2 \subseteq O_\vartheta(\langle L \rangle)$. Moreover, for each element s in $O_\vartheta(\langle L \rangle)$, we have $s^*s = \{1\} \subseteq O_\vartheta(\langle L \rangle)$; cf. Lemma 1.5.1. Thus, by Lemma 3.5.3, $O_\vartheta(\langle L \rangle) \subseteq \langle O_\vartheta(L) \rangle$.

Conversely, we know from Lemma 3.1.5(ii) that $\langle O_\vartheta(L) \rangle \subseteq O_\vartheta(\langle L \rangle)$.

(ii) If $\langle L \rangle$ is thin, so is L. If L is thin, $O_\vartheta(L) = L$; cf. Lemma 1.5.1. Thus, by (i), $\langle L \rangle = O_\vartheta(\langle L \rangle)$. From this we obtain that $\langle L \rangle$ is thin; cf. Lemma 1.5.1.

(iii) This is an immediate consequence of (i).

From Lemma 3.5.5(i) we obtain, in particular, that $O_\vartheta(\langle L \rangle)$ is a closed subset of S.

3.6 Basic Results on Coxeter Sets

Throughout this section, the letter L stands for a set of involutions of S. Instead of ℓ_L we shall write ℓ.

In accordance with Section 3.4 we shall write, for each element s in $\langle L \rangle$, $S_1(s)$ instead of $S_1(s, L)$.

Recall that L is said to satisfy the exchange condition if, for any three elements h, k in L and s in $S_1(k)$, $h \in S_1(s)$ implies $hs \subseteq sk \cup S_1(k)$.

We call L a *Coxeter set* if L is constrained and satisfies the exchange condition.

We consider Theorem 3.6.4 and Theorem 3.6.6 to be the main results of this section. The equations given in these theorems are crucial for our approach to Coxeter sets.

For the remainder of this section, we assume L to be a Coxeter set.

Lemma 3.6.1 *Let K be a subset of L, and let s be an element in $\langle K\rangle$. Then $\ell(s) = \ell_K(s)$.*

PROOF. Assume the claim to be false. Among the elements in $\langle K\rangle$ which do not satisfy the equation in question we choose s in such a way that $\ell_K(s)$ is as small as possible.

Since $\ell(s) \neq \ell_K(s)$, $1 \neq s$. Thus, by Lemma 3.1.2, there exist elements h in K and r in $\langle K\rangle$ such that $s \in hr$ and $\ell_K(s) = 1 + \ell_K(r)$.

Since $\ell(s) \neq \ell_K(s)$, $s \notin K$. Thus, as $s \in hr$ and $h \in K$, $1 \neq r$. Thus, by Lemma 3.1.2, there exist elements p in $\langle K\rangle$ and k in K such that $r \in pk$ and $\ell_K(r) = \ell_K(p)+1$. Now, by Lemma 3.4.2, there exists an element q in hp such that $s \in qk$, $\ell_K(q) = 1 + \ell_K(p)$, and $\ell_K(s) = \ell_K(q) + 1$.

Since $\ell_K(s) = \ell_K(q) + 1$, the (minimal) choice of s yields $\ell(q) = \ell_K(q)$. Similarly, as $\ell_K(s) = 1 + \ell_K(r)$ and $\ell_K(r) = \ell_K(p) + 1$, $\ell(p) = \ell_K(p)$. Thus, as $q \in hp$ and $\ell_K(q) = 1 + \ell_K(p)$, $h \in S_1(p)$.

Similarly, one obtains $p \in S_1(k)$. Thus, as L is assumed to satisfy the exchange condition, we must have $hp = pk$ or $hp \subseteq S_1(k)$.

Since $s \in hpk$, the first of these two cases yields $s \in pkk = \{p\} \cup pk$, contrary to $\ell_K(s) = \ell_K(p) + 2$. Since $q \in hp$, the second case yields $q \in S_1(k)$. Thus, as $s \in qk$, $\ell(s) = \ell(q) + 1$. (Here we use the hypothesis that L is constrained.) Thus, as $\ell(q) = \ell_K(q)$ and $\ell_K(s) = \ell_K(q)+1$, $\ell(s) = \ell_K(s)$. This contradiction finishes the proof of the lemma.

Lemma 3.6.2 *For each subset K of L, we have the following.*

(i) *The set K is a Coxeter set.*

(ii) *We have $K = L \cap \langle K\rangle$.*

PROOF. (i) This is an immediate consequence of Lemma 3.6.1.

(ii) It is clear that $K \subseteq L\cap\langle K\rangle$. Thus, we just have to show that $L\cap\langle K\rangle \subseteq K$.

In order to show that $L \cap \langle K\rangle \subseteq K$ we fix an element l in $L \cap \langle K\rangle$. Since $l \in L$, $\ell(l) = 1$. Since $l \in \langle K\rangle$, $\ell(l) = \ell_K(l)$; cf. Lemma 3.6.1. From $\ell(l) = 1$ and $\ell(l) = \ell_K(l)$ we obtain $\ell_K(l) = 1$, and that means that $l \in K$.

We are assuming that L is a Coxeter set. In particular, L is constrained. Under this hypothesis, one can modify the exchange condition, and we shall do this in the following lemma.

Formally, our modification is a consequence of the exchange condition. However, the exchange condition is obtained from our modification by setting $q = 1$.

Lemma 3.6.3 *Let h be an element in L, and let p, t be elements in $\langle L \rangle$ such that $t \in hp$ and $\ell(t) = 1 + \ell(p)$. Let k be an element in L, and let q, u be elements in $\langle L \rangle$ such that $u \in qk$ and $\ell(u) = \ell(q) + 1$.*
Assume that $t \in S_1(q)$ and $p \in S_1(u)$. Then we have $tq = pu$ or $t \in S_1(u)$.

PROOF. Assume that $t \in S_1(q)$. Then, there exists an element s in tq such that $\ell(s) = \ell(t) + \ell(q)$. Thus, as $t \in hp$ and $\ell(t) = 1 + \ell(p)$, there exists an element r in pq such that $s \in hr$, $\ell(r) = \ell(p) + \ell(q)$, and $\ell(s) = 1 + \ell(r)$; cf. Lemma 3.4.2. From $r \in pq$ and $\ell(r) = \ell(p) + \ell(q)$ we obtain $\{r\} = pq$. From $s \in hr$ and $\ell(s) = 1 + \ell(r)$ we obtain $h \in S_1(r)$.

Similarly, using $\{r\} = pq$, we conclude from $p \in S_1(u)$ that $r \in S_1(k)$. Thus, as L is assumed to satisfy the exchange condition, we may conclude that $hr = rk$ or that $hr \subseteq S_1(k)$.

Since $\{t\} = hp$, $\{r\} = pq$, and $\{u\} = qk$, the first case yields $tq = pu$.

Since $s \in hr$, the second case yields $s \in S_1(k)$. Thus, by definition, there exists an element s' in sk such that $\ell(s') = \ell(s) + 1$. Since $s' \in sk$ and

$$sk \subseteq hrk = hpqk = tu,$$

$s' \in tu$. Since $\ell(s') = \ell(s) + 1$ and

$$\ell(s) + 1 = 1 + \ell(r) + 1 = 1 + \ell(p) + \ell(q) + 1 = \ell(t) + \ell(u),$$

$\ell(s') = \ell(t) + \ell(u)$. Thus, $t \in S_1(u)$.

Recall that, for each element q in $\langle L \rangle$, $S_{-1}(q, L)$ is our notation for the set of all elements r in $\langle L \rangle$ such that there exists an element p in $\langle L \rangle$ with $r \in pq$ and $\ell(r) = \ell(p) + \ell(q)$.

In accordance with Section 3.4 we shall write $S_{-1}(s)$ instead of $S_{-1}(s, L)$.

Recall that, for each subset R of $\langle L \rangle$, $S_1(R)$ is our notation for the intersection of the sets $S_1(r)$ with $r \in \{1\} \cup R$.

Theorem 3.6.4 *For each subset K of L, we have $S_1(\langle K \rangle) = S_1(K)$.*

PROOF. Let us assume that $S_1(\langle K \rangle) \neq S_1(K)$. Then, as $S_1(\langle K \rangle) \subseteq S_1(K)$, $S_1(K) \not\subseteq S_1(\langle K \rangle)$. Among the elements in $S_1(K)$ which are not in $S_1(\langle K \rangle)$ we choose t such that $\ell(t)$ is as small as possible.

Since $t \notin S_1(\langle K \rangle)$ and $1 \in S_1(\langle K \rangle)$, $1 \neq t$. Thus, by Lemma 3.1.2, there exist elements h in L and p in $\langle L \rangle$ such $t \in hp$ and $\ell(t) = 1 + \ell(p)$.

Since $t \in hp$ and $\ell(t) = 1 + \ell(p)$, $t \in S_{-1}(p)$. Thus, as $t \in S_1(K)$, $p \in S_1(K)$; cf. Lemma 3.4.4(ii). Thus, as $\ell(t) = 1 + \ell(p)$, the (minimal) choice of t yields $p \in S_1(\langle K \rangle)$.

Since $t \notin S_1(\langle K \rangle)$, there exists an element u in $\langle K \rangle$ such that $t \notin S_1(u)$. Among the elements u in $\langle K \rangle$ satisfying $t \notin S_1(u)$ we choose u in such a way that $\ell(u)$ is minimal. Since $t \notin S_1(u)$, $1 \neq u$. Thus, by Lemma 3.1.2 together with Lemma 3.6.1, there exist elements q in $\langle K \rangle$ and k in K such that $u \in qk$ and $\ell(u) = \ell(q) + 1$.

Since $\ell(u) = \ell(q) + 1$ and $q \in \langle K \rangle$, the (minimal) choice of u yields $t \in S_1(q)$. Since $p \in S_1(\langle K \rangle)$ and $u \in \langle K \rangle$, $p \in S_1(u)$. Thus, by Lemma 3.6.3, $tq = pu$ or $t \in S_1(u)$. Thus, by the choice of u, $tq = pu$. Thus, as q, $u \in \langle K \rangle$, $t \in p\langle K \rangle$. Thus, there exists an element s in $\langle K \rangle$ such that $t \in ps$.

Since $p \in S_1(\langle K \rangle)$ and $s \in \langle K \rangle$, $p \in S_1(s)$. Thus, as $t \in ps$, $\ell(t) = \ell(p) + \ell(s)$. (Here we use the hypothesis that L is constrained.) Since $\ell(t) = 1 + \ell(p)$, this means that $\ell(s) = 1$. Thus, by Lemma 3.6.1, $s \in K$. On the other hand, as $t \in ps$ and $\ell(t) = \ell(p) + \ell(s)$, $t \in S_{-1}(s)$. Thus, $t \notin S_1(K)$, contrary to our choice of t.

Lemma 3.6.5 *Let l be an element in L, let p be an element in $S_1(l)$, and let q be the element in pl. Then $S_{-1}(p) \cap S_{-1}(l) \subseteq S_{-1}(q)$.*

PROOF. Let us define R to be the set of elements in $S_{-1}(p) \cap S_{-1}(l)$ which are not in $S_{-1}(q)$. By way of contradiction, we assume that R is not empty. We fix an element r in R, and we do this in such a way that $\ell(r)$ is as small as possible.

Since $r \in R$, $r \in S_{-1}(p)$. Thus, by definition, there exists an element u in $\langle L \rangle$ such that $r \in up$ and $\ell(r) = \ell(u) + \ell(p)$. Since $r \in S_{-1}(l)$ and $p \in S_1(l)$, $r \neq p$. Thus, as $r \in up$, $1 \neq u$. Thus, by Lemma 3.1.2, there exist elements h in L and t in $\langle L \rangle$ such that $u \in ht$ and $\ell(u) = 1 + \ell(t)$. It follows that there exists an element s in tp such that $r \in hs$, $\ell(s) = \ell(t) + \ell(p)$, and $\ell(r) = 1 + \ell(s)$; cf. Lemma 3.4.2.

Since $\ell(r) = 1 + \ell(s)$, $\ell(s) = \ell(r) - 1$. Thus, the minimal choice of r yields $s \notin R$. Moreover, as $s \in tp$ and $\ell(s) = \ell(t) + \ell(p)$, $s \in S_{-1}(p)$.

Suppose that $s \in S_{-1}(l)$. Then, as $s \notin R$ and $s \in S_{-1}(p)$, $s \in S_{-1}(q)$. On the other hand, as $r \in hs$ and $\ell(r) = 1 + \ell(s)$, $r \in S_{-1}(s)$. Thus, by Lemma 3.4.4(iii), $r \in S_{-1}(q)$, contrary to $r \in R$.

This contradiction yields $s \notin S_{-1}(l)$. Thus, by Lemma 3.4.7, $s \in S_1(l)$. Thus, as L is assumed to satisfy the exchange condition, $hs \subseteq sl \cup S_1(l)$. Thus, as $r \in hs$, $r \in sl \cup S_1(l)$. Thus, as $r \in S_{-1}(l)$, $r \in sl \subseteq tpl = tq$. It follows that,

$$\ell(r) \leq \ell(t) + \ell(q) \leq \ell(t) + \ell(p) + 1 = \ell(s) + 1 = \ell(r).$$

This yields $\ell(r) = \ell(t) + \ell(q)$. Thus, as $r \in tq$, $r \in S_{-1}(q)$, contrary to $r \in R$.

Recall that, for each subset R of $\langle L\rangle$, $S_{-1}(R)$ is our notation for the intersection of the sets $S_{-1}(r)$ with $r \in \{1\} \cup R$.

Theorem 3.6.6 *For each subset K of L, we have $S_{-1}(\langle K\rangle) = S_{-1}(K)$.*

PROOF. Let us assume that $S_{-1}(\langle K\rangle) \neq S_{-1}(K)$. Then, as $S_{-1}(\langle K\rangle) \subseteq S_{-1}(K)$, $S_{-1}(K) \not\subseteq S_{-1}(\langle K\rangle)$. Thus, there exists an element s in $S_{-1}(K)$ such that $s \notin S_{-1}(\langle K\rangle)$.

Since $s \notin S_{-1}(\langle K\rangle)$, there exists an element q in $\langle K\rangle$ such that $s \notin S_{-1}(q)$. We pick an element q in $\langle K\rangle$ with $s \notin S_{-1}(q)$, and we do this in such a way that $\ell_K(q)$ is as small as possible.

Since $s \notin S_{-1}(q)$, $1 \neq q$. Thus, Lemma 3.1.2 gives us elements p in $\langle K\rangle$ and k in K such that $q \in pk$ and $\ell_K(q) = \ell_K(p) + 1$. Now, the minimal choice of q yields $s \in S_{-1}(p)$. Thus, as $s \in S_{-1}(K) \subseteq S_{-1}(k)$, $s \in S_{-1}(q)$; cf. Lemma 3.6.5. This contradiction finishes our proof.

Lemma 3.6.7 *Let K be a subset of L, let p be an element in $\langle K\rangle$, and let q be an element in $S_{-1}(K)$. Then the following hold.*

(i) *We have $\ell(p) \leq \ell(q)$.*

(ii) *If $\ell(p) = \ell(q)$, $p = q$.*

PROOF. We are assuming that $q \in S_{-1}(K)$. Thus, by Theorem 3.6.6, $q \in S_{-1}(\langle K\rangle)$; cf. Theorem 3.6.6. Thus, as p is assumed to be an element in $\langle K\rangle$, $q \in S_{-1}(p)$. Thus, by definition, there exists an element s in $\langle L\rangle$ such that $q \in sp$ and $\ell(q) = \ell(s) + \ell(p)$.

(i) From $\ell(q) = \ell(s) + \ell(p)$ and $0 \leq \ell(s)$ we obtain $\ell(p) \leq \ell(q)$.

(ii) Assume that $\ell(q) = \ell(p)$. Then $\ell(s) = 0$. Thus, $1 = s$. Thus, as $q \in sp$, $p = q$.

The set L is called *spherical* if $S_{-1}(L)$ is not empty.

The following lemma will be needed in the beginning of Section 12.1.

Lemma 3.6.8 *The following statements are equivalent.*

(a) *The set L is spherical.*

(b) *The set $S_{-1}(L)$ has exactly one element.*

(c) *The set of all integers $\ell(s)$ with $s \in \langle L\rangle$ has a maximal element.*

PROOF. (a) $\Rightarrow$ (b) Let us assume L to be spherical, and let us pick elements p and q in $S_{-1}(L)$. We shall see that $p = q$.

From Lemma 3.6.7(i) we obtain $\ell(p) = \ell(q)$. Thus, by Lemma 3.6.7(ii), $p = q$.

(a) ⇒ (c) Assume L to be spherical. Then, by definition, $S_{-1}(L)$ is not empty. Let s be an element in $S_{-1}(L)$. Then, for each element r in $\langle L \rangle$, $\ell(r) \leq \ell(s)$; cf. Lemma 3.6.7(i).

(c) ⇒ (a) Let us assume that there exists an element s in $\langle L \rangle$ such that, for each element r in $\langle L \rangle$, $\ell(r) \leq \ell(s)$. We shall see that, for each element l in L, $s \in S_{-1}(l)$.

Let l be an element in L. Then, by Lemma 3.4.7, $s \in S_{-1}(l) \cup S_1(l)$. Assume that $s \in S_1(l)$. Then, by definition, there exists an element t in sl such that $\ell(t) = \ell(s) + 1$. From $s \in \langle L \rangle$ and $l \in L$, we obtain $t \in \langle L \rangle$, contrary to the choice of s.

4

Quotient Schemes

A subset R of S is called *naturally valenced* if each element of R has finite valency. The present chapter starts with the observation that naturally valenced schemes give rise to quotient schemes over finite closed subsets. After the definition of quotient schemes we shall always assume S to be naturally valenced.

In the first section, we compute the structure constants of quotient schemes of S in terms of those of S. We relate the complex multiplication in S to the one in quotient schemes of S, and we look at the relationship between subschemes of quotient schemes and quotient schemes of subschemes.

In the second section, we shall relate specific closed subsets of S containing a finite closed subset T to the corresponding closed subsets of the quotient scheme of S over T. Among other issues we focus on the relationship between commutators and quotient schemes. This leads naturally to the connection between the thin residue of S and the thin residue of quotient schemes of S. This relationship will be described in Theorem 4.2.8, a result which depends on Lemma 3.2.7. Theorem 4.2.8 turns out to be useful in Section 5.5 where we discuss residually thin schemes.

In the third section, we assume S to have finite valency. We investigate the arithmetic between the structure constants of S and the structure constants of the quotient schemes of S.

In the last two sections of this chapter, in which S is again assumed to have finite valency, we use the arithmetic of the structure constants of quotient schemes over closed subsets to investigate Hall subsets and Sylow subsets of S. We prove, for instance, that, if there exists a prime number p such that S is p-valenced, S possesses at least one closed subset with valency equal to the highest power of p dividing the valency of S. This is a generalization of the first of the famous theorems of Ludwig Sylow on finite groups.

We also include corresponding generalizations of Sylow's other two theorems on finite groups.

4.1 Basic Definitions

Let T be a closed subset of S. For each element s in S, we define

$$s^T := \{(yT, zT) \mid z \in yTsT\}.$$

Note that, for each element s in S, $s^T = \{(yT, zT) \mid z \in ys\}$; cf. Lemma 2.1.4.

Lemma 4.1.1 *Let p and q be elements in S, and let T be a closed subset of S. Then the conditions $\emptyset \neq p^T \cap q^T$, $TpT = TqT$, and $p^T = q^T$ are pairwise equivalent.*

PROOF. Let us first assume that $p^T \cap q^T$ is not empty. Then there exist elements y and z in X such that $(yT, zT) \in p^T \cap q^T$. From $(yT, zT) \in p^T$ we obtain $z \in yTpT$. From $(yT, zT) \in q^T$ we obtain $z \in yTqT$.

Since $z \in yTpT$, there exists an element s in TpT such that $z \in ys$. Thus, as $z \in yTqT$, $s \in TqT$. It follows that $s \in TpT \cap TqT$, so that, by Lemma 2.1.3, $TpT = TqT$.

Let us now assume that $TpT = TqT$, and let us pick elements y and z in X. By definition, we have $(yT, zT) \in p^T$ if and only if $z \in yTpT$. Similarly, we have $(yT, zT) \in q^T$ if and only if $z \in yTqT$. Thus, as we are assuming that $TpT = TqT$, we have $(yT, zT) \in p^T$ if and only if $(yT, zT) \in q^T$.

Finally, as p^T is not empty, $p^T = q^T$ implies that $p^T \cap q^T$ is not empty.

Recall that, for each closed subset T of S, X/T is our notation for the set of all sets xT with $x \in X$.

Lemma 4.1.2 *For each closed subset T of S, we have the following.*

(i) *We have $1_{X/T} = 1^T$.*

(ii) *For each element s in S, $(s^T)^* = (s^*)^T$.*

PROOF. (i) It is obvious that $1_{X/T} \subseteq 1^T$. Conversely, let y and z be elements in X such that $(yT, zT) \in 1^T$. Then $z \in yT$, so that, by Lemma 2.1.4, $yT = zT$. This means that $(yT, zT) \in 1_{X/T}$.

(ii) Let y and z be elements in X, and let s be an element in S. We have $(yT, zT) \in (s^T)^*$ if and only if $(zT, yT) \in s^T$, and $(zT, yT) \in s^T$ means that $y \in zTsT$. By Lemma 1.3.2(iii), $y \in zTsT$ is equivalent to $z \in yTs^*T$. Finally, $z \in yTs^*T$ means that $(yT, zT) \in (s^*)^T$.

Let T be a closed subset of S.

For each nonempty subset R of S, we define

$$R/\!/T := \{r^T \mid r \in R\}.$$

For any two nonempty subsets P and Q of S, we write $PQ//T$ instead of $(PQ)//T$ and $P \cap Q//T$ instead of $(P \cap Q)//T$. Moreover, for R a nonempty subset of S and U a closed subset of S, we write $R//T \cap U$ instead of $R//(T \cap U)$.

For the remainder of this section, we shall now assume S to be naturally valenced.

Assuming S to be naturally valenced, a closed subset of S is finite if and only if it has finite valency.

Theorem 4.1.3 *Let T be a finite closed subset of S. Then we have the following.*

(i) *The set $S//T$ is a scheme on X/T.*

(ii) *For any three elements p, q, and r in S, we have*

$$a_{p^T q^T r^T} n_T = \sum_{u \in TpT} \sum_{v \in TqT} a_{uvr}.$$

(iii) *For each element s in S, we have $n_{s^T} n_T = n_{TsT}$.*

Proof. (i), (ii) By Lemma 4.1.1, $S//T$ is a partition of $X/T \times X/T$. From Lemma 4.1.2(i) we know that $1_{X/T} \in S//T$. From Lemma 4.1.2(ii) we know that, for each element s in S, $(s^T)^* \in S//T$. Thus, in order to prove (i), is suffices to verify the regularity condition for $S//T$. We compute the structure constants explicitly, so that also (ii) will be proved.

Let p, q, and r be elements in S, and let y and z be elements in X such that $zT \in (yT)r^T$. Lemma 2.1.4 allows us to assume that $z \in yr$.

Since we are assuming n_T and n_p to be finite, n_{TpT} is finite; cf. Lemma 2.3.3. Thus, $yTpT$ is finite.

Define $I := yTpT \cap zTq^*T$. Then I is finite. Moreover, as $z \in yr$,

$$|I| = \sum_{u \in TpT} \sum_{v \in TqT} a_{uvr}.$$

On the other hand, with the help of Lemma 4.1.2(ii) we obtain that, for each element x in X, $x \in I$ if and only if $xT \in (yT)p^T \cap (zT)(q^T)^*$. Thus,

$$|(yT)p^T \cap (zT)(q^T)^*| n_T = \sum_{u \in TpT} \sum_{v \in TqT} a_{uvr}.$$

(iii) This follows from (ii) together with Lemma 1.1.3(i).

For each finite closed subset T of S, we call $S//T$ the *quotient scheme* of S *over* T.

Assume that S is thin. It is easy to see that, in this case, Theorem 4.1.3(i) holds for each closed subset, not only for finite such sets. The proof is almost the same as the one of Theorem 4.1.3(i).

From Theorem 4.1.3(iii) together with Lemma 2.3.3 we obtain that $S//T$ is naturally valenced for each finite closed subset T of S.

The following lemma relates the complex multiplication in S with the complex multiplication in quotient schemes of S over closed subsets of finite valency.

Lemma 4.1.4 *Let n be a positive integer, let s be an element in S, let R_1, ..., R_n be nonempty subsets of S, and let T be a finite closed subset of S. Then we have $s^T \in (R_1//T)\cdots(R_n//T)$ if and only if $s \in (TR_1T)\cdots(TR_nT)$.*

PROOF. Let us first prove the statement for $n = 1$. By definition, we have $s^T \in R_1//T$ if and only if there exists an element r in R_1 such that $s^T = r^T$. By Lemma 4.1.1, we have $s^T = r^T$ if and only if $TsT = TrT$. Thus, we have $s^T \in R_1//T$ if and only if $s \in TR_1T$.

Let us now assume that $2 \leq n$. We shall refer to the case $n = 1$.

Assume first that $s^T \in (R_1//T)\cdots(R_n//T)$. Then, by definition, there exist elements p and q in S such that $p^T \in (R_1//T)\cdots(R_{n-1}//T)$, $q^T \in R_n//T$, and $1 \leq a_{p^T q^T s^T}$. Since $1 \leq a_{p^T q^T s^T}$, there exist elements u in TpT and v in TqT such that $1 \leq a_{uvs}$; cf. Theorem 4.1.3(ii). Thus, by definition, $s \in (TpT)(TqT)$.

Since $p^T \in (R_1//T)\cdots(R_{n-1}//T)$, induction allows us to assume that $p \in (TR_1T)\cdots(TR_{n-1}T)$. Moreover, we saw earlier that $q^T \in R_n//T$ implies $q \in TR_nT$. Therefore,

$$(TpT)(TqT) \subseteq (TR_1T)\cdots(TR_nT).$$

Thus, as $s \in (TpT)(TqT)$, $s \in (TR_1T)\cdots(TR_nT)$.

Conversely, let us assume that $s \in (TR_1T)\cdots(TR_nT)$. Then there exist elements p in $(TR_1T)\cdots(TR_{n-1}T)$ and q in TR_nT such that $s \in pq$. Since $s \in pq$, $1 \leq a_{pqs}$. Thus, by Theorem 4.1.3(ii), $1 \leq a_{p^T q^T s^T}$, and that means that $s^T \in p^T q^T$.

On the other hand, by induction, we obtain from $p \in (TR_1T)\cdots(TR_{n-1}T)$ that $p^T \in (R_1//T)\cdots(R_{n-1}//T)$. Moreover, we saw in the first part of this proof that $q \in TR_nT$ is equivalent to $q^T \in R_n//T$. Thus, we have $p^T q^T \subseteq (R_1//T)\cdots(R_n//T)$. Thus, as $s^T \in p^T q^T$, $s^T \in (R_1//T)\cdots(R_n//T)$.

Lemma 4.1.5 *Let s be an element in S, and let T and U be closed subsets of S such that $T \subseteq U$. Then s^T normalizes $U//T$ if and only if $UsT \subseteq TsU$.*

PROOF. Let r be an element in S. Then, as we are assuming that $T \subseteq U$, $r^T \in (U//T)s^T$ if and only if $r \in UsT$; cf. Lemma 4.1.4. Similarly, $r^T \in s^T(U//T)$ if and only if $r \in TsU$.

Thus, we have $(U//T)s^T \subseteq s^T(U//T)$ if and only if $UsT \subseteq TsU$.

Lemma 4.1.6 *Let P and Q be nonempty subsets of S, and let T be a finite closed subset of S. Then $PTQ//T = (P//T)(Q//T)$.*

PROOF. Let s be an element in S, and let us first assume that $s^T \in PTQ//T$. Then, by definition, there exists an element r in PTQ such that $s^T = r^T$. From $r \in PTQ$ we obtain $r \in (TPT)(TQT)$. Thus, by Lemma 4.1.4, $r^T \in (P//T)(Q//T)$. Thus, as $s^T = r^T$, $s^T \in (P//T)(Q//T)$.

Let us now, conversely, assume that $s^T \in (P//T)(Q//T)$. Then, by Lemma 4.1.4, $s \in TPTQT$. Thus, there exists an element r in PTQ such that $s \in TrT$. From $r \in PTQ$ we obtain $r^T \in PTQ//T$. From $s \in TrT$ we obtain $s^T = r^T$; cf. Lemma 4.1.1. Thus, we have that $s^T \in PTQ//T$.

Lemma 4.1.7 *Let R be a nonempty subset of S, and let T be a finite closed subset of S. Then the following hold.*

(i) *Assume that $TRT \subseteq R$. Then $R//T$ is closed if and only if R is closed.*

(ii) *Assume that $R//T$ is closed, and set $U := \{s \in S \mid s^T \in R//T\}$. Then U is closed, and $R//T = U//T$.*

PROOF. (i) We are assuming that $TRT \subseteq R$. Thus, $R^*TR = R^*R$. Thus, by Lemma 4.1.6,

$$R^*R//T = (R^*//T)(R//T).$$

Thus, if R is closed, $R//T$ is closed.

Conversely, if $R//T$ is closed, the above equation yields

$$R^*R//T = R//T.$$

Let s be an element in R^*R. Then, as $R^*R//T = R//T$, $s^T \in R//T$. Thus, by Lemma 4.1.4, $s \in TRT$. Thus, as we are assuming that $TRT \subseteq R$, $s \in R$.

Since s has been chosen arbitrarily in R^*R, we have shown that $R^*R \subseteq R$, and that means that R is closed.

(ii) Let u be an element in U. Then, by definition, $u^T \in R//T$. Thus, as $R//T$ is assumed to be closed, we have that $(u^T)^* \in R//T$. Thus, by Lemma 4.1.2(ii), $(u^*)^T \in R//T$. Thus, by definition, $u^* \in U$.

Let p and q be elements in U, and let s be an element in pq. Since $p \in U$, $p^T \in R//T$. Similarly, as $q \in U$, $q^T \in R//T$. However, since $s \in pq$, $s^T \in p^T q^T$; cf. Lemma 4.1.4. Thus, as $R//T$ is assumed to be closed, $s^T \in R//T$. Thus, by definition, $s \in U$.

That $R//T = U//T$ follows right from the definition of U.

The following result describes the relationship between quotient schemes of subschemes of S and subschemes of quotient schemes of S. Note that its first part generalizes Theorem 4.1.3(i).

Theorem 4.1.8 *Let x be an element in X, and let T and U be closed subsets of S such that $T \subseteq U$ and T is finite. Then we have the following.*

(i) *The set $U_{xU}//T_{xU}$ is a scheme on xU/T_{xU}.*

(ii) *The set $(U//T)_{(xT)(U//T)}$ is a scheme on $(xT)(U//T)$.*

(iii) *We have $xU/T_{xU} = (xT)(U//T)$ and $U_{xU}//T_{xU} = (U//T)_{(xT)(U//T)}$.*

PROOF. (i) Since U is assumed to be closed, U_{xU} is a scheme on xU; cf. Theorem 2.1.8(ii). On the other hand, as T is assumed to be closed, T_{xU} is closed; cf. Lemma 2.1.9(ii). Thus, by Theorem 4.1.3(i), $U_{xU}//T_{xU}$ is a scheme on xU/T_{xU}. (Note that, for each element t in U, $n_{t_{xU}}$ is finite, and, as n_T is assumed to be finite, $n_{T_{xU}}$ is finite; cf. Theorem 2.1.8(iii).)

(ii) According to Theorem 4.1.3(i), $S//T$ is a scheme on X/T. On the other hand, as U is assumed to be closed, $U//T$ is closed; cf. Lemma 4.1.7(i). Thus, the claim follows from Theorem 2.1.8(ii).

(iii) Note that

$$xU/T_{xU} = \{y(T_{xU}) \mid y \in xU\} = \{yT \mid y \in xU\}.$$

On the other hand, $(xT)(U//T)$ is the union of the sets $(xT)u^T$ with $u \in U$, and, for each element u in U, $(xT)u^T = \{yT \mid y \in xTuT\}$. Thus,

$$(xT)(U//T) = \{yT \mid y \in xU\},$$

so that the first equation is proved.

In order to establish the second equation, we first mention that, for each element u in U,

$$(u_{xU})^{T_{xU}} = \{(y(T_{xU}), z(T_{xU})) \in xU/T_{xU} \times xU/T_{xU} \mid z \in yT_{xU}u_{xU}T_{xU}\}$$

and

$$(u^T)_{(xT)(U//T)} = \{(yT, zT) \in (xT)(U//T) \times (xT)(U//T) \mid z \in yTuT\}.$$

Thus, referring to Lemma 2.1.9(i) we obtain from the first equation that

$$(u_{xU})^{T_{xU}} = (u^T)_{(xT)(U//T)}.$$

Thus, as $U_{xU}//T_{xU} = \{(u_{xU})^{T_{xU}} \mid u \in U\}$ and

$$(U//T)_{(xT)(U//T)} = \{(u^T)_{(xT)(U//T)} \mid u \in U\},$$

we conclude that $U_{xU}//T_{xU} = (U//T)_{(xT)(U//T)}$.

Let x be an element in X, and let T and U be closed subsets of S. Assume that $T \subseteq U$ and that T has finite valency.

In the following, we shall write $(U//T)_x$ to denote the scheme $(U//T)_{(xT)(U//T)}$. According to Theorem 4.1.8(iii), $(U//T)_x$ then also denotes the scheme $U_{xU}//T_{xU}$ on xU/T_{xU}.

If $\{1\} = T$, $U//\{1\}$ is identified with U. Thus, the previous remark implies that we shall write U_x to denote the scheme U_{xU} on xU.

If $U = S$, we shall write $U//T$ instead of $(U//T)_x$.

4.2 General Facts

In this section, S is assumed to be naturally valenced. The letter T stands for a finite closed subset of S.

Lemma 4.2.1 *Let R be a nonempty subset of S, and let U be a closed subset of S such that $T \subseteq U$. Then the following hold.*

(i) *We have $R \cap U//T = R//T \cap U//T$.*

(ii) *We have $R//T \subseteq U//T$ if and only if $R \subseteq U$.*

PROOF. (i) By definition, we have $R \cap U//T \subseteq R//T \cap U//T$. Thus, it suffices to show that $R//T \cap U//T \subseteq R \cap U//T$. In order to prove this, we pick an element s in S such that $s^T \in R//T \cap U//T$. We shall be done if we succeed in showing that $s^T \in R \cap U//T$.

Since $s^T \in R//T$, there exists an element r in R such that $s^T = r^T$. Similarly, as $s^T \in U//T$, there exists an element u in U such that $s^T = u^T$. It follows that $r^T = u^T$, so that, by Lemma 4.1.1, $TrT = TuT$.

We are assuming that $T \subseteq U$. Thus, we obtain from $TrT = TuT$ that $r \in U$. It follows that $r \in R \cap U$, so that $r^T \in R \cap U//T$. Thus, as $s^T = r^T$, we conclude that $s^T \in R \cap U//T$.

(ii) It is clear that, if $R \subseteq U$, $R//T \subseteq U//T$. Therefore, we assume, conversely, that $R//T \subseteq U//T$.

In order to show that $R \subseteq U$, we fix an element r in R. Then, by definition, $r^T \in R//T$. Thus, as we are assuming that $R//T \subseteq U//T$, $r^T \in U//T$. Thus, there exists an element u in U such that $r^T = u^T$. From $r^T = u^T$ we obtain $TrT = TuT$; cf. Lemma 4.1.1. It follows that $r \in TuT \subseteq U$.

Lemma 4.2.2 *For each nonempty subset R of S, the following hold.*

(i) *If $T \subseteq \langle R \rangle$, $\langle R//T \rangle = \langle R \rangle //T$.*

(ii) *Let U be a closed subset of S such that $T \subseteq U$ and $\langle R//T \rangle = U//T$. Then $\langle R \cup T \rangle = U$.*

PROOF. (i) Without loss of generalization, we may assume that $R^* = R$.

Let s be an element in S, and let us assume that $s^T \in \langle R//T\rangle$. Then, by Lemma 3.1.1(i), there exists a non-negative integer n such that $s^T \in (R//T)^n$. Thus, by Lemma 4.1.4, $s \in (TRT)^n$. Now, as we are assuming that $T \subseteq \langle R\rangle$, $s \in \langle R\rangle$; cf. Lemma 3.1.1(i). Thus, by definition, $s^T \in \langle R\rangle//T$.

This proves that $\langle R//T\rangle \subseteq \langle R\rangle//T$.

That $\langle R\rangle//T \subseteq \langle R//T\rangle$ follows immediately from Lemma 4.1.6 together with Lemma 3.1.1(i).

(ii) We are assuming that $\langle R//T\rangle = U//T$. Thus, $\langle R \cup T//T\rangle = U//T$. Thus, by (i), $\langle R \cup T\rangle//T = U//T$, so that the claim follows from Lemma 4.2.1(ii).

Lemma 4.2.3 *For each nonempty subset R of S, we have $[R//T, T//T] = [R,T]T//T$.*

Proof. Let Q denote the union of the sets Tr^*TrT with $r \in R$. We shall show that

$$[R//T, T//T] = \langle Q//T\rangle,$$

so that our claim will follow from Lemma 4.2.2(i) together with Lemma 3.1.9(ii).

Without loss of generalization, we may assume that $R^* = R$.

Let s be an element in S such that $s^T \in [R//T, T//T]$. Then, by Lemma 3.1.1(i), there exist elements $r_1, \ldots, r_n$ in R such that

$$s^T \in (r_1^T)^* r_1^T \cdots (r_n^T)^* r_n^T.$$

Thus, for each element i in $\{1, \ldots, n\}$, there exists an element $q_i \in S$ such that $q_i^T \in (r_i^T)^* r_i^T$ and $s^T \in q_1^T \cdots q_n^T$.

Let i be an element in $\{1, \ldots, n\}$. Then, as $q_i^T \in (r_i^T)^* r_i^T$, $q_i \in Tr_i^*Tr_iT \subseteq Q$; cf. Lemma 4.1.4. Thus, by definition, $q_i^T \in Q//T$. Thus, as $s^T \in q_1^T \cdots q_n^T$, $s^T \in (Q//T)^n$. Thus, by Lemma 3.1.1(i), $s^T \in \langle Q//T\rangle$.

Conversely, let s be an element in S such that $s^T \in \langle Q//T\rangle$. Then there exists a positive integer n such that $s^T \in (Q//T)^n$; cf. Lemma 3.1.1(i). Thus, there exist elements $q_1, \ldots, q_n$ in Q such that $s^T \in q_1^T \cdots q_n^T$.

Let i be an element in $\{1, \ldots, n\}$. Then, as $q_i \in Q$, there exists an element r_i in R such that $q_i \in Tr_i^*Tr_iT$. Thus, by Lemma 4.1.4, $q_i^T \in (r_i^T)^* r_i^T$. Thus, as $s^T \in q_1^T \cdots q_n^T$,

$$s^T \in (r_1^T)^* r_1^T \cdots (r_n^T)^* r_n^T \subseteq [R//T, T//T],$$

and that finishes the proof of the theorem.

Recall that $K_S(T)$ is our notation for the set of all elements s in S which satisfy $s^*Ts \subseteq T$.

Lemma 4.2.4 *Let U and V be closed subsets of S, and let us assume that $T \subseteq U$. Then the following hold.*

(i) *If T is normal in S, $N_{V//T}(U//T) = N_V(U)//T$.*

(ii) *We have $K_{V//T}(U//T) = K_V(U)//T$.*

PROOF. (i) Let s be an element in S. Then, as T is assumed to be normal in S, we have that s^*, $s \in N_S(T)$. Thus, by Lemma 2.5.2(ii), $Ts = sT$. Thus, as $T \subseteq U$, we have $Us = UsT$ and $TsU = sU$.

From Lemma 4.1.5 we know that s^T normalizes $U//T$ if and only if $UsT \subseteq TsU$. However, $s \in N_S(U)$ means that $Us \subseteq sU$. Thus, as $Us = UsT$ and $TsU = sU$, we have $s^T \in N_{S//T}(U//T)$ if and only if $s \in N_S(U)$.

So far, we have shown that $N_{S//T}(U//T) = N_S(U)//T$. Now the claim follows from Lemma 4.2.1(i).

(ii) Let s be an element in S. By definition, we have

$$s^T \in K_{S//T}(U//T)$$

if and only if $(s^T)^*(U//T)s^T \subseteq U//T$. Since we are assuming that $T \subseteq U$, we also have that $s^*Us//T = (s^T)^*(U//T)s^T$; cf. Lemma 4.1.6. Thus, the latter condition is equivalent to $s^*Us//T \subseteq U//T$. According to Lemma 4.2.1(ii), this is equivalent to $s^*Us \subseteq U$. However, $s^*Us \subseteq U$ means that

$$s \in K_S(U).$$

Clearly, if $s \in K_S(U)$, $s^T \in K_S(U)//T$. Conversely, if $s^T \in K_S(U)//T$, there exists an element r in $K_S(U)$ such that $s^T = r^T$. From $s^T = r^T$ we obtain $s \in TrT$; cf. Lemma 4.1.1. Thus, as $T \subseteq U$, $s \in UrU$. Thus, as $r \in K_S(U)$, $s \in K_S(U)$; cf. Lemma 2.5.8(i).

So far, we have shown that $K_{S//T}(U//T) = K_S(U)//T$. Now the claim follows from Lemma 4.2.1(i).

Let U be a closed subset of S such that $T \subseteq U$.

Let V be a closed subset of S which normalizes T, and assume that $U \subseteq V$. Lemma 4.2.4(i) says, in particular, that $U//T$ is normal in $V//T$ if and only if U is normal in V.

Recall that T is called strongly normal in U if $U \subseteq K_S(T)$.

Let V be closed subsets of S satisfying $U \subseteq V$. Lemma 4.2.4(ii) says, in particular, that $U//T$ is strongly normal in $V//T$ if and only if U is strongly normal in V.

Lemma 4.2.5 *Let U be a closed subset of S such that $T \subseteq U$. Then the following hold.*

(i) *We have $O_\vartheta(U//T) = K_U(T)//T$.*

(ii) *The subset T is strongly normal in U if and only if $U//T$ is thin.*

PROOF. (i) From Lemma 2.5.9(i) we know that $O_\vartheta(U//T) = K_{U//T}(T//T)$. Thus, the claim follows from Lemma 4.2.4(ii).

(ii) By definition, T is strongly normal in U if and only if $U \subseteq K_U(T)$. By (i) (together with Lemma 4.2.1(ii)), the latter condition is equivalent to $U//T \subseteq O_\vartheta(U//T)$, and that means that $U//T$ is thin; cf. Lemma 1.5.1.

Let us now look at a distinguished series of closed subsets of S.

Let U be a closed subset of S such that $T \subseteq U$. Assume that U has finite valency. We set $(K_U)^0(T) := T$. For each positive integer n, we inductively define

$$(K_U)^n(T) := K_U((K_U)^{n-1}(T)).$$

Since U is assumed to have finite valency, we obtain from Lemma 2.5.8(ii) that $(K_U)^n(T)$ is closed for each non-negative integer n. From this we obtain

$$(K_U)^{n-1}(T) \subseteq (K_U)^n(T)$$

for each positive integer n.

The following theorem is similar to Lemma 4.2.4(ii).

Theorem 4.2.6 *Let m be a non-negative integer, let U be a closed subset of S such that $T \subseteq U$. Assume that U has finite valency, and set $V := (K_U)^m(T)$. Then, for each non-negative integer n,*

$$(K_{U//T})^n(V//T) = (K_U)^{m+n}(T)//T.$$

PROOF. The claim is obvious for $n = 0$. Therefore, we assume that $1 \leq n$. We set $W := (K_U)^{m+n-1}(T)$.

By induction, we may assume that the claim holds for $n - 1$. Thus, we have

$$(K_{U//T})^n(V//T) = K_{U//T}((K_{U//T})^{n-1}(V//T)) = K_{U//T}(W//T).$$

On the other hand, by Lemma 4.2.4(ii),

$$K_{U//T}(W//T) = K_U(W)//T = (K_U)^{m+n}(T)//T,$$

and that finishes the proof.

Let us now look how the thin residue works together with quotient schemes.

Lemma 4.2.7 *Let U be a closed subset of S such that $T \subseteq U$. Then the following hold.*

(i) *If $U//T$ is thin, $O^\vartheta(U) \subseteq T$.*

(ii) *We have* $O^\vartheta(U//T) = [U,T]T//T$.

(iii) *We have* $O^\vartheta(U)T//T \subseteq O^\vartheta(U//T)$.

(iv) *If* T *is normal in* U, $O^\vartheta(U)T//T = O^\vartheta(U//T)$.

PROOF. (i) This follows from Lemma 4.2.5(ii) together with the definition of $O^\vartheta(U)$.

(ii) From Theorem 3.2.1(ii) we know that $O^\vartheta(U//T) = [U//T, T//T]$. From Lemma 4.2.3 we know that $[U//T, T//T] = [U,T]T//T$. Thus, we have that $O^\vartheta(U//T) = [U,T]T//T$.

(iii) From Theorem 3.2.1(ii) we know that $O^\vartheta(U) = [U,1]$. Thus, $[U,1] \subseteq [U,T]$. Thus,

$$O^\vartheta(U)T//T = [U,T]T//T,$$

so that the claim follows from (ii).

(iv) Assume that T is normal in U. Then, by Lemma 2.5.6(ii), $O^\vartheta(U)T$ is strongly normal in U. Thus, by Lemma 4.2.4(ii), $O^\vartheta(U)T//T$ is strongly normal in $U//T$. Thus, by definition,

$$O^\vartheta(U//T) \subseteq O^\vartheta(U)T//T,$$

so that the claim follows from (iii).

From Theorem 3.2.1(i) we know that $O^\vartheta(T)$ is strongly normal in T. Thus, by Lemma 4.2.5(ii), $T//O^\vartheta(T)$ is thin. Thus, we obtain from Lemma 4.2.7(i) that the thin residue $O^\vartheta(T)$ of T is the uniquely defined smallest closed subset of S having a thin quotient scheme.

From Lemma 4.2.7(ii) we obtain, in particular, that $O^\vartheta(T) = [T,1]$. Thus, Lemma 4.2.7(ii) is a generalization of Theorem 3.2.1(ii).

Theorem 4.2.8 *Let* n *be a non-negative integer, and let* U *be a closed subset of* S *such that* $T \subseteq N_S(U)$. *Then we have* $(O^\vartheta)^n(TU)U//U = (O^\vartheta)^n(TU//U)$.

PROOF. Let T be a counterexample such that n_T is as small as possible. Then $1 \leq n$.

We are assuming that $T \subseteq N_S(U)$. Thus, by Lemma 2.5.1(ii), U is normal in TU. Thus, by Lemma 4.2.7(iv),

$$O^\vartheta(TU)U//U = O^\vartheta(TU//U).$$

Let us set

$$V := O^\vartheta(TU)U.$$

Then $U \subseteq V$. Thus, by Lemma 2.2.1(ii), $(T \cap V)U = TU \cap V$. Thus, as $V \subseteq TU$, $(T \cap V)U = V$.

Let us first assume that $V \neq TU$. Then $T \cap V \neq T$. Thus, the minimal choice of T yields

$$(O^\vartheta)^{n-1}(V)U//U = (O^\vartheta)^{n-1}(V//U).$$

(We write V instead of $(T \cap V)U$.) Thus, as $V//U = O^\vartheta(TU//U)$,

$$(O^\vartheta)^{n-1}(V)U//U = (O^\vartheta)^{n-1}(O^\vartheta(TU//U)) = (O^\vartheta)^n(TU//U).$$

Applying Lemma 3.2.7(i) to $O^\vartheta(TU)$ in place of T, we also obtain

$$(O^\vartheta)^{n-1}(O^\vartheta(TU))U = (O^\vartheta)^{n-1}(O^\vartheta(TU)U)U.$$

Moreover, the left hand side of this equation is equal to $(O^\vartheta)^n(TU)U$, and the right hand side is equal to $(O^\vartheta)^{n-1}(V)U$, so that

$$(O^\vartheta)^n(TU)U//U = (O^\vartheta)^{n-1}(V)U//U.$$

From this, together with $(O^\vartheta)^{n-1}(V)U//U = (O^\vartheta)^n(TU//U)$, we obtain

$$(O^\vartheta)^n(TU)U//U = (O^\vartheta)^n(TU//U),$$

contrary to the choice of T.

Let us now assume that $V = TU$. Then, as $V = O^\vartheta(TU)U$, $O^\vartheta(TU)U = TU$. Thus, by Lemma 3.2.7(ii), $(O^\vartheta)^n(TU)U = TU$. It follows that

$$(O^\vartheta)^n(TU)U//U = TU//U.$$

However, from $V = TU$ and $V//U = O^\vartheta(TU//U)$ we also obtain

$$(O^\vartheta)^n(TU//U) = TU//U,$$

so that $(O^\vartheta)^n(TU)U//U = (O^\vartheta)^n(TU//U)$. Again, this contradicts our choice of T.

Corollary 4.2.9 *Let n be a non-negative integer, and let U be a closed subset of S such that $U \subseteq (O^\vartheta)^n(T)$. Then we have $(O^\vartheta)^n(T)//U = (O^\vartheta)^n(T//U)$.*

PROOF. Since we are assuming that $U \subseteq (O^\vartheta)^n(T)$, $U \subseteq T$. Thus, $T = TU$. Thus, $(O^\vartheta)^n(TU)U = (O^\vartheta)^n(T)$, so that the claim follows from Theorem 4.2.8.

4.3 Valencies

In this section, S is assumed to have finite valency. The letter T stands for a closed subset of S.

Assuming S to have finite valency, we may speak about the quotient scheme of S over T. We also may apply all results of the previous two sections.

Lemma 4.3.1 *For each element s in S, we have the following.*

(i) *The integer n_{s^T} divides $n_T n_s$.*

(ii) *If s^*, $s \in N_S(T)$, n_{s^T} divides n_s.*

(iii) *If T is thin, n_s divides $n_{s^T} n_T$.*

PROOF. (i) From Theorem 4.1.3(iii) we know that $n_{s^T} n_T = n_{TsT}$, and from Lemma 2.3.3 that n_{TsT} divides $n_T n_s n_T$. Thus, n_{s^T} divides $n_T n_s$.

(ii) Assume that s^*, $s \in N_S(T)$. Then, by Lemma 2.5.2(ii), $Ts = sT$. From this we obtain $TsT = sT$. Thus, by Theorem 4.1.3(iii), $n_{s^T} n_T = n_{sT}$. It follows that $n_{s^T} = (n_T)^{-1} n_{sT}$, so that the claim follows from Lemma 2.3.4(iii).

(iii) Let r be an element in TsT. Then there exist elements p and q in T such that $r \in psq^*$. Since $r \in psq^*$, $n_r \leq n_{psq^*}$.

On the other hand, as p and q are thin, $\{1\} = p^*p$ and $\{1\} = q^*q$. Thus, $\{s\} = p^*psq^*q$. Thus, by Lemma 1.4.5(ii), $n_s = n_{psq^*}$.

From $n_r \leq n_{psq^*}$ and $n_s = n_{psq^*}$ we obtain $n_r \leq n_s$. Similarly, we obtain $n_s \leq n_r$, so that we have $n_r = n_s$. Thus, as r has been chosen arbitrarily in TsT, we have $n_{TsT} = |TsT| n_s$. In particular, n_s divides n_{TsT}, so that the claim follows from Theorem 4.1.3(iii).

Assume T to be normal in S, and let s be an element in S. Then, by Lemma 4.3.1(ii), n_{s^T} divides n_s. In particular, $S//T$ is thin if S is thin.

Let π be a set of prime numbers. Recall that a positive integer n is called a *π-number* if each prime divisor of n is element of π.

A nonempty subset R of S is called *π-valenced* if, for each element r in R, n_r is a π-number.

Note that, for each prime number p, a $\{p\}$-valenced nonempty subset of S is the same as a p-valenced nonempty subset of S.

Corollary 4.3.2 *Let π be a set of prime numbers, and assume that n_T is a π-number. Let U be a closed subset of S such that $T \subseteq U$. Then we have the following.*

(i) *If U is π-valenced, so is $U//T$.*

(ii) *Assume T to be thin. Then, if $U//T$ is π-valenced, so is U.*

PROOF. (i) This follows from Lemma 4.3.1(i).

(ii) This follows from Lemma 4.3.1(iii).

Lemma 4.3.3 *Let U be a closed subset S such that $T \subseteq U$. Then the following hold.*

(i) *We have $n_U = n_{U//T} n_T$.*

(ii) *Let V be a closed subset of S with $U \subseteq V$. Then $n_{V//T} = n_{V//U} n_{U//T}$.*

PROOF. (i) Let R be a transversal of (T, T) in U. Then, by Lemma 2.1.3, $\{TrT \mid r \in R\}$ is a partition of S. From this we obtain

$$n_U = \sum_{r \in R} n_{TrT}$$

and, referring to Lemma 4.1.1, that $U//T = \{r^T \mid r \in R\}$. This last equation implies

$$n_{U//T} = \sum_{r \in R} n_{r^T}.$$

Now the desired equation follows from the fact that, for each element r in R, $n_{r^T} n_T = n_{TrT}$; cf. Theorem 4.1.3(iii).

(ii) Applying (i) three times we obtain

$$n_{V//T} n_T n_U = n_V n_U = n_{V//U} n_U n_{U//T} n_T,$$

and that proves (iii).

Lemma 4.3.4 *Let U be a closed subset of S such that TU is closed. Then $n_{T//T \cap U} = n_{TU//U}$.*

PROOF. From Lemma 4.3.3(i) we know that $n_T = n_{T//T\cap U} n_{T \cap U}$ and that $n_{TU} = n_{TU//U} n_U$. From Lemma 2.3.6(i), we know that $n_T n_U = n_{TU} n_{T\cap U}$. Thus, $n_{TU//U} = n_{T//T\cap U}$.

Let π be a set of prime numbers. We shall denote by π' the set of all prime numbers not contained in π.

Lemma 4.3.5 *Let π be a set of prime numbers such that n_T is a π-number and $n_{S//T}$ a π'-number. Let s be a π-valenced element in S. Then the following hold.*

(i) *We have $\{1^T\} = s^T (s^T)^* \cap O_\vartheta(S//T)$.*

(ii) *We have $ss^* \cap K_S(T) \subseteq T$.*

PROOF. (i) From Lemma 2.5.9(iii) we know that $O_\vartheta(S//T)$ is closed. Thus, by Corollary 2.6.6,

$$\{1^T\}^{(s^T)^*} \cap O_\vartheta(S//T) = s^T (s^T)^* \cap O_\vartheta(S//T).$$

We are assuming that n_T and n_s are π-numbers. Thus, by Lemma 4.3.1(i), n_{s^T} is a π-number, too. From this we obtain that $n_{(s^T)^*}$ is a π-number; cf. Lemma 1.1.2(iii). Now recall that, by Lemma 2.3.7(ii), $n_{\{1^T\}^{(s^T)^*}}$ divides $n_{(s^T)^*}$. Thus, $n_{\{1^T\}^{(s^T)^*}}$ is a π-number.

We are assuming that $n_{S//T}$ is a π'-number. Thus, as $O_\vartheta(S//T)$ is closed, $n_{O_\vartheta(S//T)}$ must be a π'-number, too; cf. Lemma 2.3.6(ii).

Now, as $n_{\{1^T\}^{(s^T)^*}}$ is a π-number and $n_{O_\vartheta(S//T)}$ is a π'-number,

$$\{1^T\} = \{1^T\}^{(s^T)^*} \cap O_\vartheta(S//T).$$

(ii) Let r be an element in $ss^* \cap K_S(T)$. From $r \in ss^*$ we obtain $r^T \in s^T(s^T)^*$; cf. Lemma 4.1.4 (together with Lemma 4.1.2(ii)). From $r \in K_S(T)$ we obtain $r^T \in K_S(T)//T$. Thus, by Lemma 4.2.5(i), $r^T \in O_\vartheta(S//T)$.

On the other hand, we know from (i) that $\{1^T\} = s^T(s^T)^* \cap O_\vartheta(S//T)$, so that $r^T = 1^T$. Thus, by Lemma 4.1.1, $r \in T$.

4.4 Hall Subsets

In this section, S is assumed to have finite valency.

Recall that, for each set π of prime numbers, π' is our notation for the set of all prime numbers not in π.

A nonempty π-valenced subset R of S will be called a *π-subset* of S if n_R is a π-number.

Let T and U be closed subsets of S such that $T \subseteq U$. Assume that T is a π-subset of S. We call T a *Hall π-subset* of U if $n_{U//T}$ is a π'-number.

Lemma 4.4.1 *Let π be a set of prime numbers, and let T, U, and V be closed subsets of S such that T is a Hall π-subset of V, $U \subseteq V$, and $TU = UT$. Then $T \cap U$ is a Hall π-subset of U.*

PROOF. As a subset of T, the set $T \cap U$ is a closed π-subset of U; cf. Lemma 2.3.6(ii). Thus, we just have to show that $n_{U//T\cap U}$ is a π'-number.

Since $TU = UT$, TU is closed; cf. Lemma 2.1.1. Thus, by Lemma 4.3.3(ii), $n_{TU//T}$ divides $n_{V//T}$. However, as T is assumed to be a Hall π-subset of V, $n_{V//T}$ is a π'-number. Thus, $n_{TU//T}$ is a π'-number.

Recall, finally, that $n_{U//T\cap U} = n_{TU//T}$; cf. Lemma 4.3.4. Thus, $n_{U//T\cap U}$ is a π'-number.

Lemma 4.4.1 says, in particular, that, whenever a closed subset T of S contains a Hall π-subset of S, then each Hall π-subset of T is a Hall π-subset of S.

Lemma 4.4.2 *Let π be a set of prime numbers, and let T, U, and V be closed subsets of S. Assume that T is a Hall π-subset of V, that $U \subseteq V$, and that $T \subseteq N_S(U)$. Then $TU//U$ is a Hall π-subset of $V//U$.*

PROOF. From Lemma 4.3.3(i) we know that $n_T = n_{T//T\cap U} n_{T\cap U}$. Thus, as n_T is assumed to be a π-number, $n_{T//T\cap U}$ is a π-number. Moreover, we are assuming that $T \subseteq N_S(U)$. Thus, by Lemma 2.5.2(iii), TU is a closed subset of S. Thus, by Lemma 4.3.4, $n_{T//T\cap U} = n_{TU//U}$, so that $n_{TU//U}$ is a π-number.

By Lemma 4.3.3(ii), $n_{V//TU}$ divides $n_{V//T}$. Moreover, as T is assumed to be a Hall π-subset of V, $n_{V//T}$ is a π'-number. Thus, $n_{V//TU}$ is a π'-number.

It remains to prove that $TU//U$ is π-valenced. In order to do so we pick an element s in TU. We shall see that n_{s^U} is a π-number.

Since $s \in TU$, there exists an element t in T such that $s \in tU$. From $s \in tU$ we obtain $UsU = UtU$; cf. Lemma 2.1.3. Thus, by Lemma 4.1.1, $s^U = t^U$.

Since t is an element in T and T is assumed to be a Hall π-subset of V, n_t is a π-number. On the other hand, as t^*, $t \in T \subseteq N_S(U)$, n_{t^U} divides n_t; cf. Lemma 4.3.1(ii). Thus, n_{t^U} is a π-number. Thus, as $s^U = t^U$, n_{s^U} is a π-number.

4.5 Sylow Subsets

In this section, S is assumed to have finite valency. We fix a prime number and call it p.

For each element s in S, we write $\langle s\rangle$ instead of $\langle\{s\}\rangle$.

The following lemma on finite groups is commonly referred to Augustin-Louis Cauchy.

Lemma 4.5.1 *Let T be a thin closed subset of S such that p divides n_T. Then T contains a closed subset of valency p.*

Proof. Assume T to be a minimal counterexample. Then T does not contain closed subsets different from T the valency of which is divisible by p.

Let t be an element in T. Then $C_T(t)$ is closed. Thus, if $C_T(t) \neq T$, p does not divide the valency of $C_T(t)$. Thus, p divides $n_{T//C_T(t)}$.

Let us denote by $Z(T)$ the intersection of all closed subsets $C_T(t)$ with $t \in T$. Then $Z(T)$ is closed and p divides the valency of $Z(T)$. Thus, $Z(T) = T$.

Let U be a closed subsets of T such that $\{1\} \neq U \neq T$. Then, by Lemma 4.3.3(i), p divides $n_{T//U}$. Thus, as $\{1\} \neq U$, the minimal choice of T yields a closed subset V of S such that $U \subseteq V$ and $n_{V//U} = p$. Let t be an element in $V \setminus U$. Then $\langle t\rangle$ has valency p.

Let T be a closed subset of S. Note that a $\{p\}$-subset of T is the same as a p-subset of T. We also speak about *Sylow p-subsets* of T instead of Hall $\{p\}$-subsets of T.[1] Our notation for the set of all Sylow p-subsets of T will be $\mathrm{Syl}_p(T)$.

Sylow p-subsets of a closed subset T of S are particularly interesting if T is p-valenced.

[1] This is just for historical reasons.

Proposition 4.5.2 *Let V be a p-valenced closed subset of S, and let T be closed p-subset of V such that $T \notin \mathrm{Syl}_p(V)$. Then there exists a closed p-subset U of V such that $T \subseteq U \subseteq K_V(T)$ and $pn_T = n_U$.*

PROOF. We are assuming that S is p-valenced and that T is a closed p-subset of V. Thus, by Corollary 4.3.2(i), $V//T$ is p-valenced.

We are assuming that $T \notin \mathrm{Syl}_p(V)$. Thus, as T is a closed p-subset of V, p divides $n_{V//T}$. Thus, as $V//T$ is p-valenced, p divides $n_{O_\vartheta(V//T)}$. Thus, by Lemma 4.5.1, there exists a closed subset U of V such that $T \subseteq U$, $U//T \subseteq O_\vartheta(V//T)$ and $n_{U//T} = p$.

From $U//T \subseteq O_\vartheta(V//T)$ we obtain $U \subseteq K_V(T)$; cf. Lemma 4.2.5(i) (together with Lemma 4.2.1(ii)). From $n_{U//T} = p$ and $n_U = n_{U//T}n_T$ we obtain $pn_T = n_U$.

The following three theorems were first proved in [24]. A more general approach is given in [4]. The thin case was already proved in 1872 by Ludwig Sylow; cf. [36]. (Note that thin schemes are p-valenced.)

Theorem 4.5.3 *Each p-valenced closed subset of S possesses at least one Sylow p-subset.*

PROOF. Let T be a p-valenced closed subset of S. Then, Proposition 4.5.2 says that, for each power q of p which divides the valency of T, T possesses a closed p-subset of valency q. In particular, T possesses a Sylow p-subset.

Proposition 4.5.4 *Let V be a p-valenced closed subset of S, let T be a closed p-subset of V, and let U be a Sylow p-subset of V. Then there exists an element s in V such that $s^*Ts \subseteq U$.*

PROOF. Let R be a transversal of (T, U) in V. Then, by Lemma 2.1.3,

$$n_V = \sum_{r \in R} n_{TrU}.$$

Now we are assuming that n_T and n_U are powers of p. Moreover, for each element r in R, n_r is assumed to be a power of p. Thus, for each element r in R, n_{TrU} is a power of p; cf. Lemma 2.3.3.

On the other hand, for each element r in R, $n_U \leq n_{TrU}$; cf. Lemma 1.4.4(i). Thus, as n_U is the highest power of p dividing n_V, there exists an element s in R such that $n_U = n_{TsU}$. From $n_U = n_{TsU}$ we obtain $s^*Ts \subseteq U$; cf. Lemma 1.4.4(ii).

Theorem 4.5.5 *Let U be a p-valenced closed subset of S, and let T be a Sylow p-subset of U. Then $\mathrm{Syl}_p(U) = \{u^*Tu \mid u \in U,\ uu^* \subseteq T\}$.*

PROOF. Let T' be a Sylow p-subset of U. Then, by definition, $n_T = n_{T'}$. Moreover, by Proposition 4.5.4, there exists an element u in U such that $u^*Tu \subseteq T'$. Thus, $n_{u^*Tu} \leq n_{T'}$. On the other hand, we know from Lemma 1.4.5(i) that $n_T \leq n_{Tu} \leq n_{u^*Tu}$. Thus, we have $n_{u^*Tu} = n_{T'}$ and $n_T = n_{Tu}$.

From $n_{u^*Tu} = n_{T'}$ and $u^*Tu \subseteq T'$ we obtain $u^*Tu = T'$. From $n_T = n_{Tu}$ we obtain $uu^* \subseteq T$; cf. Lemma 1.4.5(ii).

Conversely, let u be an element in U such that $uu^* \subseteq T$. Then, by Lemma 2.6.1(ii), u^*Tu is a closed subset of U. Moreover, by Lemma 2.6.2(i), $n_T = n_{u^*Tu}$. Thus, as U is assumed to be p-valenced, $u^*Tu \in \mathrm{Syl}_p(U)$.

Let U be a p-valenced closed subset of S, and let T be a Sylow p-subset of U. From Theorem 4.5.5 together with Lemma 2.6.1(i) we obtain $\mathrm{Syl}_p(U) = \{T^s \mid u \in U,\ uu^* \subseteq T\}$.

Proposition 4.5.6 *Let V be a p-valenced closed subset of S, let T be a Sylow p-subset of V, and set $U := K_V(T)$. Then the following hold.*

(i) *We have $|\mathrm{Syl}_p(V)| = |\{Us \mid s \in V,\ ss^* \subseteq T\}|$.*

(ii) *The prime number p divides $n_{V//U} - 1$.*

PROOF. (i) From Lemma 2.6.3(ii) we know that

$$|\{s^*Ts \mid s \in V,\ ss^* \subseteq T\}| = |\{Us \mid s \in V,\ ss^* \subseteq T\}|,$$

so that our claim is a consequence of Theorem 4.5.5.

(ii) We are assuming that V is p-valenced. Thus, as n_T is a power of p, $V//T$ is p-valenced; cf. Corollary 4.3.2(i). On the other hand, we know from Lemma 4.2.5(i) that $O_\vartheta(V//T) = U//T$. Thus, p divides $n_{V//T} - n_{U//T}$. Thus, as $n_{V//T} = n_{V//U}n_{U//T}$, p divides $(n_{V//U} - 1)n_{U//T}$.

Since we are assuming T to be a Sylow p-subset of V, p does not divide $n_{V//T}$. Thus, as $n_{U//T}$ divides $n_{V//T}$, p does not divide $n_{U//T}$. Thus, as p divides $(n_{V//U} - 1)n_{U//T}$, p divides $n_{V//U} - 1$.

According to Theorem 4.5.3, each p-valenced closed subset of S possesses at least one Sylow p-subset. Generalizing this theorem we shall now say a little bit more about the number of Sylow p-subsets of such a closed subset of S.

Theorem 4.5.7 *The number of Sylow p-subsets of a p-valenced closed subset of S is congruent to 1 modulo p.*

PROOF. Let V be a p-valenced closed subset of S, let T be a Sylow p-subset of V, and set $U := K_V(T)$. We fix a right transversal R of U in V. Then n_V is the sum of all integers n_{Ur} with $r \in R$. Thus, as $n_V = n_{V//U}n_U$,

$$n_{V//U} = \sum_{r \in R} (n_U)^{-1} n_{Ur}.$$

Now recall that, for each element r in R, $(n_U)^{-1}n_{r^*U}$ divides n_r; cf. Lemma 2.3.4(iii). Note also that, for each element r in R, $n_{r^*U} = n_{Ur}$; cf. Lemma 1.1.2(iii) and Lemma 1.3.2(iii). Thus, for each element r in R, $(n_U)^{-1}n_{Ur}$ divides n_r. Thus, as V is assumed to be p-valenced, we conclude that, for each element r in R, $(n_U)^{-1}n_{Ur}$ is a power of p.

Let us compute the number of elements r in R such that $(n_U)^{-1}n_{Ur} = 1$.

From Lemma 1.4.5(ii) we know that, for each element s in V, $n_U = n_{Us}$ if and only if $ss^* \subseteq U$. Moreover, as V is assumed to be p-valenced, we obtain from Lemma 4.3.5(ii) that, for each element s in V, $ss^* \subseteq U$ if and only if $ss^* \subseteq T$. Thus, by the above equation, p divides

$$n_{V//U} - |\{Us \mid s \in V,\ ss^* \subseteq T\}|.$$

Now our claim follows from Proposition 4.5.6.

Let R be a subset of S such that, for each element r in R, $n_r = 2$. Then, by Corollary 3.1.7, the set $\langle R\rangle$ is 2-valenced. Thus, the three previous theorems apply to $\langle R\rangle$.

For thin schemes, the following lemma is known as a result due to William Burnside.

Lemma 4.5.8 *Let s be an element in S, let T be a Sylow p-subset of S, and let R be a nonempty subset of T such that $ss^* \subseteq T \subseteq K_S(R)$, and $T \subseteq K_S(s^*Rs)$. Then there exists an element r in $K_S(T)$ such that $r^*Rr \subseteq s^*Rs$.*

Proof. We are assuming that $ss^* \subseteq T$ and that $T \in \mathrm{Syl}_p(S)$. Thus, by Theorem 4.5.5, $s^*Ts \in \mathrm{Syl}_p(S)$. Now we have $T,\ s^*Ts \in \mathrm{Syl}_p(K_S(s^*Rs))$.[2] Thus, by Proposition 4.5.4, there exists an element q in $K_S(s^*Rs)$ such that $q^*Tq \subseteq s^*Ts$.

From $q^*Tq \subseteq s^*Ts$ we obtain

$$sq^*Tqs^* \subseteq ss^*Tss^* \subseteq T.$$

Thus, $qs^* \subseteq K_S(T)$. Thus, as $K_S(T)$ is closed, $sq^* \subseteq K_S(T)$.

From $q \in K_S(s^*Rs)$ we obtain $q^* \in K_S(s^*Rs)$. Thus, $qs^*Rsq^* \subseteq s^*Rs$. Thus, for each element r in sq^*, we have $r^*Rr \subseteq s^*Rs$.

[2] Note that, by Lemma 2.5.8(ii), $K_S(s^*Rs)$ is a closed subset of S.

5

Morphisms

Let $\tilde{X}$ be a set, let $\tilde{S}$ be a scheme on $\tilde{X}$, let ϕ be a map from $X \cup S$ to $\tilde{X} \cup \tilde{S}$, and assume that $X\phi \subseteq \tilde{X}$ and that $S\phi \subseteq \tilde{S}$. The map ϕ is called a *morphism* (or a *scheme morphism*) if, for any two elements x in X and s in S,

$$(xs)\phi \subseteq x\phi s\phi.$$

A morphism ϕ from $X \cup S$ will be called a *homomorphism* (or a *scheme homomorphism*) if, for any three elements y, z in X and s in S with $z\phi \in y\phi s\phi$, there exist elements v in X and w in vs such that $v\phi = y\phi$ and $w\phi = z\phi$.

In the first section of this chapter, we compile the most general facts about morphisms and homomorphisms.

A bijective homomorphism is called an *isomorphism* (or a *scheme isomorphism*). An isomorphism from $X \cup S$ to $X \cup S$ is called an *automorphism* (or a *scheme automorphism*) of S. The set of all automorphisms of S will be denoted by $\mathrm{Aut}(S)$.

In the second section, we shall see that the set $\mathrm{Aut}(S)$ is a group with respect to composition. We also consider subgroups of $\mathrm{Aut}(S)$ and groups which contain $\mathrm{Aut}(S)$ as a subgroup.

In the third section, we shall prove a Homomorphism Theorem and two Isomorphism Theorems for schemes of finite valency. All three of these results naturally generalize the finite versions of Emmy Noether's corresponding theorems for groups.

In Section 5.4, we define composition series of schemes having finite valency. The main result will be a generalization of a group theoretic theorem due to Camille Jordan and Otto Hölder. This generalization says that any two composition series of a closed subset of a scheme of finite valency are isomorphic.

Composition series lead naturally to the notion of a composition factor, and, according to Lemma 4.2.4(i), composition factors must be simple. Thus, the above theorem on composition series gives reason to consider simple schemes as crucial in scheme theory.

The notion of a homomorphism enables us to establish the group correspondence which has been mentioned in the beginning of the preface. The group correspondence is the content of Section 5.5.

In the last of the six sections of this chapter, we look at schemes which have finite valency and only thin composition factors.

5.1 Basic Facts

Let $\tilde{X}$ be a set, let $\tilde{S}$ be a scheme on $\tilde{X}$, let ϕ be a map from $X \cup S$ to $\tilde{X} \cup \tilde{S}$, and assume that $X\phi \subseteq \tilde{X}$ and that $S\phi \subseteq \tilde{S}$. Note that ϕ is a morphism if and only if, for any three elements y in X, s in S, and z in ys, $z\phi \in y\phi s\phi$.

Lemma 5.1.1 *Let ϕ be a morphism from $X \cup S$. Then we have the following.*

(i) *For each element s in S, $s^*\phi = (s\phi)^*$.*

(ii) *For any two elements p and q in S, $(pq)\phi \subseteq p\phi q\phi$.*

(iii) *For any two elements $\tilde{p}$ and $\tilde{q}$ in $S\phi$, $\tilde{p}\phi^{-1}\tilde{q}\phi^{-1} \subseteq (\tilde{p}\tilde{q})\phi^{-1}$.*

Proof. (i) Let s be an element in S. In order to show that $s^*\phi = (s\phi)^*$, it suffices to show that $s^*\phi \cap (s\phi)^*$ is not empty.

Let y be an element in X, and let z be an element in ys. From $z \in ys$ we obtain $y \in zs^*$. Thus, as ϕ is assumed to be a morphism, $y\phi \in z\phi s^*\phi$. On the other hand, $z \in ys$ yields $z\phi \in y\phi s\phi$, and that means that $y\phi \in z\phi(s\phi)^*$.

(ii) Let p and q be elements in S, and let s be an element in pq. We have to show that $s\phi \in p\phi q\phi$.

Let y be an element in X, and let z be an element in ys. Since $z \in ys$ and $s \in pq$, $z \in ypq$. Thus, there exists an element x in yp such that $z \in xq$. From $x \in yp$ we obtain $x\phi \in y\phi p\phi$. From $z \in xq$ we obtain $z\phi \in x\phi q\phi$. It follows that $z\phi \in y\phi p\phi q\phi$.

On the other hand, as $z \in ys$, $z\phi \in y\phi s\phi$. Thus, $s\phi \in p\phi q\phi$.

(iii) Let $\tilde{p}$ and $\tilde{q}$ be elements in $S\phi$, and let s be an element in $\tilde{p}\phi^{-1}\tilde{q}\phi^{-1}$. We have to show that $s \in (\tilde{p}\tilde{q})\phi^{-1}$.

Since $s \in \tilde{p}\phi^{-1}\tilde{q}\phi^{-1}$, there exist elements t in $\tilde{p}\phi^{-1}$ and u in $\tilde{q}\phi^{-1}$ such that $s \in tu$. Thus, by (ii),

$$s\phi \in (tu)\phi \subseteq t\phi u\phi = \tilde{p}\tilde{q},$$

whence $s \in (\tilde{p}\tilde{q})\phi^{-1}$.

Let ϕ be a morphism from $X \cup S$, and let R be a subset of S. Considering Lemma 3.1.1(i) we obtain from Lemma 5.1.1(i), (ii) that $\langle R \rangle\phi \subseteq \langle R\phi \rangle$.

Lemma 5.1.2 *Let $\tilde{X}$ be a set, let $\tilde{S}$ be a scheme on $\tilde{X}$, and let ϕ be a morphism from $X \cup S$ to $\tilde{X} \cup \tilde{S}$. Then the following hold.*

(i) *We have* $1\phi = 1_{\tilde{X}}$.

(ii) *Let* $\tilde{T}$ *be a closed subset of* $\tilde{S}$. *Then* $\tilde{T}\phi^{-1}$ *is a closed subset of* S.

PROOF. (i) Let x be an element in X. Then, as ϕ is assumed to be a morphism, $x\phi \in x\phi 1\phi$. On the other hand, as $x\phi \in \tilde{X}$, $x\phi \in x\phi 1_{\tilde{X}}$. Thus, $1\phi \cap 1_{\tilde{X}}$ is not empty, so that $1\phi = 1_{\tilde{X}}$.

(ii) We are assuming that $\tilde{T}$ is closed. Thus, $1_{\tilde{X}} \in \tilde{T}$. Thus, by (i), $1\phi \in \tilde{T}$. It follows that $1 \in \tilde{T}\phi^{-1}$. In particular, $\tilde{T}\phi^{-1}$ is not empty.

Let p and q be elements in $\tilde{T}\phi^{-1}$, and let s be an element in p^*q. Then, referring to Lemma 5.1.1(ii), (i) we obtain

$$s\phi \in (p^*q)\phi \subseteq p^*\phi q\phi = (p\phi)^* q\phi \subseteq \tilde{T}^*\tilde{T} \subseteq \tilde{T}.$$

It follows that $s \in \tilde{T}\phi^{-1}$.

Let $\tilde{X}$ be a set, and let $\tilde{S}$ be a scheme on $\tilde{X}$. For each morphism ϕ from $X \cup S$ to $\tilde{X} \cup \tilde{S}$, we set $\phi_X := \phi \cap (X \times \tilde{X})$.

Lemma 5.1.3 *Let* ϕ *be a morphism from* $X \cup S$, *and assume* ϕ_X *to be surjective. Then the following hold.*

(i) *For any two elements* $\tilde{p}$ *and* $\tilde{q}$ *in* $S\phi$, $(\tilde{p}\tilde{q})\phi^{-1} = \tilde{p}\phi^{-1}\tilde{q}\phi^{-1}$.

(ii) *The map* ϕ *is surjective.*

PROOF. (i) Let $\tilde{p}$ and $\tilde{q}$ be elements in $S\phi$, and let s be an element in $(\tilde{p}\tilde{q})\phi^{-1}$. We shall show that $s \in \tilde{p}\phi^{-1}\tilde{q}\phi^{-1}$. The claim will then follow from Lemma 5.1.1(iii).

Let y be an element in X, and let z be an element in ys. Then $z\phi \in y\phi s\phi \subseteq y\phi\tilde{p}\tilde{q}$. Thus, there exists an element $\tilde{x}$ in $y\phi\tilde{p}$ such that $z\phi \in \tilde{x}\tilde{q}$.

Since ϕ_X is assumed to be surjective, there exists an element x in X such that $x\phi = \tilde{x}$. Let p be the element in S which satisfies $x \in yp$, and let q be the element in S with $z \in xq$. From $x \in yp$ we obtain $x\phi \in y\phi p\phi$, from $z \in xq$ we obtain $z\phi \in x\phi q\phi$.

From $x\phi = \tilde{x}$, $x\phi \in y\phi p\phi$, and $\tilde{x} \in y\phi\tilde{p}$ we obtain $p\phi = \tilde{p}$. Similarly, we obtain from $x\phi = \tilde{x}$, $z\phi \in x\phi q\phi$, and $z\phi \in \tilde{x}\tilde{q}$ that $q\phi = \tilde{q}$.

Finally, as $z \in xq$ and $x \in yp$, $z \in ypq$. Thus, as $z \in ys$, $s \in pq \subseteq \tilde{p}\phi^{-1}\tilde{q}\phi^{-1}$.

(ii) Let us denote by $\tilde{X}$ the codomain of ϕ and by $\tilde{S}$ the codomain of ϕ.

In order to show that ϕ is surjective, we fix an element $\tilde{s}$ in $\tilde{S}$. We shall show that there exists an element s in S such that $s\phi = \tilde{s}$.

Since $\tilde{s} \in \tilde{S}$, there exist elements $\tilde{y}$ and $\tilde{z}$ in $\tilde{X}$ such that $\tilde{z} \in \tilde{y}\tilde{s}$. Since ϕ_X is assumed to be surjective, there exist elements y and z in X such that $y\phi = \tilde{y}$ and $z\phi = \tilde{z}$.

Let s be the uniquely determined element in S such that $z \in ys$. Then, as ϕ is assumed to be a morphism, $z\phi \in y\phi s\phi$. Thus, as $y\phi = \tilde{y}$ and $z\phi = \tilde{z}$, $\tilde{z} \in \tilde{y}s\phi$. Thus, as $\tilde{z} \in \tilde{y}\tilde{s}$, $s\phi = \tilde{s}$.

Let ϕ be a morphism from $X \cup S$. We define

$$\ker(\phi) := \{s \in S \mid s\phi = 1\phi\}.$$

The set $\ker(\phi)$ is called the *kernel* of ϕ.

From Lemma 5.1.2(ii) one obtains that $\ker(\phi)$ is closed for each morphism ϕ from $X \cup S$.

Lemma 5.1.4 *Let ϕ be a morphism from $X \cup S$, and set $T := \ker(\phi)$. Then we have the following.*

(i) *For any two elements y and z in X, we have $yT = zT$ if and only if $y\phi = z\phi$.*

(ii) *Let p and q be elements in S such that $p^T = q^T$. Then $p\phi = q\phi$.*

PROOF. (i) Let y and z be elements in X, and let us write s to denote the element in S which satisfies $z \in ys$. Since ϕ is assumed to be a morphism, we obtain from $z \in ys$ that $z\phi \in y\phi s\phi$.

Since $z \in ys$, we have $yT = zT$ if and only if $s \in T$; cf. Lemma 2.1.4. Since T stands for the kernel of ϕ, $s \in T$ is equivalent to $s\phi = 1\phi$. From $z\phi \in y\phi s\phi$ we obtain that $s\phi = 1\phi$ if and only if $y\phi = z\phi$; cf. Lemma 5.1.2(i).

(ii) We are assuming that $p^T = q^T$. Thus, by Lemma 4.1.1, $TpT = TqT$. In particular, $q \in TpT$.

Since q is not empty, there exist elements y and z in X such that $z \in yq$. Thus, as $q \in TpT$, $z \in yTpT$. Thus, there exists an element v in yT such that $z \in wpT$. Since $z \in vpT$, there exists an element w in vp such that $z \in wT$.

From $y \in vT$ we obtain $vT = yT$; from $z \in wT$ we obtain $wT = zT$.

Since ϕ is assumed to be a morphism, we obtain from $w \in vp$ that $w\phi \in v\phi p\phi$. Similarly, we obtain from $z \in yq$ that $z\phi \in y\phi q\phi$.

Since $vT = yT$, we obtain from (i) that $v\phi = y\phi$. Similarly, $wT = zT$ yields $w\phi = z\phi$. Thus, $p\phi = q\phi$.

Lemma 5.1.5 *Let $\tilde{X}$ be a set, let $\tilde{S}$ be a scheme on $\tilde{X}$, let ϕ be a morphism from $X \cup S$ to $\tilde{X} \cup \tilde{S}$, and set $T := \ker(\phi)$.*

For each element x in X, set $(xT)\bar{\phi} := x\phi$. For each element s in S, set $(s^T)\bar{\phi} := s\phi$.

Then $\bar{\phi}$ is a morphism from $X/T \cup S//T$ to $\tilde{X} \cup \tilde{S}$.

PROOF. It follows from Lemma 5.1.4 that ϕ is a (well-defined) map which maps X/T to $\tilde{X}$ and $S//T$ to $\tilde{S}$.

Let y and z be elements in X, and let s be an element in S such that $zT \in (yT)(s^T)$. Then, by definition, $z \in yTsT$. Thus, there exists an element v in yT such that $z \in vsT$. Since $z \in vsT$, there exists an element w in vs such that $z \in wT$.

From $v \in yT$ we obtain $vT = yT$, so that $v\phi = y\phi$. Similarly, $z \in wT$ yields $w\phi = z\phi$. Finally, as ϕ is a morphism, $w \in vs$ yields $w\phi \in v\phi s\phi$. Thus, we conclude that $z\phi \in y\phi s\phi$, and that means that $(zT)\bar{\phi} \in (yT)\bar{\phi}(s^T)\bar{\phi}$.

Let $\tilde{X}$ be a set, and let $\tilde{S}$ be a scheme on $\tilde{X}$. For each morphism ϕ from $X \cup S$ to $\tilde{X} \cup \tilde{S}$, we set $\phi_S := \phi \cap (S \times \tilde{S})$.

Lemma 5.1.6 *Let ϕ be a morphism from $X \cup S$. Then we have the following.*

(i) *Assume that ϕ_S is injective. Then, for any three elements y, z in X and s in S, $z\phi \in y\phi s\phi$ implies $z \in ys$.*

(ii) *Assume that, for any three elements y, z in X and s in S, $z\phi \in y\phi s\phi$ implies $z \in ys$. Then $\{1\} = \ker(\phi)$.*

(iii) *If $\{1\} = \ker(\phi)$, ϕ_X is injective.*

PROOF. (i) Let y and z be elements in X, and let s be an element in S such that $z\phi \in y\phi s\phi$. We have to show that $z \in ys$.

Let us denote by r the uniquely determined element in S satisfying $z \in yr$. Then $z\phi \in y\phi r\phi$. Thus, as $z\phi \in y\phi s\phi$, $r\phi = s\phi$. Since we are assuming ϕ to be injective, this yields $r = s$. Thus, as $z \in yr$, $z \in ys$.

(ii) Let s be an element in $\ker(\phi)$. We have to show that $s = 1$.

Since $s \in \ker(\phi)$, $s\phi = 1\phi$. Let y be an element in X, and let z be an element in ys. From $z \in ys$ we obtain $z\phi \in y\phi s\phi$. Thus, as $s\phi = 1\phi$, $z\phi \in y\phi 1\phi$. Thus, by hypothesis, $y = z$. Thus, as $z \in ys$, $s = 1$.

(iii) This is an immediate consequence of Lemma 5.1.4(i).

Recall that a morphism ϕ from $X \cup S$ is called a homomorphism if, for any three elements y, z in X and s in S with $z\phi \in y\phi s\phi$, there exist elements v in X and w in vs such that $v\phi = y\phi$ and $w\phi = z\phi$.

Let ϕ be a morphism from $X \cup S$, and assume that ϕ_S is injective. Then, by Lemma 5.1.6(i), ϕ is a homomorphism.

Lemma 5.1.7 *Let ϕ be a homomorphism from $X \cup S$, and assume that ϕ_X is injective. Then ϕ is injective.*

PROOF. Let p and q be elements in S, and let us assume that $p\phi = q\phi$. We have to show that $p = q$

Since p is not empty, there exist elements y and z in X such that $z \in yp$. From $z \in yp$ we obtain $z\phi \in y\phi p\phi$. Thus, as $p\phi = q\phi$, $z\phi \in y\phi q\phi$. Thus, as ϕ

is assumed to be a homomorphism, there exist elements v in X and w in vq such that $v\phi = y\phi$ and $w\phi = z\phi$.

Since we are assuming ϕ_X to be injective, we obtain from $v\phi = y\phi$ that $v = y$. Similarly, we obtain from $w\phi = z\phi$ that $w = z$. From $v = y$, $w = z$, and $w \in vq$ we obtain $z \in yq$. Thus, as $z \in yp$, $p = q$.

Lemma 5.1.8 *A homomorphism ϕ is injective if and only if $\{1\} = \ker(\phi)$.*

PROOF. Considering Lemma 5.1.6 this is a consequence of Lemma 5.1.7.

5.2 Isomorphisms

It follows right from the definition of a morphism that the product of two morphisms is a morphism. This is not necessarily the case with homomorphisms. However, we have the following.

Lemma 5.2.1 *Let $\tilde{X}$ be a set, let $\tilde{S}$ be a scheme on $\tilde{X}$, let ϕ be a homomorphism from $X \cup S$ to $\tilde{X} \cup \tilde{S}$, and let $\tilde{\phi}$ be a homomorphism from $\tilde{X} \cup \tilde{S}$. Assume ϕ_X to be surjective. Then $\phi\tilde{\phi}$ is a homomorphism.*

PROOF. Let y and z be elements in X, and let s be an element in S such that

$$z(\phi\tilde{\phi}) \in y(\phi\tilde{\phi})s(\phi\tilde{\phi}).$$

We have to show that there exist elements v in X and w in vs such that $v(\phi\tilde{\phi}) = y(\phi\tilde{\phi})$ and $w(\phi\tilde{\phi}) = z(\phi\tilde{\phi})$.

From $z(\phi\tilde{\phi}) \in y(\phi\tilde{\phi})s(\phi\tilde{\phi})$ we obtain

$$(z\phi)\tilde{\phi} \in (y\phi)\tilde{\phi}(s\phi)\tilde{\phi}.$$

Thus, as $\tilde{\phi}$ is assumed to be a homomorphism, there exist elements $\tilde{v}$ in $\tilde{X}$ and $\tilde{w}$ in $\tilde{v}(s\phi)$ such that $\tilde{v}\tilde{\phi} = (y\phi)\tilde{\phi}$ and $\tilde{w}\tilde{\phi} = (z\phi)\tilde{\phi}$.

Since ϕ_X is assumed to be surjective, there exist elements v' in X with $v'\phi = \tilde{v}$ and w' in X with $w'\phi = \tilde{w}$.

From $v'\phi = \tilde{v}$, $w'\phi = \tilde{w}$, and $\tilde{w} \in \tilde{v}(s\phi)$ we obtain $w'\phi \in v'\phi(s\phi)$. Thus, as ϕ is assumed to be a homomorphism, there exist elements v in X and w in vs such that $v\phi = v'\phi$ and $w\phi = w'\phi$.

From $v\phi = v'\phi$, $v'\phi = \tilde{v}$, and $\tilde{v}\tilde{\phi} = (y\phi)\tilde{\phi}$ we obtain

$$v(\phi\tilde{\phi}) = (v\phi)\tilde{\phi} = (v'\phi)\tilde{\phi} = \tilde{v}\tilde{\phi} = (y\phi)\tilde{\phi} = y(\phi\tilde{\phi}).$$

Similarly, we obtain from $w\phi = w'\phi$, $w'\phi = \tilde{w}$, and $\tilde{w}\tilde{\phi} = (z\phi)\tilde{\phi}$ that $w(\phi\tilde{\phi}) = z(\phi\tilde{\phi})$.

Recall that an isomorphism from $X \cup S$ to $X \cup S$ is called an automorphism of S. Recall also that $\mathrm{Aut}(S)$ is our notation of the set of all automorphisms of S. Recall, finally, that, for each element a in $\mathrm{Aut}(S)$, a_S is our notation for $a \cap (S \times S)$.

Let us define $\mathrm{Sch}(S)$ to be the set of all elements a in $\mathrm{Aut}(S)$ such that $a_S = \{1_S\}$.

A bijective map from a set to itself is called a *permutation.*

For each closed subset T of S, we define $\mathrm{Stc}(T)$ to be the set of all permutations σ of T such that, for any three elements p, q, and r in T, $a_{pqr} = a_{p\sigma q\sigma r\sigma}$.

Theorem 5.2.2 *We have the following.*

(i) *The set* $\mathrm{Aut}(S)$ *is a group with respect to composition.*

(ii) *The set* $\mathrm{Stc}(S)$ *is a group with respect to composition.*

(iii) *For each element a in* $\mathrm{Aut}(S)$, *we define* $a\zeta := a_S$. *Then ζ is a group homomorphism from* $\mathrm{Aut}(S)$ *to* $\mathrm{Stc}(S)$ *with kernel* $\mathrm{Sch}(S)$.

PROOF. (i) Note that the inverse map of an isomorphism is an isomorphism. Thus, the claim is an immediate consequence of Lemma 5.2.1.

(ii) This is an immediate consequence of the definition of $\mathrm{Stc}(S)$.

(iii) Let a be an automorphism of S. Then a_S is a permutation of S.

Let p, q, and r be elements in S, and let y and z be elements in X with $z \in yr$.

Since a is assumed to be an automorphism, a is a morphism. Thus, as $z \in yr$, $za \in yara$. From $z \in yr$ we also obtain $|yp \cap zq^*| = a_{pqr}$. Thus, as a is assumed to be a bijective map, $|yapa \cap za(qa)^*| = a_{pqr}$. It follows that $a_{pqr} = a_{paqara}$.

Lemma 5.2.3 *For each element σ in* $\mathrm{Stc}(S)$, *the following hold.*

(i) *For any two elements p and q in S,* $(pq)\sigma = p\sigma q\sigma$.

(ii) *We have* $1\sigma = 1$.

(iii) *For each element s in S,* $s^*\sigma = (s\sigma)^*$.

PROOF. (i) Let s be an element in $(pq)\sigma$. Then there exists an element r in pq such that $s = r\sigma$. Since $r \in pq$, $1 \leq a_{pqr}$. Thus, as $\sigma \in \mathrm{Stc}(S)$, $1 \leq a_{p\sigma q\sigma r\sigma}$. Thus, as $s = r\sigma$, $s \in p\sigma q\sigma$.

Conversely, let s be an element in $p\sigma q\sigma$. Then $1 \leq a_{p\sigma q\sigma s}$. Thus, as $\sigma \in \mathrm{Stc}(S)$, $1 \leq a_{pqs\sigma^{-1}}$. This means that $s\sigma^{-1} \in pq$. Thus, by definition, $s \in (pq)\sigma$.

(ii) From (i) we obtain $\{1\sigma\} = 1\sigma 1\sigma$. Thus, $1\sigma = 1$.

(iii) From (ii), Lemma 1.3.2(i), and (i) we obtain

$$1 = 1\sigma \in (s^*s)\sigma = s^*\sigma s\sigma.$$

Thus, the claim follows from Lemma 1.3.6(ii).

5.3 The Isomorphism Theorems

In this section, S is assumed to have finite valency.

We shall show first that each closed subset of S induces naturally a homomorphism from $X \cup S$.

Theorem 5.3.1 *Let T be a closed subset of S. For each element x in X, we set $x\phi := xT$; for each element s in S, we set $s\phi := s^T$. Then we have the following.*

(i) *The map ϕ is a surjective homomorphism from $X \cup S$ to $X/T \cup S//T$.*

(ii) *We have $T = \ker(\phi)$.*

PROOF. (i) Let y be an element in X, let s be an element in S, and let z be an element in ys. From $z \in ys$ (and $1 \in T$) we obtain $z \in yTsT$. Thus, by definition, $zT \in (yT)(s^T)$, and that means that $z\phi \in y\phi s\phi$.

Since y and z have been chosen arbitrarily in X and s arbitrarily in S, we have shown that ϕ is a morphism.

Let us now show that ϕ is a homomorphism. In order to do so we fix elements y, z in X and s in S, and we assume that $z\phi \in y\phi s\phi$.

From $z\phi \in y\phi s\phi$ we obtain $zT \in (yT)(s^T)$, and that means that $z \in yTsT$. Thus, there exists an element v in yT such that $z \in vsT$. Since $z \in vsT$, there exists an element w in vs such that $z \in wT$.

By Lemma 2.1.4, $v \in yT$ is equivalent to $vT = yT$, and that means that $v\phi = y\phi$. Similarly, we obtain from $z \in wT$ that $w\phi = z\phi$.

Thus, ϕ is a homomorphism from $X \cup S$ to $X/T \cup S//T$.

That ϕ is surjective follows right from the definition of ϕ.

(ii) Let s be an element in S. By definition, $s \in \ker(\phi)$ if and only if $s^T = 1^T$. By Lemma 4.1.1, $s^T = 1^T$ if and only if $s \in T$. Thus, $T = \ker(\phi)$.

The homomorphism ϕ in Theorem 5.3.1(i) is called the *natural homomorphism* from $X \cup S$ to $X/T \cup S//T$.

Let ϕ be a homomorphism from $X \cup S$. From Lemma 5.1.2(ii) we know that $\ker(\phi)$ is closed. Thus, we obtain from Theorem 4.1.3(i) that $S//\ker(\phi)$ is a scheme on $X/\ker(\phi)$.

The following theorem is called the *Homomorphism Theorem* for schemes.

Theorem 5.3.2 *Let ϕ be a homomorphism from $X \cup S$, and set $T := \ker(\phi)$. For each element x in X, set $(xT)\bar{\phi} := x\phi$. For each element s in S, set $(s^T)\bar{\phi} := s\phi$. Then $\bar{\phi}$ is an injective homomorphism from $X/T \cup S//T$.*

PROOF. From Lemma 5.1.5 we know that $\bar{\phi}$ is a morphism from $X/T \cup S//T$. Thus, according to Lemma 5.1.6(i), it suffices to show that $\bar{\phi}$ is injective.

That $\bar{\phi}_X$ is injective follows immediately from Lemma 5.1.4(i).

In order to show that $\bar{\phi}_S$ is injective we fix elements p and q in S, and we assume that $(p^T)\bar{\phi} = (p^T)\bar{\phi}$. Then, by definition, $p\phi = q\phi$.

Let y and z be elements in X such that $z\phi \in y\phi p\phi$. Then, as ϕ is assumed to be a homomorphism, there exist elements v in X and w in vp such that $v\phi = y\phi$ and $w\phi = z\phi$. Since $v\phi = y\phi$, $vT = yT$. Similarly, as $w\phi = z\phi$, $wT = zT$. From $w \in vp$ and $vT = yT$ we obtain $w \in yTp$. Thus, as $z \in wT$, $z \in yTpT$.

From $z\phi \in y\phi p\phi$ and $p\phi = q\phi$ we also obtain $z\phi \in y\phi q\phi$. Thus, replacing p with q in the previous paragraph, we obtain $z \in yTqT$; too. However, from $z \in yTpT$ and $z \in yTqT$ we conclude that $TpT = TqT$; cf. Lemma 2.1.3. Thus, by Lemma 4.1.1, $p^T = q^T$.

Recall that a bijective homomorphism is called an isomorphism.

Let $\tilde{X}$ be a set, and let $\tilde{S}$ be a scheme on $\tilde{X}$. The schemes S and $\tilde{S}$ are said to be *isomorphic* if there exists an isomorphism from $X \cup S$ to $\tilde{X} \cup \tilde{S}$. We shall write $S \cong \tilde{S}$ in order to indicate that S and $\tilde{S}$ are isomorphic.

The following theorem is called the *First Isomorphism Theorem* for schemes.

Theorem 5.3.3 *Let T and U be closed subsets of S such that $T \subseteq U$. Then $(S//T)//(U//T) \cong S//U$.*

PROOF. Let us denote by ϕ the natural homomorphism from $X \cup S$ to $X/T \cup S//T$ and by ψ the natural homomorphism from $X/T \cup S//T$ to

$$(X/T)/(U//T) \cup (S//T)//(U//T).$$

By Theorem 5.3.1(i) and Lemma 5.2.1, $\phi\psi$ is a surjective homomorphism from $X \cup S$ to $(X/T)/(U//T) \cup (S//T)//(U//T)$.

On the other hand, we know from Theorem 5.3.1(ii) that $U//T = \ker(\psi)$. Therefore, we have

$$\ker(\phi\psi) = \{s \in S \mid s\phi \in U//T\} = \{s \in S \mid s^T \in U//T\} = U;$$

use Lemma 4.2.1(ii) and Lemma 4.1.7(ii) for the last equation. Now the claim follows from Theorem 5.3.2.

The following theorem is called the *Second Isomorphism Theorem* for schemes.

Theorem 5.3.4 *Let T and U be closed subsets of S, and assume that $T \subseteq N_S(U)$. Then the following hold.*

(i) *The closed set $T \cap U$ is normal in T.*

(ii) *The closed set U is normal in TU.*

(iii) *For each element x in X, we have $(T//T \cap U)_x \cong (TU//U)_x$.*

Proof. (i) This follows from Lemma 2.5.1(i).

(ii) This follows from Lemma 2.5.1(ii).

(iii) For each element y in xT, we define

$$y\phi := yU.$$

For each element t in T, we define

$$(t_{xT})\phi := (t^U)_{(xU)(TU//U)}.$$

Then ϕ is a map from $xT \cup T_{xT}$ to

$$(xU)(TU//U) \cup (TU//U)_{(xU)(TU//U)}$$

which maps xT to $(xU)(TU//U)$ and T_{xT} to $(TU//U)_{(xU)(TU//U)}$.

We claim that ϕ is a morphism. In order to show this we pick elements y in xT, t in T, and z in yt.

From $z \in yt$ and $1 \in U$ we obtain $z \in yUtU$. Thus, by definition, $zU \in (yU)(t^U)$. From $y \in xT$ we similarly obtain $yU \in (xU)(TU//U)$, and from $z \in xT$ we obtain $zU \in (xU)(TU//U)$. Thus, $zU \in (yU)(t^U)_{(xU)(TU//U)}$, and that means that $z\phi = y\phi(t_{xT})\phi$.

Let us now show that ϕ is a homomorphism. In order to do so we pick elements y and z in xT and t in T such that $z\phi \in y\phi(t_{xT})\phi$.

From $z\phi \in y\phi(t_{xT})\phi$ we obtain $zU \in (yU)(t^U)$, whence, by definition, $z \in yUtU$. Since $z \in yUtU$, there exists an element v in yU such that $z \in vtU$. Since $z \in vtU$, there exists an element w in vt such that $z \in wU$.

From $w \in vt$ and $v \in yU$ we obtain $w \in yUt$. However, as we are assuming that $T \subseteq N_S(U)$, we have that $Ut \subseteq tU$. Thus, $w \in ytU$. Thus, there exists an element w' in yt such that $w \in w'U$.

From $z \in wU$ and $w \in w'U$ we obtain

$$w'\phi = w'U = zU = z\phi;$$

cf. Lemma 2.1.4. From $w' \in yt$ we obtain $w' \in yt_{xT}$. Thus, ϕ is a homomorphism.

Let us compute $\ker(\phi)$. For each element t in T, we have $(t_{xT})\phi = (1_{xT})\phi$ if and only if $t^U = 1^U$. From Lemma 4.1.1 we know that, for each element t in T, $t^U = 1^U$ is equivalent to $t \in U$. Therefore, $\ker(\phi) = (T \cap U)_{xT}$.

For each element z in xTU, there exists an element y in xT such that $yU = zU$. Therefore, ϕ_{xT} is surjective. Thus, according to Lemma 5.1.3(ii), ϕ is surjective.

Now the claim follows from Theorem 5.3.2.

Note that the Homomorphism Theorem and the First Isomorphism Theorem deal with factorizations over arbitrary closed subsets, whereas the Second Isomorphism Theorem deals with factorizations over normal closed subsets.

We shall generalize the Second Isomorphism Theorem in the next section.

The Homomorphism Theorem and the two Isomorphism Theorems were first proved in [35]. The thin case was already proved in 1929 by Emmy Noether; cf. [32; I. §2].

Let y and z be elements in X, and let T be a closed subset of S. In general, it is not true that $T_y \cong T_z$. However, from Theorem 5.3.4(iii) we immediately obtain the following.

Corollary 5.3.5 *Let T and U be closed subsets of S such that $\{1\} = T \cap U$ and $T \subseteq N_S(U)$. Then, for each element x in X, $T_x \cong (TU//U)_x$.*

Note that, by Lemma 2.5.1(ii), the hypothesis $T \subseteq N_S(U)$ in Corollary 5.3.5 says exactly that U is a normal closed subset of S.

5.4 Composition Series

In this short section, S is assumed to have finite valency. We start with a generalization of a theorem of Hans Zassenhaus; cf. [41; II. §5].

Theorem 5.4.1 *Let T, T', U, and U' be closed subsets of S. Assume that T is normal in T' and that U is normal in U'. Then the following hold.*

(i) *The set $T' \cap TU$ and $T' \cap TU'$ are closed.*

(ii) *The closed set $T' \cap TU$ is normal in $T' \cap TU'$.*

(iii) *For each element x in X, we have*

$$(T' \cap U'//T' \cap U' \cap TU)_x \cong (T' \cap TU'//T' \cap TU)_x.$$

PROOF. From Lemma 2.2.1(i) we know that

$$T' \cap TU = T(T' \cap U)$$

and that

$$T' \cap TU' = T(T' \cap U').$$

Since $T' \cap U \subseteq N_S(T)$, $T(T' \cap U)$ is closed; cf. Lemma 2.5.2(iii). Thus, as $T' \cap TU = T(T' \cap U)$, $T' \cap TU$ is closed.

Similarly, one obtains from $T' \cap TU' = T(T' \cap U')$ that $T' \cap TU'$ is closed.

Since $T' \cap TU'$ is closed, $(T' \cap U')(T' \cap TU) \subseteq T' \cap TU'$. Conversely, as $T' \cap TU' = (T' \cap U')T$, $T' \cap TU' \subseteq (T' \cap U')(T' \cap TU)$. It follows that

$$(T' \cap U')(T' \cap TU) = T' \cap TU'.$$

Note also that, by Lemma 2.5.1(i),

$$T' \cap U' \subseteq N_{T'}(TU) \subseteq N_S(T' \cap TU).$$

Thus, the theorem follows from Theorem 5.3.4(i), (iii) (with $T' \cap U'$ in place of T and $T' \cap TU$ in place of U).

Let T be a closed subset of S, and let $\mathcal{U}$ be a set of closed subsets of T such that $\{1\} \in \mathcal{U}$ and $T \in \mathcal{U}$. Let us assume that, for any two elements V and W in $\mathcal{U}$, $V \subseteq W$ or $W \subseteq V$.

For each element U in $\mathcal{U} \setminus \{T\}$, we define $U^{\mathcal{U}}$ to be the intersection of all elements V of $\mathcal{U} \setminus \{U\}$ which contain U as a subset. (Since S is assumed to have finite valency, we have $U^{\mathcal{U}} \in \mathcal{U}$.) The set $\mathcal{U}$ is called a *subnormal series* of T if, for each element U in $\mathcal{U}$, U is normal in $U^{\mathcal{U}}$. A maximal subnormal series of T is called a *composition series* of T.

Two composition series $\mathcal{U}$ and $\mathcal{V}$ of T are called *isomorphic*, if there exists a bijective map η from $\mathcal{U} \setminus \{T\}$ to $\mathcal{V} \setminus \{T\}$ such that, for any two elements x in X and U in $\mathcal{U} \setminus \{T\}$,

$$(U^{\mathcal{U}}//U)_x \cong (U^{\eta\mathcal{V}}//U^{\eta})_x.$$

The following theorem is [35; Theorem 4.2]. The thin case was already proved in 1889 by Otto Hölder; cf. [26; §10].

Theorem 5.4.2 *Let T be a closed subset of S. Then any two composition series of T are isomorphic.*

PROOF. Let $\mathcal{U}$ and $\mathcal{V}$ be composition series of T, let us fix an element U in $\mathcal{U}\setminus\{T\}$, and let us denote by U^{ν} the uniquely determined element V in $\mathcal{V}\setminus\{T\}$ which satisfies $U^{\mathcal{U}} \subseteq UV^{\mathcal{V}}$ and $U^{\mathcal{U}} \not\subseteq UV$.

Since $U^{\mathcal{U}} \subseteq UU^{\nu\mathcal{V}}$,

$$U^{\mathcal{U}} = U^{\mathcal{U}} \cap UU^{\nu\mathcal{V}}.$$

Thus, by Theorem 5.4.1(ii), $U^{\mathcal{U}} \cap UU^{\nu}$ is normal in $U^{\mathcal{U}}$.

Since $U^{\mathcal{U}} \not\subseteq UU^{\nu}$, $U^{\mathcal{U}} \cap UU^{\nu} \neq U^{\mathcal{U}}$. Moreover, U is normal in $U^{\mathcal{U}} \cap UU^{\nu}$, and $U^{\mathcal{U}} \cap UU^{\nu}$ is normal in $U^{\mathcal{U}}$. Thus, as $\mathcal{U}$ is assumed to be a composition series of T, we conclude that

$$U = U^{\mathcal{U}} \cap UU^{\nu}.$$

Using the last two equations we obtain

$$(U^{\mathcal{U}} \cap U^{\nu\mathcal{V}}//U^{\mathcal{U}} \cap U^{\nu\mathcal{V}} \cap UU^{\nu})_x \cong (U^{\mathcal{U}}//U)_x$$

for each element x in X; cf. Theorem 5.4.1(iii).

Let us now define a map υ from $\mathcal{V} \setminus \{T\}$ to $\mathcal{U} \setminus \{T\}$ analogously to ν. Then $U^{\nu\mathcal{V}} \subseteq U^{\nu}U^{\nu\upsilon\mathcal{U}}$ and $U^{\nu\mathcal{V}} \not\subseteq U^{\nu}U^{\nu\upsilon}$.

From the above given isomorphism we obtain $U^{\nu\mathcal{V}} \not\subseteq UU^{\nu}$. Therefore, $U^{\nu\mathcal{V}} \not\subseteq U^{\nu}U$. Thus, by definition of υ, $U \subseteq U^{\nu\upsilon}$.

On the other hand, as above, the fact that $\mathcal{V}$ is a composition series of T yields $U^{\nu} = U^{\nu\mathcal{V}} \cap U^{\nu}U^{\nu\upsilon}$. Therefore, $U^{\mathcal{U}} \subseteq U^{\nu\upsilon}$ would lead to $U^{\mathcal{U}} \cap U^{\nu\mathcal{V}} \subseteq U^{\nu}$, and that contradicts the above given isomorphism. Thus, $U^{\mathcal{U}} \not\subseteq U^{\nu\upsilon}$.

From $U \subseteq U^{\nu\upsilon}$ and $U^{\mathcal{U}} \not\subseteq U^{\nu\upsilon}$ we obtain $U = U^{\nu\upsilon}$.

Now we have shown that ν and υ are inverse to each other. In particular, ν is a bijective map from $\mathcal{U} \setminus \{T\}$ to $\mathcal{V} \setminus \{T\}$, and υ is a bijective map from $\mathcal{V} \setminus \{T\}$ to $\mathcal{U} \setminus \{T\}$. Thus, the above given isomorphism and its analogue yield that, for each element x in X, $(U^{\mathcal{U}}//U)_x \cong (U^{\nu\mathcal{V}}//U^{\nu})_x$.

Let T be a closed subset of S. A scheme $\tilde{S}$ is called a *composition factor* of T if there exists a composition series $\mathcal{U}$ of T, an element U in $\mathcal{U} \setminus \{T\}$, and an element x in X such that $\tilde{S} \cong (U^{\mathcal{U}}//U)_x$.

A closed subset T of S is called *simple* if T has exactly two normal closed subsets, namely $\{1\}$ and T.

It follows immediately from Lemma 4.2.4(i) that composition factors are simple.

5.5 The Group Correspondence

Let G be a set. A map from $G \times G$ to G is called an *operation on* G.

Let us fix an operation on G and call it μ. For any two elements e and f in G, we write ef instead of $(e, f)\mu$.

The operation μ is called *associative* if, for any three elements d, e, and f in G, $d(ef) = (de)f$.

An element n of G is called a *neutral* element of G if, for each element g in G, $gn = g = ng$.

Assume that n and n' are neutral elements of G. Then, by definition, $n' = n'n = n$. Thus, G possesses at most one neutral element.

Let us assume that G possesses a neutral element n, and let e and f be elements of G. The element f is called an *inverse* of e if $ef = n = fe$.

The set G is called a *group* (with respect to μ) if μ is associative, if G possesses a neutral element and, for each of its elements, an inverse.

Let G be a group. Then, by definition, there exists an operation μ on G.

If, for any two elements e and f in G, we write ef instead of $(e, f)\mu$ (as we did so far), then we say that G is written *multiplicatively*. In this case, the neutral element of G is denoted by 1, and it is also called the *identity element* of G. Moreover, for each element g in G, the inverse of g is denoted by g^{-1}.

Usually, we write groups multiplicatively. However, sometimes we write, for any two elements e and f in G, $e + f$ instead of $(e, f)\mu$. In this case, we say that G is written *additively*, we denote the neutral element of G by 0, and, for each element g in G, the inverse of g will be denoted by $-g$.

Assume that an element d in G has inverse elements e and f. Then, by definition,

$$e = e1 = e(df) = (ed)f = 1f = f.$$

Thus, d possesses at most one inverse element.

Recall that, if S is thin, S^γ is our notation for the set of all sets $\{s\}$ with $s \in S$.

Theorem 5.5.1 *Assume S to be thin. Then S^γ is a group with respect to the restriction of the complex multiplication in S to S^γ and with $\{1\}$ as identity element.*

PROOF. From Lemma 2.5.9(i), (ii) together with Lemma 1.5.2 we obtain that the restriction of the complex multiplication in S to S^γ is an operation on S^γ. From Lemma 1.3.1 we know that this operation is associative.

It follows right from the definition of the complex multiplication in S that $\{1\}$ is an identity element of S^γ. Finally, as S is assumed to be thin, we deduce from Lemma 1.3.2(i) that, for each element s in S, $s^*s = \{1\} = ss^*$, so that

$$\{s\}^{-1} = \{s^*\}.$$

This finishes the proof of the theorem.

Let G be a group. Recall that, for each element g in G, g^τ is our notation for the set of all pairs (e, f) of elements of G satisfying $eg = f$. Recall also that G^τ is our notation for the set of all sets g^τ with $g \in G$.

Theorem 5.5.2 *Let G be a group. Then G^τ is a thin scheme on G.*

PROOF. It is clear that $1^\tau = 1_G$ and that, for each element g in G,

$$(g^\tau)^* = (g^{-1})^\tau.$$

In order to verify the regularity condition we pick elements b, c, d, e, and f in G. We assume that $(b, c) \in f^\tau$. Thus, by definition, we have that $bf = c$. It follows that $f = b^{-1}c$.

Let us now assume that $bd^\tau \cap c(e^\tau)^*$ is not empty. Then, as $(e^\tau)^* = (e^{-1})^\tau$, $bd^\tau \cap c(e^{-1})^\tau$ is not empty. Thus, as $\{bd\} = bd^\tau$ and $\{ce^{-1}\} = c(e^{-1})^\tau$, $bd = ce^{-1}$. From $bd = ce^{-1}$ we obtain $de = b^{-1}c$.

From $f = b^{-1}c$ and $de = b^{-1}c$ we obtain $de = f$.

Note, finally, that, for each element g in G, $\{1^\tau\} = (g^\tau)^* g^\tau$. Thus, G^τ is thin.

Theorem 5.5.3 *If S is thin, $S^{\gamma\tau} \cong S$.*

PROOF. We set $G := S^\gamma$, and we fix an element v in X.

For each element x in X, we define $x\phi$ to be the uniquely determined element g in G which satisfies $x \in vg$. For each element s in S, we set $s\phi := \{s\}^\tau$.

Since S is assumed to be thin, ϕ is a bijective map from $X \cup S$ to $G \cup G^\tau$.

Let y and z be elements in X, and let s be the uniquely defined element in S satisfying $z \in ys$. By Lemma 5.1.6(i), we shall be done if we succeed in showing that $z\phi \in y\phi s\phi$.

Let p be the uniquely defined element in S such that $y \in vp$, and let q be the uniquely defined element in S satisfying $z \in vq$. From $z \in ys$ and $y \in vp$ we obtain $z \in vps$. Thus, as $z \in vq$, $q \in ps$. Thus, by Lemma 1.5.2, $\{q\} = \{p\}\{s\}$. Thus, by definition, $\{q\} \in \{p\}\{s\}^\tau$.

On the other hand, as $y \in vp$, $\{p\} = y\phi$. Similarly, as $z \in vq$, $\{q\} = z\phi$. Thus, as $s\phi = \{s\}^\tau$, $z\phi \in y\phi s\phi$.

Let G and $\tilde{G}$ be groups. A map from G to $\tilde{G}$ is called a *group homomorphism* or simply a *homomorphism* if, for any two elements e and f in G, $(ef)\phi = e\phi f\phi$.

Two groups G and $\tilde{G}$ are called *isomorphic* if there exists a bijective group homomorphism from G to $\tilde{G}$. If two groups G and $\tilde{G}$ are isomorphic, one writes $G \cong \tilde{G}$.

Theorem 5.5.4 *For each group G, $G^{\tau\gamma} \cong G$.*

PROOF. For each element g in G, we define $g\phi := \{g^\tau\}$. Then ϕ is a bijective map from G to $G^{\tau\gamma}$.

Let e and f be elements in G, and set $g := ef$. We shall be done if we succeed in showing that $g\phi = e\phi f\phi$.

First of all, since $g = ef$, $g \in ef^\tau$. Moreover, we also have $e \in 1e^\tau$ and $g \in 1g^\tau$.

From $g \in ef^\tau$ and $e \in 1e^\tau$ we obtain $g \in 1e^\tau f^\tau$. Thus, as $g \in 1g^\tau$, $g^\tau \in e^\tau f^\tau$. But, by Theorem 5.5.2, $1 = n_{e^\tau}$ and $1 = n_{f^\tau}$. Thus, by Lemma 1.5.2, $\{g^\tau\} = e^\tau f^\tau$, and that means that $\{g^\tau\} = \{e^\tau\}\{f^\tau\}$. It follows that $g\phi = e\phi f\phi$.

The four theorems which have been proved in this section show that we may view thin schemes as groups. In particular, the class of all groups can be viewed as a subclass of the class of all schemes.

5.6 Residually Thin Schemes

Throughout this section, S is assumed to have finite valency. The letter T will stand for a closed subset of S.

We define $O^{\Theta}(T)$ to be the intersection of the sets $(O^{\vartheta})^n(T)$ with n a non-negative integer.

Since S is assumed to have finite valency, there exists a non-negative integer m such that, for each integer n with $m \leq n$, $O^{\Theta}(T) = (O^{\vartheta})^n(T)$.

We call T *residually thin* if $\{1\} = O^{\Theta}(T)$.

Theorem 5.6.1 *The set T is residually thin if and only if each composition factor of T is thin.*

PROOF. Define $\mathcal{T}$ to be the set of all sets $(O^{\vartheta})^n(T)$ such that n is a non-negative integer. Then, by Theorem 3.2.1(i) and Lemma 2.5.5, $\{\{1\}\} \cup \mathcal{T}$ is a subnormal series of T. Let $\mathcal{U}$ be a composition series of T which contains $\{\{1\}\} \cup \mathcal{T}$.

Let us first assume T to be residually thin, and let us pick an element U in $\mathcal{U} \setminus \{T\}$.

Since T is assumed to be residually thin, $\{1\} = O^{\Theta}(T)$. Thus, as $U \in \mathcal{U}$, there exists a non-negative integer n such that $(O^{\vartheta})^{n+1}(T) \subseteq U$ and $U^{\mathcal{U}} \subseteq (O^{\vartheta})^n(T)$.[1] By Lemma 2.5.6(iii), this implies that U is strongly normal in $U^{\mathcal{U}}$. Thus, by Lemma 4.2.5(ii), $U^{\mathcal{U}}/\!/U$ is thin. Thus, for each element x in X, $(U^{\mathcal{U}}/\!/U)_x$ is thin; cf. Theorem 2.1.8(iii). Thus, as U has been chosen arbitrarily in $\mathcal{U} \setminus \{T\}$, we have proved that each composition factor of T is thin; cf. Theorem 5.4.2.

Let us now assume that T is not residually thin. We shall show that T has a composition factor which is not thin.

Since T is assumed not to be residually thin, $\{1\} \neq O^{\Theta}(T)$. Thus, as $O^{\Theta}(T) \in \mathcal{U}$, there exists an element U in $\mathcal{U}$ such that $U^{\mathcal{U}} = O^{\Theta}(T)$.

Since $U^{\mathcal{U}} = O^{\Theta}(T)$, $O^{\vartheta}(U^{\mathcal{U}}) = U^{\mathcal{U}}$. Therefore, by Lemma 4.2.5(ii), $U^{\mathcal{U}}/\!/U$ is not thin. Thus, for each element x in X, $(U^{\mathcal{U}}/\!/U)_x$ is not thin; cf. Theorem 2.1.8(iii).

Lemma 5.6.2 *Let U be a closed subsets of S such that $T \subseteq U$. Then, if U is residually thin, so is T.*

PROOF. This is an immediate consequence of Lemma 3.2.6.

Lemma 5.6.3 *For each closed subset U of S, the following hold.*

[1] Recall that $U^{\mathcal{U}}$ stands for the intersection of all elements V of $\mathcal{U} \setminus \{U\}$ which contain U.

(i) *Assume that* $T \subseteq N_S(U)$ *and that* $O^{\Theta}(TU) \subseteq U$. *Then* $TU//U$ *is residually thin.*

(ii) *Assume that* T *is normal in* U. *Then, if* U *is residually thin, so is* $U//T$.

PROOF. (i) This is an immediate consequence of Theorem 4.2.8.

(ii) This is a particular case of (i).

Lemma 5.6.2 and Lemma 5.6.3(ii) say that the property of being residually thin is inherited by closed subsets and by quotient schemes over normal closed subsets. Here is a partial converse.

Theorem 5.6.4 *Let* U *be a closed subset of* S *such that* $T \subseteq N_S(U)$. *Then, if* T *and* U *are residually thin, so is* TU.

PROOF. We are assuming that $T \subseteq N_S(U)$. Thus, by Theorem 5.3.4(i), $T \cap U$ is normal in T.

Let us assume that T is residually thin. Then, as $T \cap U$ is normal in T, we obtain from Lemma 5.6.3(ii) that $T//T \cap U$ is residually thin.

On the other hand, from our hypothesis that $T \subseteq N_S(U)$ we also obtain

$$(T//T \cap U)_x \cong (TU//U)_x$$

for each element x in X; cf. Theorem 5.3.4(iii). Thus, as $T//T \cap U$ is residually thin, $TU//U$ is residually thin.

Since $TU//U$ is residually thin, $O^{\Theta}(TU) \subseteq U$; cf. Theorem 4.2.8. Thus, as U is assumed to be residually thin, Lemma 5.6.2 yields $\{1\} = O^{\Theta}(TU)$, and that means that TU is residually thin.

From Theorem 5.6.4 we obtain that S possesses a uniquely determined biggest residually thin normal closed subset.

Let us now introduce a second operator on the set of all closed subsets of S. This operator is dual to O^{Θ}.

Let T be a closed subset of S. We shall write $O_{\Theta}(T)$ to denote the union of the sets $(K_T)^n(\{1\})$ such that n is a non-negative integer.

Note that, by Lemma 2.5.8(ii), $O_{\Theta}(T)$ is a closed subset of T.

Note also that there exists a non-negative integer m such that, for each integer n with $m \leq n$, $(K_T)^n(\{1\}) = O_{\Theta}(T)$. (Recall that S is assumed to have finite valency.)

Lemma 5.6.5 *The following statements hold.*

(i) *We have* $O_{\vartheta}(T) \subseteq O_{\Theta}(T)$.

(ii) *We have* $O_{\Theta}(T//O_{\vartheta}(T)) = O_{\Theta}(T)//O_{\vartheta}(T)$.

PROOF. (i) From Lemma 2.5.9(i) we know that $O_\vartheta(S) = K_S(\{1\})$. Thus, the claim is an immediate consequence of the definition of $O_\Theta(T)$.

(ii) We fix a non-negative integer and call it n. Applying Theorem 4.2.6 to $O_\vartheta(T)$, T, and 1 instead of T, U, and m, we obtain

$$(K_{T//O_\vartheta(T)})^n(K_T(O_\vartheta(T))//O_\vartheta(T)) = (K_T)^n(O_\vartheta(T))//O_\vartheta(T).$$

On the other hand, by Lemma 4.2.4(ii),

$$K_{T//O_\vartheta(T)}(O_\vartheta(T)//O_\vartheta(T)) = K_T(O_\vartheta(T))//O_\vartheta(T),$$

so that

$$(K_{T//O_\vartheta(T)})^n(K_{T//O_\vartheta(T)}(O_\vartheta(T)//O_\vartheta(T))) = (K_T)^n(O_\vartheta(T))//O_\vartheta(T).$$

Now recall that, by Lemma 2.5.9(i), $K_T(\{1\}) = O_\vartheta(T)$. Thus,

$$(K_{T//O_\vartheta(T)})^{n+1}(O_\vartheta(T)//O_\vartheta(T)) = (K_T)^{n+1}(\{1\})//O_\vartheta(T).$$

Now the claim follows from the fact that n has been chosen arbitrarily.

Theorem 5.6.6 *The set $O_\Theta(T)$ is residually thin.*

PROOF. We set $U := O_\Theta(T)$. Then, as S is assumed to have finite valency, there exists a positive integer n such that $(K_T)^n(\{1\}) = U$. It follows that

$$(O^\vartheta)^0(U) = (K_T)^n(\{1\}).$$

Let us now prove that, for each element i in $\{1, \ldots, n\}$,

$$(O^\vartheta)^i(U) \subseteq (K_T)^{n-i}(\{1\}).$$

By induction, we may (and shall) assume that

$$(O^\vartheta)^{i-1}(U) \subseteq (K_T)^{n-i+1}(\{1\}).$$

By definition, $(K_T)^{n-i}(\{1\})$ is strongly normal in $(K_T)^{n-i+1}(\{1\})$. Thus, referring to Theorem 3.2.1(iii) we obtain

$$(O^\vartheta)^i(U) = O^\vartheta((O^\vartheta)^{i-1}(U)) \subseteq O^\vartheta((K_T)^{n-i+1}(\{1\})) \subseteq (K_T)^{n-i}(\{1\}).$$

It follows that $\{1\} = (O^\vartheta)^n(U)$, and that means that U is residually thin.

Theorem 5.6.7 *Let p be a prime number, and let us assume that n_T is a power of p. Then the following conditions are equivalent.*

(a) *We have $O_\Theta(T) = T$.*

(b) *The set T is residually thin.*

(c) *The set T is a closed p-subset of S.*

PROOF. (a) $\Rightarrow$ (b) This follows from Theorem 5.6.6.

(b) $\Rightarrow$ (c) Let T be a counterexample of minimal valency. Then T is residually thin. Thus, there exists a positive integer n such that $\{1\} = (O^{\vartheta})^{n+1}(T)$ and $\{1\} \neq (O^{\vartheta})^{n}(T)$. We set

$$U := (O^{\vartheta})^{n}(T).$$

Then $\{1\} \neq U \subseteq O_{\vartheta}(T)$ and, by Corollary 4.2.9, $U//U = (O^{\vartheta})^{n}(T//U)$. The latter equation implies that $T//U$ is residually thin.

On the other hand, we are assuming that n_T is a power of p. Thus, by Lemma 4.3.3(i), $n_{T//U}$ is a power of p. Thus, the (minimal) choice of T forces $T//U$ to be p-valenced. Thus, as U is thin, T is p-valenced; cf. Corollary 4.3.2(ii).

(c) $\Rightarrow$ (a) Let T be a counterexample of minimal valency. Then T is a closed p-subset of S. Thus, by Lemma 2.3.9, $\{1\} \neq O_{\vartheta}(T)$.

On the other hand, the hypothesis that T is p-valenced implies also that $T//O_{\vartheta}(T)$ is p-valenced; cf. Corollary 4.3.2(i). Moreover, as n_S is assumed to be a power of p, $n_{T//O_{\vartheta}(T)}$ is a power of p; cf. Lemma 4.3.3(i). Thus, the (minimal) choice of T leads to

$$O_{\Theta}(T//O_{\vartheta}(T)) = T//O_{\vartheta}(T).$$

From Lemma 5.6.5(ii) we also know that

$$O_{\Theta}(T//O_{\vartheta}(T)) = O_{\Theta}(T)//O_{\vartheta}(T).$$

Thus, by Lemma 4.2.1(ii), $O_{\Theta}(T) = T$, contrary to the choice of T.

Corollary 5.6.8 *Let p be a prime number. Then T is a closed p-subset of S if and only if each composition factor of T is thin and has valency p.*

PROOF. Assume first that T is a closed p-subset of S. Then, T is p-valenced and n_T is a power of p. Thus, by Theorem 5.6.7, T is residually thin. Thus, by Theorem 5.6.1, each composition factor of S is thin.

Conversely, let us assume that each composition factor of T is thin and has valency p. We may assume that $\{1\} \neq T$. Then T possesses a thin closed subset U of valency p. Thus, by induction, $T//U$ is a closed p-subset of $S//U$. Thus, by Corollary 4.3.2(i), T is a closed p-subset of S.

Corollary 5.6.9 *Let p be a prime number, assume that T is a closed p-subset of S, and let U be a closed p-subset of S such that $T \subseteq N_S(U)$. Then TU is a closed p-subset of S.*

PROOF. By hypothesis, n_T and n_U are powers of p. Thus, by Lemma 2.3.6(i), n_{TU} is a power of p.

We assume T to be a closed p-subset of S. Thus, by Theorem 5.6.7, T is residually thin. Similarly, as U is assumed to be a closed p-subset of S, U is residually thin. Thus, by Theorem 5.6.4, TU is residually thin. Thus, as n_{TU} is a power of p, TU is a closed p-subset of S; cf. Theorem 5.6.7.

6

Faithful Maps

A map χ from a subset Y of X to X is called *faithful* if, for any three elements v, w in Y and s in S, $w \in vs$ implies $w\chi \in v\chi s$.

The first section of this chapter is a collection of basic facts about faithful maps.

Let Y and Z be subsets of X such that $Y \subseteq Z$, let χ be a faithful map from Y to X, and let $\bar{\chi}$ be a faithful map from Z to X. We say that χ *extends faithfully* to $\bar{\chi}$ if, for each element y in Y, $y\chi = y\bar{\chi}$.

Let T and U be closed subsets of S, and assume that $T \subseteq U$. We say that T is *faithfully embedded in* U if, for any two elements y in X and z in yT, each faithful map χ from $\{y, z\}$ to yU extends faithfully to a bijective (faithful) map from yT to $y\chi T$.

The second section of this chapter deals with faithfully embedded closed subsets of S. We mainly discuss the question to which extent the property of being faithfully embedded is inherited from given quotient schemes of closed subsets of S to other quotient schemes of closed subsets of S.

A closed subset of S will be called *schurian* if it is faithfully embedded in itself.

Let x be an element in X, and let T be a closed subset of S. It is easy to see that the set of all bijective faithful maps from xT to xT is a group with respect to composition. We call this group the *Schur group* of T *with respect to* x. If $T = S$, we just speak about the Schur group of S.

Recall that $\mathrm{Sch}(S)$ is defined to be the set of all elements a in $\mathrm{Aut}(S)$ such that $a_S = \{1_S\}$. Referring to this notation it is clear that the Schur group of S is just the set of all elements a_S with $a \in \mathrm{Sch}(S)$.

The Schur group is the subject of the third section of this chapter. We shall see, in particular, that S is schurian if and only if S is isomorphic to a quotient scheme of a thin scheme, and this is the case if and only if, modulo the group correspondence, S is isomorphic to the quotient scheme of its own Schur group over one of the 'one-point stabilizers'; cf. Theorem 6.3.1.

In Section 6.4, we use previously obtained results about Schur groups in order to establish a recognition theorem for certain schemes of finite valency all elements of which have valency at most 2. The theorem is related to one of the most significant results in finite group theory, to George Glauberman's Z^*-Theorem.

In Section 6.5, we assume S to have finite valency. We shall prove that closed subsets of S which are generated by a single symmetric element of valency 2 are faithfully embedded in S. We also establish the corresponding recognition theorem. After that we shall look at closed subsets of S in which each non-identity element has valency 2.

Faithful maps are particularly interesting in connection with constrained sets. Section 6.6 provides a few results about this relationship. They will turn out to be useful in Chapter 11.

In the last of the seven sections of this chapter, we investigate closed subsets T of S which have finite valency and satisfy $O^{\vartheta}(T) \subseteq O_{\vartheta}(T)$. We shall give a sufficient criterion for T to be faithfully embedded in S.

6.1 Basic Facts

Recall that a map χ from a subset Y of X to X is called faithful if, for any three elements v, w in Y and s in S, $w \in vs$ implies $w\chi \in v\chi s$.

Note that faithful maps are necessarily injective.

Lemma 6.1.1 *Let x be an element in X, let T be a closed subset of S, and let χ be a map from xT to X. Then the following hold.*

(i) *The map χ is faithful if and only if, for any two elements w in xT and t in T, $wt\chi \subseteq w\chi t$.*

(ii) *Assume T to have finite valency. Then χ is faithful if and only if, for any two elements w in xT and t in T, $wt\chi = w\chi t$.*

(iii) *Assume T to have finite valency and χ to be faithful. Then χ is bijective map from xT to $x\chi T$, and its inverse is faithful.*

Proof. (i) This is just the definition of a faithful map rewritten under the hypothesis that the domain of χ contains wt.

(ii) Assume χ to be faithful. Then χ is injective. Thus, as wt and $w\chi t$ have the same (finite) number of elements, $\chi|_{wt}$ is a bijective map from wt to $w\chi t$.

(iii) This follows from (ii).

Lemma 6.1.2 *Let x be an element in X, let T and U be closed subsets of S such that $T \subseteq U$, and let χ be a faithful map from xU to X. For each element $w \in xU$, we define*

$$(wT)\chi^T := (w\chi)T.$$

Then χ^T is a faithful map from $(xT)(U/\!/T)$ to X/T.

Proof. Let v and w be elements in xU such that $vT = wT$. Since $vT = wT$, there exists an element t in T such that $w \in vt$. Thus, as χ is assumed to be faithful, $w\chi \in (v\chi)t$. In particular, $(v\chi)T = (w\chi)T$. Thus, by definition, $(vT)\chi^T = (wT)\chi^T$. Thus, as v and w have been chosen arbitrarily in xU such that $vT = wT$, we have shown that χ^T is a map.

In order to show that χ^T is faithful, we pick two elements y and z in xU, and we denote by u the uniquely determined element in S which satisfies $z \in yu$. Since $z \in yu$, $zT \in (yT)u^T$.

On the other hand, as χ is assumed to be faithful, $z \in yu$ yields $z\chi \in (y\chi)u$. From $z\chi \in (y\chi)u$ we obtain $(z\chi)T \in ((y\chi)T)u^T$. Thus, by definition, $(zT)\chi^T \in (yT)\chi^T u^T$. Thus, as y and z have been chosen arbitrarily in xU, we have shown that χ^T is faithful.

Lemma 6.1.3 *Let y and z be elements in X, and let T and U be closed subsets of S. Assume that T is strongly normal in U. Let χ be a faithful map from $\{y, z\}$ to X. Then, if $z \in yU$ and $y\chi \in yT$, $z\chi \in zT$.*

PROOF. Let us assume that $z \in yU$. Then there exists an element u in U such that $z \in yu$.

From $u \in U$ and $U \subseteq K_S(T)$ we obtain $u^*Tu \subseteq T$. Since $z \in yu$, $y \in zu^*$ and, as χ is assumed to be faithful, $z\chi \in y\chi u$. Thus, assuming that $y\chi \in yT$, we obtain

$$z\chi \in yTu \subseteq zu^*Tu \subseteq zT,$$

and that finishes the proof of the lemma.

Lemma 6.1.4 *Let T, U, and V be closed subsets of S such that $T \subseteq U \subseteq V$. Assume that T is normal in V and that, for each element s in $V \backslash U$, $\{s\} = sT$. Let x be an element in X, and let χ be a map from xV to X such that, for each element w in xV, $\chi|_{wU}$ is faithful and $w\chi \in wT$. Then χ is faithful.*

PROOF. Let y and z be elements in xV. Then, by Lemma 2.1.4, $z \in yV$. Thus, there exists an element s in V such that $z \in ys$. We have to show that $z\chi \in y\chi s$.

Since $\chi|_{yU}$ is assumed to be faithful, there is nothing to show if $z \in yU$. Thus, we may assume that $z \notin yU$. From $z \notin yU$ and $z \in ys$ we obtain $s \in V \setminus U$. Thus, by hypothesis, $\{s\} = sT$. On the other hand, we are assuming that T

is normal in V. Thus, by Lemma 2.5.2(ii), $Ts = sT$. Thus, we obtain from $\{s\} = sT$ that $\{s\} = TsT$.

We are assuming that $y\chi \in yT$. Thus, by Lemma 2.1.4, $y \in y\chi T$. Thus, as $z\chi \in zT$ and $z \in ys$,

$$z\chi \in ysT \subseteq y\chi TsT.$$

Thus, as $\{s\} = TsT$, $z\chi \in y\chi s$.

The following lemma shows in which way generating subsets of S and faithful maps work together.

Lemma 6.1.5 *Let R be a subset of S, and assume that R has finite valency. Assume that $p = q$ for any two elements p and q in $\langle R\rangle$ with $\ell_R(p) = \ell_R(q)$. Then the following hold.*

(i) *There exists a symmetric element r in R such that $\{r\} = R$.*

(ii) *Let x be an element in X, let r be the element in R, and let χ be an injective map from $x\langle R\rangle$ to X such that, for each element y in $x\langle R\rangle$, $yr\chi \subseteq y\chi r$. Then χ is faithful.*

PROOF. (i) This follows immediately from the hypothesis that $\langle R\rangle$ contains only one element of length 1 (with respect to R).

(ii) Let us denote by Q the set of all elements s in $\langle R\rangle$ for which there exists an element y in $x\langle R\rangle$ with $ys\chi \not\subseteq y\chi s$. By way of contradiction, we assume that Q is not empty.

Let us write ℓ instead of ℓ_R, and let us fix an element q in Q with $\min \ell(Q) = \ell(q)$.

Since $1 \notin Q$ and $q \in Q$, $1 \neq q$. Thus, referring to (i) and Lemma 3.1.2 we find an element p in $\langle R\rangle$ such that $q \in pr$ and $\ell(q) = \ell(p) + 1$.

Since $q \in Q$, there exists an element y in $x\langle R\rangle$ with

$$yq\chi \not\subseteq y\chi q.$$

Thus, there exists an element z in yq with $z\chi \notin y\chi q$. Since $z \in yq$ and $q \in pr$, $z \in ypr$. Thus, there exists an element w in yp such that $z \in wr$.

Since $\ell(p) = \ell(q) - 1$, the (minimal) choice of q forces $p \notin Q$. Thus, as $p \in \langle R\rangle$ and $y \in x\langle R\rangle$,

$$yp\chi \subseteq y\chi p.$$

Thus, as $w \in yp$, $w\chi \in y\chi p$. On the other hand, we obtain from $z \in wr$ and $wr\chi \subseteq w\chi r$ that $z\chi \in w\chi r$. Thus, $z\chi \in y\chi pr$, so that there exists an element t in pr with $z\chi \in y\chi t$.

From $z\chi \in y\chi t$ and $z\chi \notin y\chi q$ we obtain $t \neq q$.

Since $t \in pr$, $\ell(t) \leq \ell(p) + 1$. Thus, as $\ell(q) = \ell(p) + 1$, $\ell(t) \leq \ell(q)$.

From $t \neq q$ and our hypothesis we obtain $\ell(t) \neq \ell(q)$. Thus, as $\ell(t) \leq \ell(q)$, we must have that $\ell(t) \leq \ell(q) - 1$. Thus, the (minimal) choice of q forces $yt\chi \subseteq y\chi t$. However, we are assuming χ to be injective. Thus, as $|yt| = |y\chi t|$, $yt\chi = y\chi t$.

From $z\chi \in y\chi t$ and $yt\chi = y\chi t$ we obtain $z\chi \in yt\chi$. Thus, there exists an element z' in yt such that $z\chi = z'\chi$. Thus, as χ is assumed to be injective, we obtain $z = z'$. Thus, as $z' \in yt$, $z \in yt$. Thus, as $z \in yq$, $q = t$, contradiction.

6.2 Faithfully Embedded Closed Subsets

Let Y and Z be subsets of X such that $Y \subseteq Z$, let χ be a faithful map from Y to X, and let $\bar{\chi}$ be a faithful map from Z to X. Recall that we say that χ extends faithfully to $\bar{\chi}$ if, for each element y in Y, $y\chi = y\bar{\chi}$.

Let T and U be closed subsets of S such that $T \subseteq U$. Recall that T is said to be faithfully embedded in U if, for any two elements y in X and z in yT, each faithful map χ from $\{y, z\}$ to yU extends faithfully to a bijective map from yT to $y\chi T$.

Lemma 6.2.1 *Thin closed subsets of S are faithfully embedded in S.*

Proof. Let us fix a thin closed subset of S and call it T. Let y be an element in X, let z be an element in yT, and let χ be a faithful map from $\{y, z\}$ to X. We have to show that χ extends faithfully to a bijective map from yT to $y\chi T$.

For each element x in yT, we define $x\bar{\chi}$ to be the uniquely defined element in $y\chi t$, where t stands for the uniquely defined element in T with $x \in yt$. Then $\bar{\chi}$ is a bijective map from yT to $y\chi T$ satisfying $y\chi = y\bar{\chi}$ and $z\chi = z\bar{\chi}$.

We claim that $\bar{\chi}$ is faithful.

In order to show this we pick two elements v and w in yT, and we denote by t the uniquely determined element in T satisfying $w \in vt$. We have to show that $w\bar{\chi} \in v\bar{\chi}t$.

Let us denote by p the element in T which satisfies $v \in yp$ and by q the element in T which satisfies $w \in yq$.

From $v \in yp$ we obtain $y \in vp^*$. Thus, as $w \in yq$, $w \in vp^*q$. Thus, as $w \in vt$, $t \in p^*q$. However, we know from Lemma 1.5.2 that $|p^*q| = 1$. Thus, $\{t\} = p^*q$.

From $v \in yp$ we obtain $v\bar{\chi} \in y\bar{\chi}p$, and from $w \in yq$ we obtain $w\bar{\chi} \in y\bar{\chi}q$. Thus, $w\bar{\chi} \in v\bar{\chi}p^*q$. Thus, as $\{t\} = p^*q$, $w\bar{\chi} \in v\bar{\chi}t$.

The following two propositions show that the property of being faithfully embedded is inherited to closed subsets and to quotient schemes.

Proposition 6.2.2 *Let T, U, V, and W be closed subsets of S such that $T \subseteq V$, $U \subseteq W$, $T \subseteq U$, and $V \subseteq W$. Then, if U is faithfully embedded in W, T is faithfully embedded in V.*

PROOF. Let y be an element in X, let z be an element in yT, and let χ be a faithful map from $\{y, z\}$ to yV. We have to show that there exists a bijective faithful map $\bar{\chi}$ from yT to $y\chi T$ such that $y\chi = y\bar{\chi}$ and $z\chi = z\bar{\chi}$.

Since $z \in yT$ and $T \subseteq U$, $z \in yU$. Since $V \subseteq W$ and χ is a faithful map from $\{y, z\}$ to yV, χ is a faithful map from $\{y, z\}$ to yW. Thus, as U is assumed to be faithfully embedded in W, there exists a bijective faithful map $\bar{\chi}$ from yU to $y\chi U$ such that $y\chi = y\bar{\chi}$ and $z\chi = z\bar{\chi}$.

Since $\bar{\chi}$ is faithful, we have $(yT)\bar{\chi} \subseteq y\bar{\chi}T$. Thus, as $y\chi = y\bar{\chi}$, $(yT)\bar{\chi} \subseteq y\chi T$. Thus, as $\bar{\chi}|_{yT}$ is a faithful map from yT to $y\chi T$. Finally, as $\bar{\chi}$ is a bijective from yU to $y\chi U$, $\bar{\chi}|_{yT}$ is a bijective from yT to $y\chi T$.

Proposition 6.2.3 *Let T, U, and V be closed subsets of S such that $T \subseteq U \subseteq V$. If U is faithfully embedded in V, $U/\!/T$ is faithfully embedded in $V/\!/T$.*

Proof. Let y and z be elements in X such that $z \in yU$, and let ϕ be a faithful map from $\{yT, zT\}$ to $(yT)(V/\!/T)$. We have to show that there exists a bijective faithful map $\bar{\phi}$ from $(yT)(U/\!/T)$ to $(yT)\phi(U/\!/T)$ such that $(yT)\phi = (yT)\bar{\phi}$ and $(zT)\phi = (zT)\bar{\phi}$.

Since $z \in yU$, there exists an element u in U such that $z \in yu$. In particular, $zT \in (yT)u^T$. Thus, as ϕ is assumed to be faithful, $(zT)\phi \in (yT)\phi u^T$. Thus, by definition, $(zT)\phi \subseteq (yT)\phi TuT$. Thus, there exist elements v in $(yT)\phi$ and w in $(zT)\phi$ such that $w \in vu$. Thus, as U is assumed to be faithfully embedded in V, we obtain from $v \in yV$ a bijective faithful map χ from yU to vU such that $y\chi = v$ and $z\chi = w$.

For each element $x \in yU$, we define

$$(xT)\bar{\phi} := (x\chi)T.$$

By Lemma 6.1.2, $\bar{\phi}$ is a faithful map from $(yT)(U/\!/T)$ to $(vT)(U/\!/T)$. Moreover, we have that

$$(yT)\phi = vT = (y\chi)T = (yT)\bar{\phi}$$

and, similarly, $(zT)\phi = (zT)\bar{\phi}$.

Since χ is a bijective map from yU to vU, we obtain from $(yT)\phi = vT$ that $\bar{\phi}$ is a bijective map from $(yT)(U/\!/T)$ to $(yT)\phi(U/\!/T)$.

For the remainder of this section, we shall assume S to have finite valency.

Proposition 6.2.2 says that the property of being faithfully embedded is inherited to closed subsets. Given closed subsets T and U in S with $T \subseteq U$, we shall now ask ourselves under which hypotheses the property of being faithfully embedded in S can be lifted from T to U.

Lemma 6.2.4 *Let T, U, and V be closed subsets of S such that $T \subseteq U \subseteq V \subseteq K_S(T)$ and, for each element s in $V \setminus U$, $\{s\} = sT$. Assume that U is faithfully embedded in S.*

Let y, z be elements in X such that $z \in yV$, and let χ be a faithful map from $\{y, z\}$ to X such that $y\chi \in yT$ and $z\chi \in zT$. Then χ extends faithfully to a bijective map from yV to $y\chi V$.

PROOF. Since U is assumed to be faithfully embedded in S, there exists a bijective faithful map $\bar{\chi}$ from $yU \cup zU$ to $y\chi U \cup z\chi U$ such that $\chi|_{yU}$ and $\chi|_{zU}$ are faithful. (It might be that $yU = zU$, it also might be that $yU \neq zU$.) Thus, for each element x in $yU \cup zU$, $x\chi \in xT$; cf. Lemma 6.1.3. For each element x in $yV \setminus (yU \cup zU)$ we define $x\chi := x$. Then, by Lemma 6.1.4, χ must be faithful.

Lemma 6.2.5 *Let T, U, and V be closed subsets of S such that $T \subseteq U \subseteq V \subseteq K_S(T)$ and, for each element s in $V \setminus U$, $\{s\} = sT$. Assume that U is faithfully embedded in S.*

Let x be an element in X. Then, for each faithful map ψ from $(xT)(V//T)$ to X/T, there exists a faithful map ϕ from xV to X such that, for each element w in xV, $w\phi \in (wT)\psi$.

PROOF. By Lemma 2.1.4, we find a subset Z of xV such that, for each element w in xV, $|Z \cap wU| = 1$. Let ψ be a faithful map from $(xT)(V//T)$ to X/T.

For each element u in Z, we pick an element in $(uT)\psi$ and call it $u\phi$. Since U is assumed to be faithfully embedded in S, there exists, for each element u in Z, a bijective faithful map ϕ_u from uU to $u\phi U$ such that $u\phi_u = u\phi$. We extend ϕ from Z to xV by setting $w\phi := w\phi_u$ for any two elements u in Z and w in uU.

Let us first prove that, for each element w in xV, $w\phi \in (wT)\psi$.

In order to do so we pick an element w in xV, and we denote by u the uniquely determined element in $Z \cap wU$. Since $u \in wU$, $w \in uU$; cf. Lemma 2.1.3. Thus, there exists an element p in U such that $w \in up$. Thus, as ϕ is faithful on uU, $w\phi \in u\phi p$. On the other hand, $u\phi$ is defined to be an element of $(uT)\psi$. Thus,

$$w\phi \in (uT)\psi p.$$

By Lemma 4.2.5(ii), $V//T$ is thin. Thus, there exists an element v in X such that

$$\{vT\} = (uT)\psi p^T.$$

Thus, by Lemma 4.1.4, $vT = (uT)\psi T p T$. Thus, as $w\phi \in (uT)\psi p$,

$$w\phi \in vT.$$

From $w \in up$ we also obtain $wT \in (uT)p^T$. Thus, as ψ is assumed to be faithful, we must have

$$(wT)\psi \in (uT)\psi p^T.$$

Thus, as $\{vT\} = (uT)\psi p^T$, $(wT)\psi = vT$. Thus, as $w\phi \in vT$, $w\phi \in (wT)\psi$.

Let us now show that ϕ is faithful. In order to do so we pick two elements y and z in xV. From y, $z \in xV$ we obtain $z \in yV$; cf. Lemma 2.1.4. Thus, there exists an element q in V such that $z \in yq$. We have to show that $z\phi \in y\phi q$.

If $q \in U$, we are done by the definition of ϕ. Thus, we may assume that $q \notin U$.

Since $z \in yq$, $zT \in (yT)q^T$. Thus, as ψ is assumed to be faithful, we must have

$$(zT)\psi \in (yT)\psi q^T.$$

From $y\phi \in (yT)\psi$ we obtain $y\phi T = (yT)\psi$. Similarly, as $z\phi \in (zT)\psi$, $z\phi T = (zT)\psi$. Thus, as $(zT)\psi \in (yT)\psi q^T$, $z\phi T \in (y\phi T)q^T$. By Lemma 4.1.4, this means that $z\phi \in y\phi TqT$.

On the other hand, we are assuming that $q \notin U$. Thus, by hypothesis, $\{q\} = qT$. Thus, as we are assuming that $V \subseteq K_S(T)$, we obtain from Lemma 2.5.5 and Lemma 2.5.2(ii) that $Tq = qT$. Thus, as we are assuming that $\{q\} = qT$, $\{q\} = TqT$. Thus, as $z\phi \in y\phi TqT$, $z\phi \in y\phi q$.

The following theorem is a partial inverse of Proposition 6.2.2. It is similar to [25; Theorem A].

Theorem 6.2.6 *Let T, U, and V be closed subsets of S such that $T \subseteq U \subseteq V$, $V \subseteq K_S(T)$, and, for each element s in $V \setminus U$, $\{s\} = sT$. Then, if U is faithfully embedded in S, so is V.*

PROOF. Let y, z be elements in X such that $z \in yV$, and let χ be a faithful map from $\{y, z\}$ to X. We have to show that χ extends faithfully to a bijective map from yV to $y\chi V$.

Since $z \in yV$, there exists an element r in V such that $z \in yr$. Since χ is assumed to be faithful, we have $z\chi \in y\chi r$.

Since $z \in yr$, $zT \in (yT)(r^T)$. Similarly, as $z\chi \in y\chi r$, $z\chi T \in (y\chi T)(r^T)$. Thus, there exists a faithful map ψ from $(yT)(V//T)$ to X/T with $(yT)\psi = y\chi T$ and $(zT)\psi = z\chi T$; cf. Lemma 6.2.1 together with Lemma 4.2.5(ii). Thus, by Lemma 6.2.5, there exists a faithful map ϕ from yV to X such that, for each element x in yV, $x\phi \in (xT)\psi$.

From $y\phi \in (yT)\psi$ and $(yT)\psi = y\chi T$ we obtain $y\phi \in y\chi T$. Thus, by Lemma 2.1.4, $y\chi \in y\phi T$. Similarly, we obtain from $z\phi \in (zT)\psi$ and $(zT)\psi = z\chi T$ that $z\chi \in z\phi T$.

Finally, as $y\chi \in y\phi T$ and $z\chi \in z\phi T$, there exists a faithful map ϕ' from $y\chi V$ to X such that $y\phi\phi' = y\chi$ and $z\phi\phi' = z\chi$; cf. Lemma 6.2.4.

Let T and U be closed subsets of S such that $T \subseteq U$.

Proposition 6.2.3 says that the property of being faithfully embedded is inherited from U to $U//T$. We shall now ask ourselves under which hypotheses the property of being faithfully embedded in S can be transferred from $U//T$ to U.

Let x be an element in X, and let us denote by $\mathcal{F}$ (respectively $\mathcal{F}^T$) the set of all faithful maps from xU (respectively $(xT)(U//T)$) to X (respectively X/T). For each element χ in $\mathcal{F}$, we define $\lambda_x(\chi) := \chi^T$. Then by Lemma 6.1.2, λ_x is a map from $\mathcal{F}$ to $\mathcal{F}^T$.

We shall say that U *covers* T if, for each element x in X, the map λ_x is surjective.

Lemma 6.2.7 *Let T and U be closed subsets of S such that $T \subseteq U$. Assume that U contains a closed subset V of S which covers T. Assume further that $\{u \in U \mid 1 \neq |TuT|\} \subseteq V$. Then U covers T.*

Proof. Let x be an element in X, let ϕ be a faithful map from $(xT)(U//T)$ to X/T. We have to find a faithful map χ from xU to X such that $\chi^T = \phi$.

Let Y be a subset of xU such that, for each element y in xU, $|Y \cap yV| = 1$. Let w be an element in Y. Then $\phi|_{(wT)(V//T)}$ is faithful. Thus, as V is assumed to cover T, there exists a faithful map χ_w from wV to X such that $(\chi_w)^T = \phi|_{(wT)(V//T)}$.

For each element y in xU, we define $y\chi := y\chi_w$, where w denotes the uniquely determined element w in Y which satisfies $y \in wV$. Then χ is a map from xU to X satisfying $\chi^T = \phi$. We claim that χ is faithful.

In order to prove that χ is faithful, we pick two elements y and z in xU, and we denote by u the uniquely determined element in U which satisfies $z \in yu$. We have to show that $z\chi \in y\chi u$.

Let w denote the uniquely determined element in $Y \cap yV$. Then, if $u \in V$, $z \in yu \subseteq yV = wV$. Thus, in this case, $z\chi = z\chi_w \in y\chi_w u = y\chi u$. Thus, we may (and we shall) assume that $u \notin V$.

Since $w \in Y \cap yV$,

$$(yT)\phi = (yT)\phi|_{(wT)(V//T)} = (yT)(\chi_w)^T = (y\chi_w)T = (y\chi)T.$$

Similarly, $(zT)\phi = (z\chi)T$.

From $z \in yu$ we obtain $zT \in (yT)u^T$. Thus, as ϕ is assumed to be faithful,

$$(zT)\phi \in (yT)\phi u^T.$$

Thus, the above two equations yield $(z\chi)T \in ((y\chi)T)u^T$. Thus, by definition, $z\chi \in y\chi TuT$. On the other hand, we are assuming that $u \notin V$. Thus, by hypothesis, $\{u\} = TuT$. Thus, $z\chi \in y\chi u$.

In our last result, we do not need the full strength of the notion of a faithfully embedded closed subset. In order to weaken this concept we fix closed subsets

T and U such that $T \subseteq U$. We say that U is *faithfully T-embedded* in U if, for any two elements y in X and z in yU, each faithful map χ from $\{y, z\}$ to X with $y\chi \in yT$ and $z\chi \in zT$ extends faithfully to a bijective (faithful) map from yU to $y\chi U$.

Here is a partial inverse of Proposition 6.2.3.

Theorem 6.2.8 *Let T and U be closed subsets of S such that $T \subseteq U$. Assume that U contains a closed subset V of S which covers T. Assume further that $\{h \in U \mid 1 \neq |ThT|\} \subseteq V$ and that V is faithfully T-embedded in S. Then, if $U//T$ is faithfully embedded in $S//T$, U is faithfully embedded in S.*

PROOF. Let y and z be elements in X such that $z \in yU$, and let χ be a faithful map from $\{y, z\}$ to X. We have to show that χ extends faithfully to a bijective map from yU to $(y\chi)U$.

Since $z \in yU$, there exists an element u in U such that $z \in yu$. Thus, as χ is assumed to be faithful, $z\chi \in y\chi u$.

From $z \in yu$, we obtain $zT \in (yT)u^T$. Similarly, as $z\chi \in y\chi u$, $(z\chi)T \in (y\chi)Tu^T$. Thus, as $U//T$ is assumed to be faithfully embedded in $S//T$, there exists a faithful map ϕ from $(yT)(U//T)$ to X/T such that $(y\chi)T = (yT)\phi$ and $(z\chi)T = (zT)\phi$.

We are assuming that V covers T and that $\{u \in U \mid 1 \neq |TuT|\} \subseteq V$. Thus, by Lemma 6.2.7, U covers T. Thus, by definition, there exists a faithful map ζ from yU to X such that $\zeta^T = \phi$. From $\zeta^T = \phi$ we obtain

$$(y\zeta)T = (yT)\zeta^T = (yT)\phi = (y\chi)T.$$

In particular, $y\zeta \in y\chi T$. Similarly, $z\zeta \in z\chi T$.

Set $y\zeta\eta := y\chi$ and $z\zeta\eta := z\chi$. Then $y\zeta\eta \in y\zeta T$ and $z\zeta\eta \in z\zeta T$. Moreover, as $z\chi \in y\chi u$ and $z\zeta \in y\zeta u$, η is faithful. Thus, as we are assuming that U is faithfully T-embedded in S, η extends faithfully to a faithful map $\bar{\chi}$ from $(y\zeta)U$ to $(y\zeta\eta)U$. Thus, as ζ is a faithful map from yU to $(y\zeta)U$, $\bar{\chi}$ is faithful. This finishes the proof, because we have $\bar{\chi}|_{\{y,z\}} = \chi$.

6.3 The Schur Group of a Closed Subset

Let x be an element in X, and let T be a closed subset of S.

Recall that the group of all bijective faithful maps from xT to xT is called the Schur group of T with respect to x. The identity on xT is the identity element of the Schur group of T with respect to x.

Let G be a subgroup of the Schur group of T with respect to x.

Let Y be a subset of xT. Assume that, for any two elements y in Y and g in G, $yg \in Y$. One says that the group G *acts transitively* on Y if, for any two elements v and w in Y, there exists an element g in G such that $vg = w$.

We set

$$G_x := \{g \in G \mid xg = x\}.$$

The set G_x is called the *stabilizer* of x in G.

Recall that a closed subset of S is called schurian if it is faithfully embedded in itself.

Theorem 6.3.1 *Let x be an element in X, and let T be a closed subset of S. Then the following statements are equivalent.*

(a) *The Schur group G of T with respect to x acts transitively on xT and, for each element t in T, G_x acts transitively on xt.*

(b) *Let G be the Schur group of T with respect to x. Then $T_x \cong G^\tau//(G_x)^\tau$.*

(c) *The scheme T_x is a quotient scheme of a thin scheme.*

(d) *The scheme T_x is schurian.*

PROOF. (a) $\Rightarrow$ (b) We set $H := G_x$.

We pick an element w in xT. For any two elements e and f in G with $w = xe^{-1}$ and $w = xf^{-1}$, we have $xe^{-1}f = x$. That means that $e^{-1}f \in H$. Thus, by Lemma 1.3.3(ii) and Lemma 2.1.4, $eH = fH$, and this implies $eH^\tau = fH^\tau$.

For any two elements w in xT and g in G with $w = xg^{-1}$, we define

$$w\phi := gH^\tau.$$

Let e and f be elements in G such that $xe^{-1} \in xt$ and $xf^{-1} \in xt$. Then, by (a), there exists an element h in H such that $xe^{-1}h = xf^{-1}$. It follows that $e^{-1}hf \in H$, so that $e \in HfH$. Thus, $HeH = HfH$, and this yields $(e^\tau)^{H^\tau} = (f^\tau)^{H^\tau}$.

For any two elements t in T and g in G with $xg^{-1} \in xt$, we define

$$(t_{xT})\phi := (g^\tau)^{H^\tau}.$$

Note that ϕ is a map from $xT \cup T_{xT}$ to $G/H^\tau \cup G^\tau//H^\tau$ which maps xT to G/H^τ and T_{xT} to $G^\tau//H^\tau$. Let us now show that ϕ is an isomorphism from $xT \cup T_{xT}$ to $G/H^\tau \cup G^\tau//H^\tau$.

Let w be an element in xT, and let t be an element in T such that $w \in xt$. Since G is assumed to act transitively on xT, we shall be done if we succeed in showing that $w\phi \in x\phi t_{xT}\phi$.

Since G acts transitively on xT, there exists an element g in G such that $w = xg^{-1}$. Thus, by definition, $w\phi = gH^\tau$.

From $w = xg^{-1}$ and $w \in xt$ we obtain $xg^{-1} \in xt$. Thus, by definition, $(t_{xT})\phi = (g^\tau)^{H^\tau}$.

Since $g \in 1H^\tau g^\tau H^\tau$, $gH^\tau \in (1H^\tau)(g^\tau)^{H^\tau}$. Thus, as $x\phi = 1H^\tau$, $w\phi = gH^\tau$ and $(t_{xT})\phi = (g^\tau)^{H^\tau}$, $w\phi \in x\phi t_{xT}\phi$.

(b) $\Rightarrow$ (c) This is obvious.

(c) $\Rightarrow$ (d) Considering Lemma 6.2.1 this is an immediate consequence of Proposition 6.2.3.

(d) $\Rightarrow$ (a) Let y and z be elements in xT. Since T_x is assumed to be schurian, T_{xT} is faithfully embedded in T_{xT}. Thus, there exists a bijective faithful map g from xT to xT such that $yg = z$.

Let t be an element in T, and let y and z be elements in xt. Then, as T_{xT} is assumed to be faithfully embedded in T_{xT}, there exists a bijective faithful map from xT to xT such that $yg = z$. From this we obtain (a).

Corollary 6.3.2 *The scheme S is schurian if and only if S is a quotient scheme of a thin scheme.*

PROOF. This follows immediately from Theorem 6.3.1.

Theorem 6.3.3 *A closed subset T of S is schurian if and only if, for each element x in X, the Schur group G of T with respect to x acts transitively on xT and, for each element t in T, G_x acts transitively on xt.*

PROOF. Let T be a closed subset of S, and let us first assume T to be schurian. We fix an element x in X. In order to show that the Schur group G of T with respect to x acts transitively on xT, we fix elements y and z in xT. The map χ from $\{y\}$ to $\{z\}$ with $y\chi := z$ is faithful. Thus, by definition, there exists an element g in G such that $yg = z$. This shows that G acts transitively on xT.

Let t be an element in T, and let y and y' be elements in xt. Define $x\chi := x$ and $y\chi := y'$. Then χ is a faithful map from $\{x, y\}$ to xT. Thus, as T is assumed to be schurian, there exists an element g in G_x such that $yg = y'$.

Let us now assume that, for each element x in X, the Schur group G of T with respect to x acts transitively on xT and that, for each element t in T, G_x acts transitively on xt. We have to show that T is faithfully embedded in T.

Let y be an element in X, let z be an element in yT, and let χ be a faithful map from $\{y, z\}$ to yT.

Since $z \in yT$, there exists an element t in T such that $z \in yt$. Thus, as χ is assumed to be faithful, $z\chi \in y\chi t$. We are assuming that G acts transitively on yT. Thus, there exists an element e in G such that $ye = y\chi$. We are assuming that G_{ye} acts transitively on yet. Thus, there exists an element f in G_{ye} such that $zef = z\chi$.

Theorem 6.3.1, together with Proposition 6.2.2, also shows that, if S is schurian, each closed subset of S is faithfully embedded in S. Lemma 6.2.1 says that thin closed subsets are faithfully embedded in S. The set of all closed subsets of S which are faithfully embedded in S can be viewed as a measure how close S is to be schurian.

The following lemma will be needed in Section 6.5 and in Section 12.3.

Lemma 6.3.4 *Let x be an element in X, let R be a subset of S, and let G be a subgroup of the Schur group of $\langle R\rangle$ with respect to x. Assume that, for any two elements y and z in $x\langle R\rangle$ with $z \in y(R^* \cup R)$, there exists an element g in G with $yg = z$. Then G acts transitively on $x\langle R\rangle$.*

PROOF. This follows from Lemma 3.1.1(i) by induction.

We shall now look at closed subsets T of S in which each element has valency at most 2.

The following lemma is [23; Lemma 3.5]. Note that we do not assume S to have finite valency.

Lemma 6.3.5 *Let x be an element in X, let T be a closed subset of S, and let G denote the Schur group of T with respect to x. Assume that T is not thin and that, for each element t in T, $n_t \leq 2$. Then $2 \leq |G_x|$.*

PROOF. Let x be an element in X, and let w be an element in xT. Since $w \in xT$, there exists an element t in T such that $w \in xt$. If $n_t = 1$, we set $wh := w$. If $n_t = 2$, we define wh to be the uniquely determined element in $xt \setminus \{w\}$.

Note that $h \in G_x$. Note also that h is a bijective map from xT to xT. Since T is assumed to be not thin, $1 \neq h$. Thus, we just have to show that h is faithful.

In order to show that h is faithful we pick elements y and z in xT. From $y \in xT$ and $z \in xT$ we obtain $z \in yT$; cf. Lemma 2.1.4. Thus, there exists an element t in T such that $z \in yt$. We now have to show that $zh \in yht$.

Let us denote by p the uniquely determined element in T such that $y \in xp$. Then, as $z \in yt$, $z \in xpt$. Thus, there exists an element q in pt such that $z \in xq$. From $z \in xq$ we obtain $zh \in xq$.

Let us first assume that $n_p = 1$. In this case, we have $\{y\} = xp$ and $yh = y$. Thus, as $zh \in xq$ and $q \in pt$,

$$zh \in xpt = yt = yht,$$

so that we are done in this case.

The case where $n_q = 1$ is treated similarly. Let us, therefore, assume that $n_p = 2$ and that $n_q = 2$.

Since $y \in xp$ and $z \in xq \cap yt$, $1 \leq a_{qt^*p}$. Thus, as $yh \in xp$, there exists an element w in $xq \cap yht$. Since $w \in xq = \{z, zh\}$ we now must have $w = z$ or $w = zh$.

If $w = zh$, $w \in yht$ leads to $zh \in yht$, and we are done.

If $w = z$, $\{y, yh\} \subseteq xp \cap zt^*$. Thus, as $z \in xq$, $2 \leq a_{ptq}$. Thus, as $zh \in xq$, $2 \leq |xp \cap zht^*|$. Thus, as $xp = \{y, yh\}$, $\{y, yh\} \subseteq zht^*$. It follows that $zh \in yht$; again, we are done.

The following lemma is [23; Theorem 3.7(iv)]. Note that, again, we do not require S to have finite valency.

Lemma 6.3.6 *Let x be an element in X, let T be a closed subset of S, and let G denote the Schur group of T with respect to x. Assume that $O^\vartheta(T)$ does not have valency 2 and that, for each element t in T, $n_t \leq 2$. Then $|G_x| \leq 2$.*

PROOF. Assume, by way of contradiction, that $3 \leq |G_x|$. Then there exists an element h in $G_x \setminus \{1\}$ which fixes an element in $x(T \setminus O_\vartheta(T))$.

We define P to be the set of all elements t in $T \setminus O_\vartheta(T)$ such that h fixes none of the two elements in xt, and we set $Q := T \setminus P \setminus O_\vartheta(T)$.

Since $1 \neq h$, P is not empty. Recall also that h fixes at least one element in $x(T \setminus O_\vartheta(T))$. Thus, Q is not empty. Thus, by Lemma 1.5.5, we shall be done if we succeed in showing that, for any two elements p in P and q in Q, $|p^*q| = 1$. (Recall that we are assuming that $O^\vartheta(T)$ does not have valency 2.)

Let p be an element in P, let q be an element in Q, let v be an element in xp, and let w be an element in xq. Then $vh \neq v$ and $wh = w$.

Since $v \in xp$, $x \in vp^*$. Thus, as $w \in xq$, $w \in vp^*q$. From $w \in vp^*q$ we obtain an element r in p^*q such that $w \in vr$. It follows that $v \in xp \cap wr^*$.

Since $v \in xp$ and $xh = x$,

$$vh \in xph \subseteq xhp = xp.$$

Similarly, as $wh = w$ and $v \in wr^*$, $vh \in wr^*$. Thus, $vh \in xp \cap wr^*$.

From $\{vh, v\} \subseteq xp \cap wr^*$, $w \in xq$, and $vh \neq v$, we conclude that $2 \leq a_{prq}$. Thus, as $n_p = 2$, $a_{prq} = n_p$. Thus, by Lemma 1.4.3, $\{r\} = p^*q$.

6.4 Elements of Valency 2

In Lemma 6.3.5 and Lemma 6.3.6 we investigated the Schur group of closed subsets of S in which each element has valency at most 2. From Lemma 1.5.6(ii) we know that, for each element s in S with $n_s = 2$, $n_{s^*s} \in \{2, 3\}$. In this section, we focus on closed subsets T of S in which each element t satisfies $n_t \leq 2$ and $n_{t^*t} \neq 2$.

We shall first prove that a closed subset T of S is schurian if each element t of T satisfies $n_t \leq 2$ and $n_{t^*t} \neq 2$. After that, we shall see which finite groups are Schur groups of these closed subsets. This way we shall establish a recognition theorem for these schemes.

The following proposition is [31; (4.1)]. Note that we do not require the closed subset T of S to have finite valency.

Proposition 6.4.1 *Let T be a closed subset of S, and assume that, for each element t in T, $n_t \leq 2$ and $n_{t^*t} \neq 2$. Then T is faithfully embedded in S.*

PROOF. Let y and y' be elements in X, let s be an element in T, let z be an element in ys, and let z' be an element in $y's$. We have to show that there exists a bijective faithful map χ from yT to $y'T$ such that $y\chi = y'$ and $z\chi = z'$.

Let us first assume that $n_s = 2$. In this case, we define χ in the following way.

Let x be an element in yT. Then there exist elements p and q in T such that $x \in yp \cap zq^*$. Thus, as $z \in ys$, $1 \leq a_{pqs}$. On the other hand, we are assuming that $n_p \leq 2$, $n_q \leq 2$, $n_s = 2$, and $n_{p^*p} \neq 2$. Thus, by Lemma 1.5.4, $a_{pqs} \leq 1$, so that $a_{pqs} = 1$. Thus, as $z' \in y's$, $y'p \cap z'q^*$ contains exactly one element. We define $x\chi$ to be the uniquely determined element in $y'p \cap z'q^*$.

It follows from the definition of χ that χ is a surjective map from yT to $y'T$, that $y\chi = y'$, and that $z\chi = z'$.

In order to show that χ is faithful we fix elements v in yT and r in T. We have to show that $vr\chi \subseteq v\chi r$; cf. Lemma 6.1.1(i).

Let us assume, by way of contradiction, that $vr\chi \not\subseteq v\chi r$. Then there exists an element w in vr such that $w\chi \notin v\chi r$. Since $w\chi \notin v\chi r$, there exists an element r' in T such that $r' \neq r$ and $w\chi \in v\chi r'$.

Let p and q be elements in S such that $v \in yp \cap zq^*$, and let t and u be elements in S such that $w \in yt \cap zu^*$. From the definition of χ we obtain $v\chi \in y\chi p \cap z\chi q^*$ and $w\chi \in y\chi t \cap z\chi u^*$.

Since $w\chi \in v\chi r'$ and $v\chi \in y\chi p$, $w\chi \in y\chi pr'$. Thus, as $w\chi \in y\chi t$, $t \in pr'$. Thus, as $w \in yt$, $w \in ypr'$. Thus, there exists an element v' in yp such that $w \in v'r'$. From $w \in v'r'$, $w \in vr$, and $r' \neq r$ we obtain $v' \neq v$.

Similarly, as $z\chi \in w\chi u$ and $w\chi \in v\chi r'$, $z\chi \in v\chi r'u$. Thus, as $z\chi \in v\chi q$, $q \in r'u$. Thus, as $z \in vq$, $z \in vr'u$. Thus, there exists an element w' in vr' such that $z \in w'u$. From $w' \in vr'$, $w \in vr$, and $r' \neq r$ we obtain $w' \neq w$.

Let us denote by t' the uniquely determined element in T satisfying $w' \in yt'$. If $t' = t$, $\{w, w'\} \subseteq yt \cap zu^*$. Thus, by Lemma 1.5.4, $w' = w$, contradiction. Thus, $t' \neq t$.

From $w' \in yt'$ we obtain $w'\chi \in y\chi t'$. Thus, as $w\chi \in y\chi t$ and $t' \neq t$, $w'\chi \neq w\chi$.

From $w' \in zu^*$ we obtain $w'\chi \in z\chi u^*$. Thus, as $w\chi \in z\chi u^*$, $w'\chi \neq w\chi$ and $n_{u^*} \leq 2$, $z\chi u^* = \{w\chi, w'\chi\}$.

Since $z \in vq$ and $w \in vr \cap zu^*$, $1 \leq a_{ruq}$. Thus, as $z\chi \in v\chi q$, there exists an element x in $v\chi r \cap z\chi u^*$. From $x \in v\chi r$, $w\chi \in v\chi r'$, and $r' \neq r$ we obtain $x \neq w\chi$. Thus, as $x \in z\chi u^* = \{w\chi, w'\chi\}$, $x = w'\chi$. Thus, as $x \in v\chi r$, $w'\chi \in v\chi r$.

Since $w'\chi \in y\chi t'$ and $v\chi \in y\chi p \cap w'\chi r^*$, $1 \leq a_{prt'}$. Thus, as $w' \in yt'$, there exists an element x in $yp \cap w'r^*$. From $x \in w'r^*$, $v \in w'r'^*$, and $r' \neq r$ we obtain $x \neq v$. Thus, as $x \in yp = \{v, v'\}$, $x = v'$. Thus, as $w' \in xr$, $w' \in v'r$.

Since $v \in yp$ and $w \in yt \cap vr$, $1 \leq a_{tr^*p}$. Thus, as $v' \in yp$, there exists an element x in $yt \cap v'r$.

Similarly, as $v' \in yp$ and $w \in yt \cap v'r'$, $1 \leq a_{tr'^*p}$. Thus, as $v \in yp$, there exists an element x' in $yt \cap vr'$.

Since $x \in v'r$, $w \in v'r'$, and $r' \neq r$, $x \neq w$. Since $x' \in vr'$, $w \in vr$, and $r' \neq r$, $x' \neq w$. Thus, as $\{w, x, x'\} \subseteq yt$, $x' = x$. Thus, as $x' \in vr'$, $x \in vr'$. It follows that $\{x, w'\} \subseteq vr' \cap v'r$.

Since $v' \in yp$ and $y \in vp^*$, $v' \in vp^*p$. Thus, as $v' \neq v$, there exists an element m in $p^*p \setminus \{1\}$ such that $v' \in vm$. Since $m \in p^*p \setminus \{1\}$, $\{1, m\} = p^*p$; cf. Lemma 1.5.6(i). Thus, as we are assuming that $n_{p^*p} \neq 2$, $n_m = 2$. Thus, by Lemma 1.5.4, $a_{r'r^*m} = 1$.

From $a_{r'r^*m} = 1$, $v' \in vm$, and $\{w', x\} \subseteq vr' \cap v'r$ we obtain $w' = x$. Thus, as $w' \in yt'$ and $x \in yt$, $t' = t$, contradiction. This contradiction proves that χ is faithful.

Let us now assume that $n_s = 1$. By Lemma 6.2.1, we may assume that T is not thin. Thus, there exists an element r in T such that $n_r = 2$. Let x be an element in yr^*. Then, by the previous case (applied to x and y instead of y and z), we obtain a bijective faithful map χ from yT to $y'T$ such that $y\chi = y'$. Since s is assumed to be thin, $z\chi = z'$.

Let p and q be elements in S. If S is thin, one usually writes q^p instead of $\{q\}^p$.

Recall that, for each element s in S, we write $\langle s \rangle$ instead of $\langle \{s\} \rangle$.

The following lemma shows in which way thin schemes provide the situation required in Proposition 6.4.1.

Lemma 6.4.2 *Assume S to be thin, let s be an element in S, and let l be an involution of S. Then the following hold.*

(i) *We have $n_{s^{\langle l \rangle}} \leq 2$.*

(ii) *Assume that $n_{s^{\langle l \rangle}} = 2$. Then $n_{(s^{\langle l \rangle})^* s^{\langle l \rangle}} = 2$ if and only if $ll^s = l^s l$.*

Proof. (i) Since l is assumed to be an involution, $n_{\langle l \rangle s \langle l \rangle} \leq 4$.

On the other hand, we know from Theorem 4.1.3(iii) that $n_{s^{\langle l \rangle}} n_{\langle l \rangle} = n_{\langle l \rangle s \langle l \rangle}$. Thus, as $n_{\langle l \rangle} = 2$, $n_{s^{\langle l \rangle}} \leq 2$.

(ii) Let us assume that $n_{s^{\langle l \rangle}} = 2$. Then, by Lemma 1.5.6(i), there exists an element r in S such that $1^{\langle l \rangle} \neq r^{\langle l \rangle}$ and $\{1^{\langle l \rangle}, r^{\langle l \rangle}\} = (s^{\langle l \rangle})^* s^{\langle l \rangle}$.

From $1^{\langle l \rangle} \neq r^{\langle l \rangle}$ we obtain $r \notin \langle l \rangle$; cf. Lemma 4.1.1.

From $r^{\langle l \rangle} \in (s^{\langle l \rangle})^* s^{\langle l \rangle}$ we obtain $r \in \langle l \rangle s^* \langle l \rangle s \langle l \rangle$; cf. Lemma 4.1.4. Thus, as $r \in S \setminus \langle l \rangle$, $r \in \langle l \rangle l^s \langle l \rangle$. Thus, $lr = rl$ if and only if $ll^s = l^s l$.

However, we have $lr = rl$ if and only if $n_{r^{\langle l \rangle}} = 1$; cf. Lemma 4.2.5(i). Moreover, as $\{1^{\langle l \rangle}, r^{\langle l \rangle}\} = (s^{\langle l \rangle})^* s^{\langle l \rangle}$, this is the case if and only if $n_{(s^{\langle l \rangle})^* s^{\langle l \rangle}} = 2$.

Assume S to be thin, and let l be an involution of S. If S has finite valency, the local condition that, for each element s in $S \setminus C_S(l)$, $ll^s \neq l^s l$ can be expressed globally.[1] This fact, namely that condition (c) of the following theorem follows from condition (a) or condition (b), is due to George Glauberman; cf. [15; Theorem 1].

The proof of Glauberman's theorem requires Richard Brauer's Second Main Theorem on modular representations of finite groups; cf. [5]. A proof of Glauberman's theorem is beyond the scope of this monograph.

From Lemma 2.5.2(iv) and Lemma 2.3.6(i) we obtain that a thin scheme possesses a uniquely determined maximal normal closed subset of odd valency. We shall denote this closed subset by $O(S)$.

Theorem 6.4.3 *Assume S to be thin and to have finite valency. Then, for each involution l of S, the following conditions are equivalent.*

(a) *For each element s in $S \setminus C_S(l)$, $l^s l$ has odd order.*

(b) *For each element s in $S \setminus C_S(l)$, $ll^s \neq l^s l$.*

(c) *We have $[O(S), l]C_S(l) = S$.*

PROOF. (a) $\Rightarrow$ (b) Let us assume, by way of contradiction, that $S \setminus C_S(l)$ contains an element s with $ll^s = l^s l$. Then $(l^s l)^2 = l^s l l^s l = l^s l^s l l = 1$. Thus, $l^s l$ has order at most 2. Thus, as $s \notin C_S(l)$, $l^s l$ has order 2.

(b) $\Rightarrow$ (a) Let us assume, by way of contradiction, that $S \setminus C_S(l)$ contains an element s such that $l^s l$ has even order. Then there exists an integer n such that $(l^s l)^n$ is an involution. Thus, $(ll^s)^n = (l^s l)^n$. Set $k := (l^s l)^{n-1} l^s$. Then $lk = kl$. Moreover, if n is even and $2m + 2 = n$, $k = l^{l^s (ll^s)^m}$. If n is odd and $2m + 1 = n$, then $k = (l^s)^{(ll^s)^m}$.

(a) $\Rightarrow$ (c) This is the content of [15; Theorem 1].

(c) $\Rightarrow$ (a) Assume that $[O(S), l]C_S(l) = S$, and let s be an element in $S \setminus C_S(l)$. Then there exist elements p in $[O(S), l]$ and q in $C_S(l)$ such that $qp = s$. Then $l^p = l^s$. Thus, $l^s l = l^p l = p^{-1} p^l \in O(S)$, so that $l^s l$ has odd order.

Assume that S is simple and satisfies condition (c) in Theorem 6.4.3. Then $\langle l \rangle = S$ or $[O(S), l] = S$. In the second case, $O(S) = S$, and that means that S has odd valency. However, $\langle l \rangle$ has even valency, so that, according to Lemma 2.3.6(ii), S cannot have odd valency. This contradiction shows that S cannot

[1] Recall that $C_S(l)$ is an abbreviation for $C_S(\{l\})$.

be simple if S satisfies one of the three conditions (a), (b), or (c) of Theorem 6.4.3.

The following theorem is the main result of this section. It was first proved in [31; (5.1)].

Theorem 6.4.4 *Assume that S is not thin and that S has finite valency. Assume that, for each element s in S, $n_s \leq 2$ and $n_{s^*s} \neq 2$. Then there exists a finite thin scheme $\bar{S}$ and an involution $\bar{l}$ in $\bar{S}$ such that $[O(\bar{S}), \bar{l}]C_{\bar{S}}(\bar{l}) = \bar{S}$ and $S \cong \bar{S}//\langle \bar{l} \rangle$.*

PROOF. Let us denote by G the Schur group of S, and let us fix an element x in X. Then, by Lemma 6.3.5 and Lemma 6.3.6, there exists an element h in $G \setminus \{1\}$ such that $\{1, h\} = G_x$.

Let g be an element in $G \setminus C_G(h)$. We shall prove that $hh^g \neq h^g h$.

Let us denote by s the uniquely determined element in S satisfying $xg \in xs$. Then

$$xgh \in xsh \subseteq xhs = xs,$$

so that $\{xg, xgh\} \subseteq xs$.

Assume that $xgh = xg$. Then $xghg^{-1} = x$. Thus, $ghg^{-1} \in G_x$. Thus, as $\{1, h\} = G_x$, $ghg^{-1} = h$. It follows that $g \in C_G(h)$, contrary to the choice of g. This contradiction shows that $xgh \neq xg$.

From $\{xg, xgh\} \subseteq xs$ and $xgh \neq xg$ we obtain $\{xg, xgh\} = xs$. In particular,

$$xgs^* \cup xghs^* = xss^*.$$

Since $xg \in xs$, $x \in xsg^{-1} \subseteq xg^{-1}s$. Thus, $xg^{-1} \in xs^*$, so that

$$xh^g = xg^{-1}hg \in xs^*hg \subseteq xhgs^* = xgs^*.$$

On the other hand, we obtain from $xg \in xs$ also that $x \in xgs^*$. Thus, as $x \neq xh^g$,

$$\{x, xh^g\} = xgs^*.$$

From this we obtain

$$\{x, xh^g h\} = xghs^*.$$

Thus, as $xgs^* \cup xghs^* = xss^*$, $\{x, xh^g, xh^g h\} = xss^*$. Thus, as we are assuming that $n_{ss^*} \neq 2$, we conclude that $xh^g \neq xh^g h$. In particular, we have $hh^g \neq h^g h$.

We set $\bar{S} := G^\tau$ and $\bar{l} := h^\tau$.

For each element g in $G \setminus C_G(h)$, we have $hh^g \neq h^g h$. Thus, for each element $\bar{s}$ in $\bar{S} \setminus C_{\bar{S}}(\bar{l})$, $\bar{l}\bar{l}^{\bar{s}} \neq \bar{l}^{\bar{s}}\bar{l}$. Thus, by Theorem 6.4.3, $[O(\bar{S}), \bar{l}]C_{\bar{S}}(\bar{l}) = \bar{S}$.

Note finally that, by Theorem 6.3.1 and Proposition 6.4.1, $S \cong \bar{S}//\langle \bar{l} \rangle$.

Note that each element s in S satisfies $n_{ss^*} \neq 2$ if $O_\vartheta(O^\vartheta(S))$ has odd valency. Thus, the conclusion of Theorem 6.4.4 remains valid if we assume that all elements of S have valency at most 2 and $O^\vartheta(S)$ has odd valency.

Here is the converse of Theorem 6.4.4. It makes Theorem 6.4.4 to be one of our recognition theorems.

Theorem 6.4.5 *Assume S to be thin and to have finite valency. Let l be an involution of S such that $[O(S), l]C_S(l) = S$. Then $S//\langle l \rangle$ has finite valency and, for each element s in S, $n_{s^{\langle l \rangle}} \leq 2$ and $n_{(s^{\langle l \rangle})^* s^{\langle l \rangle}} \neq 2$.*

PROOF. Considering Theorem 6.4.3 we obtain the claim from Lemma 6.4.2.

Note that Theorem 6.4.4, in the form as we stated it, requires the above-mentioned theorem of George Glauberman. In order to avoid the use of Glauberman's theorem one could, of course, replace the condition that

$$[O(\bar{S}), \bar{l}]C_{\bar{S}}(\bar{l}) = \bar{S}$$

with one of the equivalent conditions (a) or (b) in Lemma 6.4.3.

Contrary to this, the following corollary of Theorem 6.4.4 relies essentially on Glauberman's theorem.

Corollary 6.4.6 *Assume that S has finite valency and that, for each element s in S, $n_s \leq 2$ and $n_{ss^*} \neq 2$. Then $O^\vartheta(S)O_\vartheta(S) = S$, and $O^\vartheta(S)$ has odd valency.*

PROOF. The statement is obvious if S is thin. If S is not thin, we obtain from Theorem 6.4.4 a finite thin scheme $\bar{S}$ and an involution $\bar{l}$ in $\bar{S}$ such that $[O(\bar{S}), \bar{l}]C_{\bar{S}}(\bar{l}) = \bar{S}$ and $S \cong \bar{S}//\langle \bar{l} \rangle$.

From $[O(\bar{S}), \bar{l}]C_{\bar{S}}(\bar{l}) = \bar{S}$ we obtain

$$([O(\bar{S}), \bar{l}]//\langle \bar{l} \rangle)(C_{\bar{S}}(\bar{l})//\langle \bar{l} \rangle) = \bar{S}//\langle \bar{l} \rangle;$$

cf. Lemma 4.1.6.

From $S \cong \bar{S}//\langle \bar{l} \rangle$ we obtain $O^\vartheta(S) \cong O^\vartheta(\bar{S}//\langle \bar{l} \rangle)$. On the other hand, we know from Lemma 4.2.3 that $O^\vartheta(\bar{S}//\langle \bar{l} \rangle) = [\bar{S}, \langle \bar{l} \rangle]\langle \bar{l} \rangle//\langle \bar{l} \rangle$. Thus, as $[\bar{S}, \langle \bar{l} \rangle] = [O(\bar{S}), \bar{l}]$,

$$O^\vartheta(S) \cong [O(\bar{S}), \bar{l}]\langle \bar{l} \rangle//\langle \bar{l} \rangle.$$

From $S \cong \bar{S}//\langle \bar{l} \rangle$ we also obtain $O_\vartheta(S) \cong O_\vartheta(\bar{S}//\langle \bar{l} \rangle)$. Moreover, by Lemma 4.2.5(i), $O_\vartheta(\bar{S}//\langle \bar{l} \rangle) = C_{\bar{S}}(\bar{l})//\langle \bar{l} \rangle$. Thus,

$$O_\vartheta(S) \cong C_{\bar{S}}(\bar{l})//\langle \bar{l} \rangle.$$

Now $O^\vartheta(S)O_\vartheta(S) = S$ follows from $S \cong \bar{S}//\langle \bar{l} \rangle$.

That $O^\vartheta(S)$ has odd valency follows from $O^\vartheta(S) \cong [O(\bar{S}), \bar{l}]\langle \bar{l} \rangle//\langle \bar{l} \rangle$.

6.5 More About Elements of Valency 2

In this section, S is assumed to have finite valency. We shall first prove that a closed subset of S which is generated by a single symmetric element of valency 2 is faithfully embedded in S. After that we shall look at closed subsets in which each non-identity element has valency 2.

Lemma 6.5.1 *Let s be a symmetric element of S, assume that s has valency 2, and let y and z be elements in X. Then there exists a bijective faithful map χ from $y\langle s\rangle$ to $z\langle s\rangle$ such that $y\chi = z$.*

PROOF. By Lemma 3.3.2 and Lemma 6.1.5(ii), we shall be done if we succeed in showing that there exists a map χ from $y\langle s\rangle$ to X such that $y\chi = z$ and $ys\chi \subseteq y\chi s$. (As for the bijectivity we refer to Lemma 6.1.1(iii).)

Let us write ℓ instead of $\ell_{\{s\}}$. For each non-negative integer n, we define R_n to be the set of all elements r in $\langle s\rangle$ such that $\ell(r) \leq n$.

Let us assume, by way of contradiction, that there does not exist a map χ from $y\langle s\rangle$ to X with $y\chi = z$ and $ys\chi \subseteq y\chi s$. Then, by Lemma 3.1.1(i), there exists a non-negative integer n such that there does not exist a map χ from yR_n to X such that $y\chi = z$ and, for any two elements u and v in yR_n, $v \in us$ implies $v\chi \in u\chi s$. We pick n as small as possible.

Clearly, $1 \leq n$. Thus, the minimal choice of n gives us a map χ from yR_{n-1} to X such that $y\chi = z$ and, for any two elements u and v in yR_{n-1}, $v \in us$ implies $v\chi \in u\chi s$.

From Lemma 3.3.2 we know that $|R_n \setminus R_{n-1}| = 1$. Let us denote by q the uniquely determined element in $R_n \setminus R_{n-1}$.

From $q \in R_n$ we obtain $q \in \langle s\rangle$. Thus, by Lemma 3.3.1, $n_q \leq 2$. Thus, there exist elements v and w in yq such that $\{v, w\} = yq$.

Let us first assume that $v \neq w$.

Since $q \in R_n \setminus R_{n-1}$, $\ell(q) = n$. Thus, there exists an element p in $\langle s\rangle$ such that $q \in ps$ and $\ell(q) = \ell(p) + 1$; cf. Lemma 3.1.2.

Since $v \in yq$ and $q \in ps$, $v \in yps$. Thus, there exists an element v' in yp such that $v \in v's$. Similarly, we find an element w' in yp such that $w \in w's$.

We define $v\chi$ to be the unique element in $v'\chi s \cap y\chi R_n$. Similarly, we define $w\chi$ to be the unique element in $w'\chi s \cap y\chi R_n$.

Note that v' is the only element in $vs \cap yR_{n-1}$. Otherwise we would have $2 \leq a_{psq}$. Then, as $a_{psq}n_q = a_{qs^*p}n_p$, $n_q = 1$, and then $v = w$. Also, if $w \in vs$, $w\chi \in v\chi s$. Thus, we are done in this case.

Let us now assume that $v = w$. In this case, we have $1 = |yR_n \setminus yR_{n-1}|$, so that $n_q = 1$. Thus, $|y\chi R_n \setminus y\chi R_{n-1}| = 1$. We denote by $v\chi$ the uniquely determined element in $y\chi R_n \setminus y\chi R_{n-1}$.

Since $n_q = 1$, there exist elements t and u in $vs \cap yR_{n-1}$ such that $t \neq u$. Thus, $a_{psq} = 2$. Thus, $v\chi \in t\chi s$ and $v\chi \in u\chi s$. Again, we are done.

Proposition 6.5.2 *Let s be a symmetric element of S, and assume that s has valency 2. Then $\langle s\rangle$ is faithfully embedded in S.*

PROOF. Let y be an element in X, let z be an element in $y\langle s\rangle$, and let χ be a faithful map from $\{y, z\}$ to X. We have to show that χ extends faithfully to a bijective map from $y\langle s\rangle$ to $y\chi\langle s\rangle$.

From Lemma 6.5.1 we obtain a bijective faithful map ϕ from $y\langle s\rangle$ to X such that $y\phi = y\chi$.

Since $z \in y\langle s\rangle$, there exists an element q in $\langle s\rangle$ such that $z \in yq$.

Since χ is faithful, we obtain from $z \in yq$ that $z\chi \in y\chi q$.

Since ϕ is faithful, we obtain from $z \in yq$ that $z\phi \in y\phi q$. Thus, as $y\phi = y\chi$, $z\phi \in y\chi q$.

By Lemma 3.3.1, each element in $\langle s\rangle$ has valency at most 2. Thus, by Lemma 6.3.5, there exists a (bijective) faithful map ψ from $y\chi\langle s\rangle$ to $y\chi\langle s\rangle$ such that ($y\chi\psi = y\chi$ and) $z\phi\psi = z\chi$. Now $\phi\psi$ is a faithful map from $y\langle s\rangle$ to $y\chi\langle s\rangle$ which coincides with χ on $\{y, z\}$.

For any two elements p and q in S, we write $\langle p, q\rangle$ instead of $\langle\{p, q\}\rangle$.

Theorem 6.5.3 *Let s be a symmetric element of S such that $n_s = 2$ and $\langle s\rangle = S$. Then there exists a finite thin scheme $\bar{S}$ containing involutions $\bar{h}$ and $\bar{k}$ with $\bar{h}\bar{k} \neq \bar{k}\bar{h}$, $\langle \bar{h}, \bar{k}\rangle = \bar{S}$, and $S \cong \bar{S}//\langle\bar{k}\rangle$.*

PROOF. Let us denote by G the Schur group of S, and let us fix an element x in X.

It is easy to see that our claim holds if $n_{ss^*} = 2$. Let us, therefore, assume that $n_{ss^*} \neq 2$. Then $O^\vartheta(S)$ does not have valency 2. Thus, referring to Lemma 3.3.1 we obtain from Lemma 6.3.5 and Lemma 6.3.6 that G possesses an involution k such that $\langle k\rangle = G_x$.

We set $\bar{S} := G^\tau$ and $\bar{k} := k^\tau$. Then $\bar{S}$ is a finite thin scheme, and, by Theorem 6.3.1 and Proposition 6.5.2, $S \cong \bar{S}//\langle\bar{k}\rangle$.

From $S \cong \bar{S}//\langle\bar{k}\rangle$ and $\langle s\rangle = S$ we obtain an element $\bar{h}$ in $\bar{S}$ such that $\bar{h}^{\langle\bar{k}\rangle}$ has valency 2 and $\langle\bar{h}^{\langle\bar{k}\rangle}\rangle = \bar{S}//\langle\bar{k}\rangle$.

Since $\bar{h}^{\langle\bar{k}\rangle}$ has valency 2, $\bar{h}\bar{k} \neq \bar{k}\bar{h}$; cf. Lemma 4.2.5(i). Since $\langle\bar{h}^{\langle\bar{k}\rangle}\rangle = \bar{S}//\langle\bar{k}\rangle$, $\langle\bar{h}, \bar{k}\rangle = \bar{S}$; cf. Lemma 4.2.2(i).

Here is the converse of Theorem 6.5.3. It makes Theorem 6.5.3 to be one of our recognition theorems.

Theorem 6.5.4 *Assume S to be thin. Let h and k be involutions of S such that $hk \neq kh$ and $\langle h, k\rangle = S$. Then $h^{\langle k\rangle}$ is symmetric, $h^{\langle k\rangle}$ has valency 2, and we have $\langle h^{\langle k\rangle}\rangle = S//\langle k\rangle$.*

PROOF. Applying Lemma 4.2.2(i) to $\{h, k\}$ instead of R, we obtain $\langle h^{\langle k\rangle}\rangle = S//\langle k\rangle$.

From Lemma 6.4.2(i) we know that $n_{h^{\langle k\rangle}} \leq 2$. Since $hk \neq kh$, $2 \leq n_{h^{\langle k\rangle}}$; cf. Lemma 4.2.5(i). Thus, $n_{h^{\langle k\rangle}} = 2$.

From Lemma 4.1.2(ii) we obtain that $h^{\langle k\rangle}$ is symmetric.

Note the similarity between Theorem 6.4.4 and Theorem 6.5.3.

The first theorem says that groups with a 'Glauberman involution' provide the only examples for non-thin schemes of finite valency in which each element s satisfies $n_s \leq 2$ and $n_{ss^*} \neq 2$.

The second theorem says that finite dihedral groups provide the only examples for schemes of finite valency which are generated by a single symmetric element.

We now list a few consequences of Theorem 6.5.3.

Corollary 6.5.5 *Let s be a symmetric element of S, and assume that s has valency 2. Set $n := n_{\langle s\rangle}$. Then the following hold.*

(i) *If n is odd, $\{1\} = O_\vartheta(\langle s\rangle)$ and $O^\vartheta(\langle s\rangle) = \langle s\rangle$.*

(ii) *Assume that n is even. Then $O^\vartheta(\langle s\rangle) \neq \langle s\rangle$ and there exists an element r in $\langle s\rangle$ such that $\ell_{\{s\}}(r) = \max \ell_{\{s\}}(\langle s\rangle)$ and $\{1, r\} = O_\vartheta(\langle s\rangle)$.*

(iii) *Assume that n is even and that 4 does not divide n. Then we have $\{1\} = O^\vartheta(\langle s\rangle) \cap O_\vartheta(\langle s\rangle)$ and $O^\vartheta(\langle s\rangle)O_\vartheta(\langle s\rangle) = \langle s\rangle$.*

(iv) *If 4 divides n, $O_\vartheta(\langle s\rangle) \subseteq O^\vartheta(\langle s\rangle)$.*

PROOF. All these statements follow from Theorem 6.5.3.

The second part of the following theorem is due to Mitsugu Hirasaka. It is [22; Theorem 4.9].

Theorem 6.5.6 *Let T be a closed subset of S in which each element has valency at most 2. Then we have the following.*

(i) *The set $O^\vartheta(T)$ is schurian.*

(ii) *If each element in $T \setminus \{1\}$ has valency 2, T is schurian.*

PROOF. (i) Let x be an element in X, set $U := O^\vartheta(T)$, and let us write G to denote the Schur group of U with respect to x.

Each element in T is assumed to have valency at most 2. Thus, as $U \subseteq T$, each element in U has valency at most 2. Thus, we obtain from Lemma 6.3.5 that, for any two elements w in xU and u in U, G_w acts transitively on (the two elements of) wu. Thus, we just have to show that G acts transitively on xU; cf. Theorem 6.3.1 and Theorem 6.3.3.

In order to show this latter condition we fix elements y and z in xU. According to Lemma 6.3.4, we may assume that there exists an element t in T such that $z \in yt^*t$. Since $z \in yt^*t$, there exists an element w in yt^* such that $z \in wt$.

From $y \in wt$ and $z \in wt$ we now obtain an element g in the Schur group of T with respect to x which (fixes w) and maps y to z. From $yg = z$ and $z \in yU$ we obtain that the restriction of g to xU is in G.

(ii) Referring to Theorem 3.3.5 this follows immediately from (i).

6.6 Constrained Sets of Involutions

Let L be a set of involutions, and let us write ℓ instead of ℓ_L.

Recall that, for each element q in $\langle L\rangle$, $S_1(q, L)$ is our notation for the set of all elements p in $\langle L\rangle$ such that pq contains an element r with $\ell(r) = \ell(p) + \ell(q)$.

In accordance with Section 3.4 we shall write, for each element s in $\langle L\rangle$, $S_1(s)$ instead of $S_1(s, L)$.

Recall that L is called constrained if, for any two elements q in $\langle L\rangle$ and p in $S_1(q)$, $1 = |pq|$.

Throughout this section, the letter L stands for a constrained set of involutions.

Lemma 6.6.1 *Let q be an element in $\langle L\rangle$, and let p be an element in $S_1(q)$. Let x be an element in X, let y be an element in xp, and let z be an element in yq. Let χ be a map from $\{x, y, z\}$ to X. Then, if $\chi|_{\{x,y\}}$ and $\chi|_{\{y,z\}}$ are faithful, so is χ.*

Proof. Let us denote by s the uniquely determined element in S which satisfies $z \in xs$. We have to prove that $z\chi \in x\chi s$.

Since $z \in yq$ and $y \in xp$, $z \in xpq$. Thus, as $z \in xs$, $s \in pq$. Thus, as we are assuming that $p \in S_1(q)$ and that L is constrained, $\{s\} = pq$.

Assume that $\chi|_{\{x,y\}}$ is faithful. Then, as $y \in xp$, $y\chi \in x\chi p$. Assume that $\chi|_{\{y,z\}}$ is faithful. Then, as $z \in yq$, $z\chi \in y\chi q$. From $z\chi \in y\chi q$ and $y\chi \in x\chi p$ we obtain $z\chi \in x\chi pq$. Thus, as $\{s\} = pq$, $z\chi \in x\chi s$.

Let y and z be elements in X, and let n be the smallest non-negative integer n with $z \in yL^n$. We write $D(y, z)$ to denote the union of the sets $yL^i \cap zL^j$ which satisfy $i + j = n$.

Lemma 6.6.2 *Let y be an element in X, let z be an element in $y\langle L\rangle$, let v be an element in $D(y,z)$, and let w be an element in $D(v,z)$. Let χ be a map from $\{v,w,y,z\}$ to X. Then, if $\chi|_{\{y,v,z\}}$ and $\chi|_{\{y,w,z\}}$ are faithful, so is χ.*

Proof. Let us denote by p the uniquely determined element in S which satisfies $v \in yp$, by t the one which satisfies $w \in vt$. Then, we have $w \in ypt$. Thus, there exists an element s in pt such that $w \in ys$.

Since $w \in ys$, $w\chi \in y\chi s$. (We are assuming that $\chi|_{\{y,w,z\}}$ is faithful.) Thus, as $s \in pt$, $w\chi \in y\chi pt$. Thus, there exists an element x in $y\chi p$ such that $w\chi \in xt$.

Let us denote by u the uniquely determined element in S which satisfies $z \in wu$. Then, as $w \in vt$, $z \in vtu$. Thus, there exists an element q in tu such that $z \in vq$. From $q \in tu$ and $w \in D(v,z)$ we obtain $\ell(q) = \ell(t) + \ell(u)$. Thus, $t \in S_1(u)$. Thus, as L is assumed to be constrained, we obtain from $q \in tu$ that $\{q\} = tu$.

Since $z \in wu$, $z\chi \in w\chi u$. (Again, we use the hypothesis that $\chi|_{\{y,w,z\}}$ is faithful.) Thus, as $w\chi \in xt$, $z\chi \in xtu$. Thus, as $\{q\} = tu$, $z\chi \in xq$.

From $z \in vq$ and $v \in yp$ we obtain $z \in ypq$. Thus, there exists an element r in pq such that $z \in yr$. From $r \in pq$ and $v \in D(y,z)$ we obtain $\ell(r) = \ell(p) + \ell(q)$. Thus, by Lemma 3.5.2, $a_{pqr} = 1$.

Since $z \in yr$, $z\chi \in y\chi r$. Since $v \in yp \cap zq^*$, $v\chi \in y\chi p \cap z\chi q^*$. (This time, we use that $\chi|_{\{y,v,z\}}$ is assumed to be faithful.) On the other hand, we also have $x \in y\chi p \cap z\chi q^*$. Thus, as $a_{pqr} = 1$, $v\chi = x$. Thus, as $w\chi \in xt$, $w\chi \in v\chi t$.

Lemma 6.6.3 *Let x be an element in X, and let χ be a map from $x\langle L\rangle$ to X. Assume that, for any two elements w in $x\langle L\rangle$ and l in L, $wl\chi \subseteq w\chi l$. Then χ is faithful.*

PROOF. Let us assume that χ is not faithful, and let us define R to be the set of all elements s in $\langle L\rangle$ such that $xs\chi \not\subseteq x\chi s$. Then R is not empty. We fix an element r in R such that $\ell(r)$ is as small as possible.

Among the elements in R we pick r as small as possible. Then $1 \neq r$. Thus, by Lemma 3.1.2, there exist elements q in $\langle L\rangle$ and l in L such that $r \in ql$ and $\ell(r) = \ell(q) + 1$.

From $q \in \langle L\rangle$ and $l \in L$ we obtain $q \in \langle L\rangle$. Thus, as $\ell(q) = \ell(r) - 1$, the minimal choice of r yields $q \notin R$. That means that $xq\chi \subseteq x\chi q$. By hypothesis, we have that, for each element w in $x\langle L\rangle$, $wl\chi \subseteq w\chi l$. Since L is assumed to be constrained, we also have $\{r\} = ql$. Thus,

$$xr\chi = x(ql)\chi = (xq)l\chi \subseteq (xq\chi)l \subseteq (x\chi q)l = x\chi(ql) = x\chi r,$$

contradiction.

Let x be an element in X, and let G be a subgroup of the Schur group of $\langle L\rangle$ with respect to x. For any two elements y and z in X, we define G_{yz} to be the intersection of G_y and G_z.

Lemma 6.6.4 *Let x be an element in X, and let G be a subgroup of the Schur group of $\langle L\rangle$ with respect to x. Let q be an element in $\langle L\rangle$, let p be an element in $S_1(q)$. Then we have the following.*

(i) *If G_x acts transitively on xpq, G_x acts transitively on xp.*

(ii) *Assume that G_x acts transitively on xp and that, for each element w in xp, G_{xw} acts transitively on wq. Then G_x acts transitively on xpq.*

PROOF. (i) Let y and z be elements in xp. We shall prove that there exists an element g in G_x such that $yg = z$.

We are assuming that $p \in S_1(q)$. Thus, as L is assumed to be constrained, there exists an element r in pq such that $\{r\} = pq$.

Let v be an element in yq, and let w be an element in zq. From $v \in yq$ and $y \in xp$ we obtain $v \in xpq$. Thus, as $\{r\} = pq$, $v \in xr$. Similarly, we obtain from $w \in zq$ and $z \in xp$ that $w \in xr$. Thus, as we are assuming that G_x acts transitively on xpq, there exists an element g in G_x such that $vg = w$.

Since $y \in xp \cap vq^*$, $yg \in xp \cap vgq^* = xp \cap wq^*$. Thus, as $\{r\} = pq$, $yg = z$, cf. Lemma 3.5.2.

(ii) Let y and z be elements in xpq. We shall prove that there exists an element g in G_x such that $yg = z$.

Since $y \in xpq$, there exists an element v in xp such that $y \in vq$. Since $z \in xpq$, there exists an element w in xp such that $z \in wq$.

We are assuming that G_x acts transitively on xp. Thus, as v, $w \in xp$, there exists an element e in G_x such that $ve = w$. We are also assuming that G_{xw} acts transitively on wq. Thus, there exists an element f in G_{xw} such that $yef = z$. Finally, as $e \in G_x$ and $f \in G_{xw}$, $ef \in G_x$.

Lemma 6.6.5 *Let x be an element in X, and let G be the Schur group of $\langle L\rangle$ with respect to x. Assume that, for any four elements l in L, s in $S_1(l)$, y in $x\langle L\rangle$, and z in ys, G_{yz} acts transitively on zl. Then, $\langle L\rangle_x$ is schurian.*

PROOF. That G acts transitively on $x\langle L\rangle$ follows from the hypothesis with $s = 1$.

From Lemma 6.6.4(ii) we obtain by induction that, for each element s in $\langle L\rangle$, G_x acts transitively on xs. Thus, the claim follows from Theorem 6.3.1.

6.7 Thin Thin Residues

In this section, the letter T stands for a closed subset of S which has finite valency and satisfies

$$O^{\vartheta}(T) \subseteq O_{\vartheta}(T).$$

We shall prove that, under a certain additional condition, T is faithfully embedded in S.

Recall from Section 2.6 that, for each element s in S, $D_T(s)$ is our notation for the set of all elements t in T satisfying $t^*t \subseteq s^*s$.

Lemma 6.7.1 *For each element t in T, the following hold.*

(i) *We have $t^*t \subseteq O_{\vartheta}(T)$.*

(ii) *We have $O^{\vartheta}(T) \subseteq D_T(t)$.*

(iii) *We have $\{t\} = tt^*t$.*

(iv) *The set t^*t is closed.*

(v) *We have $O^{\vartheta}(T) \subseteq K_S(t^*t)$.*

(vi) *We have $n_t = n_{t^*t}$.*

PROOF. (i) From Theorem 3.2.1(ii) we know that, for each element t in T, $t^*t \subseteq O^{\vartheta}(T)$. Thus, as we are assuming that $O^{\vartheta}(T) \subseteq O_{\vartheta}(T)$, we must have $t^*t \subseteq O_{\vartheta}(T)$.

(ii) It follows right from the definition of $D_T(t)$ that thin elements of T belong to $D_T(t)$. Thus, by Lemma 1.5.1, $O_{\vartheta}(T) \subseteq D_T(t)$. Thus, the claim follows from our hypothesis that $O^{\vartheta}(T) \subseteq O_{\vartheta}(T)$.

(iii) From (i) we know that $t^*t \subseteq O_{\vartheta}(T)$. Thus, by Corollary 2.6.6 (applied to t, $\{1\}$ and T in the roles of s, T and U), $1^t = t^*t$. Thus, by definition, $tt^*t \subseteq \{t\}$. It follows that $\{t\} = tt^*t$.

(iv) From (iii) we obtain $t^*tt^*t \subseteq t^*t$, and this implies that t^*t is closed.

(v) From (i) we know that $t^*t \subseteq O_{\vartheta}(T)$, from (ii) we know that $O^{\vartheta}(T) \subseteq D_T(t^*)$. Clearly, we also have that

$$tO^{\vartheta}(T) \subseteq tO^{\vartheta}(T)t^*t \subseteq O^{\vartheta}(T)t.$$

Thus, $O^{\vartheta}(T) \subseteq O^{\vartheta}(T)^t$. Thus, by Lemma 2.6.8, $O^{\vartheta}(T) \subseteq K_S(t^*t)$.

(vi) Considering Lemma 1.4.4(ii) this follows from (iii).

Corollary 6.7.2 *Let π be a set of primes such that, for each element t in T, n_t is a π-number. Then the valency of $O^{\vartheta}(T)$ is a π-number.*

PROOF. From Theorem 3.2.1(ii) we know that $O^{\vartheta}(T)$ is generated by the subsets t^*t with $t \in T$. From Lemma 6.7.1(iv), (v) we know that, for each element t in T, t^*t is a normal closed subset of T, and Lemma 6.7.1(vi) says that, for each element t in T, $n_t = n_{t^*t}$.

Thus, as we are assuming that, for each element t in T, n_t is a π-number, $O^\vartheta(T)$ is generated by normal closed subsets of valency a π-number. Thus, considering Lemma 2.1.1 the claim follows from Lemma 2.3.6(i).

Assume that there exists a prime p such that $n_t = p$ for each element t in T. Then, via the group correspondence, $O^\vartheta(T)$ is an elementary abelian p-group; cf. Corollary 6.7.2.

We shall now focus on a condition which turns out to be sufficient for T to be faithfully embedded in S. We shall assume that, for any two elements p and q in T, $p^*p \subseteq q^*q$ or $q^*q \subseteq p^*p$.

Lemma 6.7.3 *Assume that $\{t^*t \mid t \in T\}$ is linearly ordered with respect to set theoretic inclusion. Then, for each element t in T, $tt^* = t^*t$.*

PROOF. We are assuming that T has finite valency. Thus, by Lemma 1.1.2(iii), $n_{t^*} = n_t$. Thus, by Lemma 6.7.1(vi), $n_{tt^*} = n_{t^*t}$. Thus, as we are assuming that $t^*t \subseteq tt^*$ or $tt^* \subseteq t^*t$, $tt^* = t^*t$.

Lemma 6.7.4 *Assume that $\{t^*t \mid t \in T\}$ is linearly ordered with respect to set theoretic inclusion. Then, for each element t in T, $D_T(t)$ is closed.*

PROOF. Let p and q be elements in $D_T(t)$, and let s be an element in p^*q. We have to show that $s \in D_T(t)$. Thus, we shall be done if we succeed in showing that $s^*s \subseteq t^*t$.

Since $\{t^*t \mid t \in T\}$ is assumed to be linearly ordered with respect to set theoretic inclusion, we have $pp^* \subseteq qq^*$ or $qq^* \subseteq pp^*$.

Let us first assume that $pp^* \subseteq qq^*$. Then, as $s \in p^*q$, $s^*s \subseteq q^*pp^*q \subseteq q^*qq^*q$. However, according to Lemma 6.7.1(iv), q^*q is closed, so that $q^*qq^*q \subseteq q^*q$. Thus, $s^*s \subseteq q^*q$. On the other hand, we have picked q in $D_T(t)$, and, therefore, we have $q^*q \subseteq t^*t$. It follows that $s^*s \subseteq t^*t$.

Let us now assume that $qq^* \subseteq pp^*$. Then, as $s \in p^*q$,

$$ss^* \subseteq p^*qq^*p \subseteq p^*pp^*p.$$

However, we know from Lemma 6.7.1(iv) that p^*p is closed. Thus, $ss^* \subseteq p^*p$. Thus, as p has been chosen from $D_T(t)$, $p^*p \subseteq t^*t$. It follows that $ss^* \subseteq t^*t$. Thus, by Lemma 6.7.3, $s^*s \subseteq t^*t$.

The following theorem is [25; Theorem B].

Theorem 6.7.5 *Assume that $\{t^*t \mid t \in T\}$ is linearly ordered with respect to set theoretic inclusion. Then, T is faithfully embedded in S.*

PROOF. Let us abbreviate $V := O^\vartheta(T)$. By Lemma 6.2.1, we may assume that T is not thin. Thus, by Lemma 4.2.5(ii), $\{1\} \neq V$. In particular, there exists an element t in T such that $t^*t \neq V$.

Let us denote by U the set of all elements t in T which satisfy $t^*t \neq V$. Since we are assuming $\{t^*t \mid t \in T\}$ to be linearly ordered with respect to set theoretic inclusion, there exists an element s in T such that $U = D_T(s)$. Thus, by Lemma 6.7.4, U is closed. Note also that $U \neq T$.

We are assuming that $O^\vartheta(T) \subseteq O_\vartheta(T)$. Thus, by Theorem 3.2.1(iii), we also have $O^\vartheta(U) \subseteq O_\vartheta(U)$. Thus, by induction, U is faithfully embedded in S.

Recall that $V \subseteq O_\vartheta(T)$ and $U = D_T(s)$. Thus, as $O_\vartheta(T) \subseteq D_T(s)$, $V \subseteq U$.

According to the definition of U, we have, for each element t in $T \setminus U$, $t^*t = V$. Thus, for each element t in $T \setminus U$, we have $\{t\} = tV$; cf. Lemma 6.7.1(iii). Thus, by Theorem 6.2.6, T is faithfully embedded in S.

Corollary 6.7.6 *If $O^\vartheta(T)$ is simple, T is faithfully embedded in S.*

PROOF. From Lemma 6.7.1(iv), (v) we know that, for each element t in T, t^*t is a normal closed subset of $O^\vartheta(T)$. Thus, as we are assuming $O^\vartheta(T)$ to be simple, we have that, for each element t in T, $\{1\} = t^*t$ or $t^*t = O^\vartheta(T)$. Thus, the claim follows from Theorem 6.7.5.

Lemma 6.7.7 *Assume that $\{t^*t \mid t \in T\}$ is linearly ordered with respect to set theoretic inclusion. Then the following hold.*

(i) *For any two elements p and q in T, we have $p^*p \subseteq q^*q$ if and only if $D_T(p) \subseteq D_T(q)$.*

(ii) *The set $\{D_T(t) \mid t \in T\}$ is linearly ordered with respect to set theoretic inclusion.*

(iii) *Let p and q be elements in T such that $q \notin D_T(p)$. Then $|p^*q| = 1$.*

PROOF. (i) Let p and q be elements in T, and let us first assume that $p^*p \subseteq q^*q$.

Let s be an element in $D_T(p)$. Then, by definition, $s^*s \subseteq p^*p$. Thus, as we are assuming that $p^*p \subseteq q^*q$, we conclude that $s^*s \subseteq q^*q$. Thus, by definition, $s \in D_T(q)$.

Since s has been chosen arbitrarily in $D_T(p)$, we have shown that $D_T(p) \subseteq D_T(q)$.

Let us now, conversely assume that $D_T(p) \subseteq D_T(q)$. Then $p \in D_T(q)$. Thus, by definition, $p^*p \subseteq q^*q$.

(ii) This is an immediate consequence of (i).

(iii) Let t and u be elements in p^*q. We shall see that $t = u$.

Assume that $t \in D_T(p)$. From Lemma 6.7.4 we know that $D_T(p)$ is closed. Thus, as $p \in D_T(p)$, $pt \subseteq D_T(p)$.

On the other hand, we obtain from $t \in p^*q$ that $q \in pt$; cf. Lemma 1.3.3(ii). Thus, $q \in D_T(p)$, contradiction.

Thus, $t \notin D_T(p)$, so that, according to (i), $p \in D_T(t)$. Thus, by definition, $p^*p \subseteq t^*t$. Thus, by Lemma 6.7.3, $p^*p \subseteq tt^*$.

From $u \in p^*q$ and $q \in pt$ we obtain $u \in p^*pt$. From $u \in p^*pt$ and $p^*p \subseteq tt^*$ we obtain $u \in tt^*t$. Now recall that, by Lemma 6.7.1(iii), $\{t\} = tt^*t$. Thus, $t = u$.

7

Products

In this chapter we investigate various types of products arising naturally in scheme theory. We define direct products of closed subsets of S, direct products of schemes, quasi-direct products of schemes, and semidirect products of schemes.

7.1 Direct Products of Closed Subsets

Let n be a positive integer, let $T_1, \ldots, T_n$ be closed subsets of S, and let us write T to denote the closed subset generated by the union of the sets T_i with $i \in \{1, \ldots, n\}$.

For each element i in $\{1, \ldots, n\}$, we define $\hat{T}_i$ to be the closed subset generated by the union of the sets T_j with $j \in \{1, \ldots, n\} \setminus \{i\}$.

Assume that, for each element i in $\{1, \ldots, n\}$, $\{1\} = T_i \cap \hat{T}_i$. We call T the *direct product* of the closed subsets $T_1, \ldots, T_n$ if, for any two elements i and j in $\{1, \ldots, n\}$, $T_i \subseteq N_S(T_j)$.

We shall write

$$T_1 \times \ldots \times T_n = T$$

in order to indicate that the closed subset T is the direct product of the closed subsets $T_1, \ldots, T_n$.

Assume that $T_1 \times \ldots \times T_n = T$. Then, by Lemma 2.1.1, $T_1 \cdots T_n = T$. Moreover, for any two elements i and j in $\{1, \ldots, n\}$ with $i \neq j$, $T_i \subseteq C_S(T_j)$; cf. Lemma 2.5.3.

Theorem 7.1.1 *Let n be a positive integer, and let $T_1, \ldots, T_n$ be closed subsets of S such that, for any two elements i and j in $\{1, \ldots, n\}$, $T_i \subseteq N_S(T_j)$. Then, for each closed subset T of S, the following conditions are equivalent.*

(a) *We have $T_1 \times \ldots \times T_n = T$.*

(b) *Let t be an element in T. Then there exists, for each element i in $\{1, \ldots, n\}$, a uniquely determined element t_i in T_i such that $t \in t_1 \cdots t_n$.*

PROOF. The claim is obvious for $n = 1$. Therefore, we assume that $2 \leq n$.

(a) $\Rightarrow$ (b) Since $t \in \hat{T}_n T_n$, there exist uniquely determined elements s in $\hat{T}_n$ and t_n in T_n such that $t \in st_n$; cf. Lemma 2.1.2. By induction, there exists, for each element i in $\{1, \ldots, n-1\}$, a uniquely determined element t_i in T_i such that $s \in t_1 \cdots t_{n-1}$.

From $t \in st_n$ and $s \in t_1 \cdots t_{n-1}$ we obtain $t \in t_1 \cdots t_n$.

(b) $\Rightarrow$ (a) We have to show that, for each element i in $\{1, \ldots, n\}$, $\{1\} = T_i \cap \hat{T}_i$. We are assuming that, for any two elements i and j in $\{1, \ldots, n\}$, $T_i \subseteq N_S(T_j)$. Thus, for any two elements i and j in $\{1, \ldots, n\}$, $T_i T_j = T_j T_i$. Therefore it suffices to prove that $\{1\} = T_n \cap \hat{T}_n$.

Let s be an element in $T_n \cap \hat{T}_n$. Since $\hat{T}_n = T_1 \cdots T_{n-1}$, we obtain from $s \in \hat{T}_n$ that $s \in T_1 \cdots T_{n-1}$. Thus, there exists, for each element i in $\{1, \ldots, n-1\}$, an element t_i in T_i such that $s \in t_1 \cdots t_{n-1}$.

On the other hand, we have $s \in T_n$. Thus, the uniqueness assumed in (b) yields $s = 1$.

Lemma 7.1.2 *Let n be a positive integer, let $T_1, \ldots, T_n$ be closed subsets of S, and let T be a closed subset of S such that $T_1 \times \ldots \times T_n = T$. Then we have the following.*

(i) *For each element t in T, $t \in (T_1 \cap t\hat{T}_1) \cdots (T_n \cap t\hat{T}_n)$.*

(ii) *Let U be a closed subset of S such that $U \subseteq T$. Assume that*

$$(T_1 \cap U\hat{T}_1) \cdots (T_n \cap U\hat{T}_n) = T.$$

Then, for each element i in $\{1, \ldots, n\}$, $U\hat{T}_i = T$.

PROOF. (i) Let t be an element in T. Since we are assuming that $T_1 \cdots T_n = T$, there exists, for each element i in $\{1, \ldots, n\}$, an element t_i in T_i such that $t \in t_1 \cdots t_n$.

Let i be an element in $\{1, \ldots, n\}$. Then, for each element j in $\{1, \ldots, n\} \setminus \{i\}$, $\{1\} = T_i \cap T_j$, $T_i \subseteq N_S(T_j)$, and $T_j \subseteq N_S(T_i)$. Thus, for each element j in $\{1, \ldots, n\} \setminus \{i\}$, $t_i t_j = t_j t_i$; cf. Lemma 2.5.3. Therefore, $t \in t_i \hat{T}_i$. Thus, by Lemma 2.1.4, $t_i \in t\hat{T}_i$. It follows that $t_i \in T_i \cap t\hat{T}_i$.

Since $i \in \{1, \ldots, n\}$ has been chosen arbitrarily, we now obtain from $t \in t_1 \cdots t_n$ that $t \in (T_1 \cap t\hat{T}_1) \cdots (T_n \cap t\hat{T}_n)$.

(ii) Let i be an element in $\{1, \ldots, n\}$, and let t be an element in T_i. Then $t \in T$, so that, by hypothesis, $t \in (T_1 \cap U\hat{T}_1) \cdots (T_n \cap U\hat{T}_n)$.

Since $t \in (T_1 \cap U\hat{T_1}) \cdots (T_n \cap U\hat{T_n})$, there exists, for each element j in $\{1, \ldots, n\}$, an element t_j in $T_j \cap U\hat{T_j}$ such that $t \in t_1 \cdots t_n$. Thus, as $t \in T_i$, $t = t_i$; cf. Theorem 7.1.1. It follows that $t \in U\hat{T_i}$.

Since $t \in T_i$ has been chosen arbitrarily, we have proved that $T_i \subseteq U\hat{T_i}$. It follows that $T = T_i\hat{T_i} \subseteq U\hat{T_i}$.

A closed subset T of S is called *decomposable* if there exist closed subsets U and V of S such that $\{1\} \neq U$, $\{1\} \neq V$, and $U \times V = T$. A closed subset of S different from $\{1\}$ is called *indecomposable* if it is not decomposable.

Theorem 7.1.3 *Let T be a closed subset of S such that $\{1\} \neq T$. Then we have the following.*

(i) *Let U and U' be closed subsets of S such that $U \times U' = T$. Let V and W be closed subsets of S such that $V \times W = U'$. Then $U \times V \times W = T$.*

(ii) *If S has finite valency, T is the direct product of indecomposable closed subsets of S.*

Proof. (i) Let s be an element in $V \cap UW$. Since $s \in UW$, there exist elements u in U and w in W such that $s \in uw$. Thus, by Lemma 1.3.3(i), $u \in sw^* \subseteq VW$. It follows that $u \in U \cap U' = \{1\}$. Thus, as $s \in uw$, $s = w$. Thus, we obtain from $s \in V$, $w \in W$, and $V \cap W = \{1\}$ that $s = 1$.

Since s has been chosen arbitrarily in $V \cap UW$, we have shown that $\{1\} = V \cap UW$. Similarly, one obtains $\{1\} = W \cap UV$.

From $V \times W = U'$ we obtain $V \subseteq N_S(W)$ and $W \subseteq N_S(V)$. Since $U \times U' = T$, $VW = U' \subseteq N_S(U)$. Referring to Lemma 2.5.3 we also obtain

$$U \subseteq C_S(U') \subseteq C_S(V) \subseteq N_S(V).$$

Similarly, we obtain $U \subseteq N_S(W)$.

(ii) This follows from (i) by induction.

Proposition 7.1.4 *Let T be a closed subset of S. Assume that T has finite valency and that, for any two elements p and q in T, $qp = pq$.*

Let m and n be positive integers, and let $U_1, \ldots, U_m$ and $V_1, \ldots, V_n$ be indecomposable closed subsets of S such that

$$U_1 \times \ldots \times U_m = T = V_1 \times \ldots \times V_n.$$

Then, for each element i in $\{1, \ldots, m\}$, there exists an element j in $\{1, \ldots, n\}$ such that $U_i \times \hat{V_j} = T = V_j \times \hat{U_i}$.

Proof. Let i be an element in $\{1, \ldots, m\}$, and set

$$\tilde{U_i} := (V_1 \cap U_i\hat{V_1}) \cdots (V_n \cap U_i\hat{V_n}).$$

For each element j in $\{1, \ldots, n\}$, we define

$$\tilde{V}_j := (U_1 \cap V_j\hat{U}_1) \cdots (U_m \cap V_j\hat{U}_m).$$

Let us first assume that $\tilde{U}_i \neq T$.

By Lemma 7.1.2(i), $U_i \subseteq \tilde{U}_i$. Thus, by Lemma 2.2.1(i), $U_i(\tilde{U}_i \cap \hat{U}_i) = \tilde{U}_i$. On the other hand, as $U_i \cap \hat{U}_i = \{1\}$, $U_i \cap (\tilde{U}_i \cap \hat{U}_i) = \{1\}$. Thus,

$$U_i \times (\tilde{U}_i \cap \hat{U}_i) = \tilde{U}_i.$$

(Recall that we are assuming that, for any two elements p and q in T, $qp = pq$.)

By definition, we also have

$$(V_1 \cap U_i\hat{V}_1) \times \ldots \times (V_n \cap U_i\hat{V}_n) = \tilde{U}_i.$$

With the help of Theorem 7.1.3(i) we may refine both of the above representations of $\tilde{U}_i$ to direct products of indecomposable closed subsets of S. Thus, as we are assuming that $\tilde{U}_i \neq T$, we obtain, by induction, an element j in $\{1, \ldots, n\}$ and an indecomposable closed subset W of $V_j \cap U_i\hat{V}_j$ such that

$$W \times (\tilde{U}_i \cap \hat{U}_i) = \tilde{U}_i.$$

From $W \subseteq \tilde{U}_i$ and $W \cap (\tilde{U}_i \cap \hat{U}_i) = \{1\}$ we conclude that $W \cap \hat{U}_i = \{1\}$. On the other hand, we have that

$$W\hat{U}_i = W(\tilde{U}_i \cap \hat{U}_i)\hat{U}_i = \tilde{U}_i\hat{U}_i = U_i(\tilde{U}_i \cap \hat{U}_i)\hat{U}_i = T.$$

Therefore,

$$W \times \hat{U}_i = T.$$

From $W \cap \hat{U}_i = \{1\}$ we conclude that $W \cap (V_j \cap \hat{U}_i) = \{1\}$.

From $W\hat{U}_i = T$ and $W \subseteq V_j$ we obtain $W(V_j \cap \hat{U}_i) = V_j \cap W\hat{U}_i = V_j$; cf. Lemma 2.2.1(i).

Thus,

$$W \times (V_j \cap \hat{U}_i) = V_j.$$

However, by hypothesis, V_j is indecomposable. Therefore, $V_j = W$. Thus, as $W \times \hat{U}_i = T$,

$$V_j \times \hat{U}_i = T.$$

From $V_j = W$ and $W \subseteq U_i\hat{V}_j$ we obtain $V_j \subseteq U_i\hat{V}_j$. Thus, $T = V_j\hat{V}_j \subseteq U_i\hat{V}_j$. It follows that $U_i\hat{V}_j = T$.

On the other hand, as $U_i \times \hat{U}_i = T = V_j \times \hat{U}_i$, Lemma 2.3.6(i) yields $n_{U_i} = n_{V_j}$. (Recall that T is assumed to have finite valency.) Therefore, referring to Lemma 2.3.6(i) once again, we obtain from $U_i\hat{V}_j = T$ that

$$U_i \times \hat{V}_j = T.$$

This proves the proposition in the case where $\tilde{U}_i \neq T$.

Let us now assume that $\tilde{U}_i = T$. Then, for each element j in $\{1, \ldots, n\}$, $U_i\hat{V}_j = T$; cf. Lemma 7.1.2(ii). We shall now prove that there exists an element j in $\{1, \ldots, n\}$ such that $V_j\hat{U}_i = T$.

Assume, by way of contradiction, that, for each element j in $\{1, \ldots, n\}$, $V_j\hat{U}_i \neq T$. Then, for each element j in $\{1, \ldots, n\}$, $\tilde{V}_j \neq T$; cf. Lemma 7.1.2(ii). Thus, similar to the first case, we find, for each element l in $\{1, \ldots, n\}$, an element k in $\{1, \ldots, m\}$ such that

$$U_k \times \hat{V}_l = T = V_l \times \hat{U}_k.$$

Thus, if $k = i$, we are done. Let us, therefore, assume that $k \neq i$.

From $U_k \times \hat{V}_l = T$ we obtain

$$V_1U_k//U_k \times \ldots \times V_{l-1}U_k//U_k \times V_{l+1}U_k//U_k \times \ldots \times V_nU_k//U_k = T//U_k.$$

Thus, as we are assuming that $k \neq i$, we find, by induction, an element j in $\{1, \ldots, n\}$ such that

$$V_jU_k//U_k \times \widehat{U_iU_k//U_k} = T//U_k.$$

From

$$U_1//U_k \times \ldots \times U_{k-1}//U_k \times U_{k+1}//U_k \times \ldots \times U_m//U_k = T//U_k$$

we obtain

$$\widehat{U_iU_k//U_k} = (\hat{U}_i \cap \hat{U}_k)U_k//U_k.$$

From this we obtain

$$V_jU_k(\hat{U}_i \cap \hat{U}_k) = T.$$

On the other hand, as $k \neq i$, $U_k \subseteq \hat{U}_i$. Thus, by Lemma 2.2.1(i),

$$\hat{U}_i = U_k(\hat{U}_i \cap \hat{U}_k).$$

It follows that $V_j\hat{U}_i = T$, contrary to our hypothesis.

Thus, we have shown that there exists an element j in $\{1, \ldots, n\}$ such that

$$U_i\hat{V}_j = T = V_j\hat{U}_i.$$

Similar to the reasoning in the first case we obtain from this that

$$U_i \times \hat{V}_j = T = V_j \times \hat{U}_i;$$

cf. Lemma 2.3.6(i).

The following theorem is due to Pamela Ferguson and Alexandre Turull; cf. [11; Theorem 3.11].

Theorem 7.1.5 *Assume that S has finite valency and that, for any two elements p and q in S, $qp = pq$.*
Let m and n be positive integers, and let T_1, ..., T_m and U_1, ..., U_n be indecomposable closed subsets of S satisfying

$$T_1 \times \ldots \times T_m = S = U_1 \times \ldots \times U_n.$$

Then the following hold.

(i) *We have $m = n$.*

(ii) *There exists a permutation π of the set $\{1, \ldots, n\}$ such that, for any two elements x in X and i in $\{1, \ldots, n\}$, $(T_i)_x \cong (U_{i\pi})_x$.*

Proof. Let i be a positive integer such that $i \leq m$. Then, by Proposition 7.1.4, there exists an element j in $\{1, \ldots, n\}$ such that

$$T_i \times \hat{U}_j = S = U_j \times \hat{T}_i.$$

Thus, as $T_i \times \hat{T}_i = S$,

$$T_i \times \hat{U}_j = T_i \times \hat{T}_i$$

and

$$T_i \times \hat{T}_i = U_j \times \hat{T}_i.$$

From the second of these equations we obtain

$$(T_i)_x \cong S//\hat{T}_i \cong (U_j)_x$$

for each element x in X; cf. Corollary 5.3.5. Thus, setting $i\pi := j$ we obtain $(T_i)_x \cong (U_{i\pi})_x$.
From the first of the above two equations we obtain

$$(\hat{T}_i)_x \cong S//T_i \cong (\hat{U}_j)_x$$

for each element x in X; cf. Corollary 5.3.5. Thus, we are done by induction.

7.2 Quasidirect Products of Schemes

Let n be a positive integer. For each an element i in $\{1, \ldots, n\}$, we fix a nonempty set X_i and a scheme S_i on X_i.
For each element $(s_1, \ldots, s_n)$ in $S_1 \times \ldots \times S_n$, we define $(s_1, \ldots, s_n)_\varsigma$ to be the set of all pairs $((y_1, \ldots, y_n), (z_1, \ldots, z_n))$ which satisfy, for each element i in $\{1, \ldots, n\}$, $z_i \in y_i s_i$.

Lemma 7.2.1 *Let n be a positive integer. For each element i in $\{1,\ldots,n\}$, we fix a nonempty set X_i and a scheme S_i on X_i. Then the following hold.*

(i) *The set $(S_1 \times \ldots \times S_n)\varsigma$ is a partition of $X_1 \times \ldots \times X_n$.*

(ii) *We have $(1_{X_1},\ldots,1_{X_n})\varsigma = 1_{X_1\times\ldots\times X_n}$.*

PROOF. For each element i in $\{1,\ldots,n\}$, we fix elements y_i and z_i in X_i.

(i) For each element i in $\{1,\ldots,n\}$, there exists a uniquely determined element s_i in S_i such that $z_i \in y_i s_i$. Thus, $(s_1,\ldots,s_n)\varsigma$ is the uniquely determined element in $(S_1 \times \ldots \times S_n)\varsigma$ which satisfies

$$((y_1,\ldots,y_n),(z_1,\ldots,z_n)) \in (s_1,\ldots,s_n)\varsigma.$$

(ii) We have $((y_1,\ldots,y_n),(z_1,\ldots,z_n)) \in (1_{X_1},\ldots,1_{X_n})\varsigma$ if and only if, for each element i in $\{1,\ldots,n\}$, $y_i = z_i$. The latter condition is equivalent to $(y_1,\ldots,y_n) = (z_1,\ldots,z_n)$. Thus, $(1_{X_1},\ldots,1_{X_n})\varsigma = 1_{X_1\times\ldots\times X_n}$.

Lemma 7.2.2 *Let n be a positive integer. For each element i in $\{1,\ldots,n\}$, we fix a nonempty set X_i, a scheme S_i on X_i, and an element s_i in S_i. Then the following hold.*

(i) *We have $((s_1,\ldots,s_n)\varsigma)^* = (s_1^*,\ldots,s_n^*)\varsigma$.*

(ii) *Let us fix, for each element i in $\{1,\ldots,n\}$, elements y_i and z_i in X_i and elements p_i and q_i in S_i. Then, if $((y_1,\ldots,y_n),(z_1,\ldots,z_n)) \in (s_1,\ldots,s_n)\varsigma$,*

$$|(y_1,\ldots,y_n)(p_1,\ldots,p_n)\varsigma \cap (z_1,\ldots,z_n)((q_1,\ldots,q_n)\varsigma)^*| = a_{p_1q_1s_1}\cdots a_{p_nq_ns_n}.$$

PROOF. (i) We have

$$((y_1,\ldots,y_n),(z_1,\ldots,z_n)) \in (s_1,\ldots,s_n)\varsigma$$

if and only if, for each element i in $\{1,\ldots,n\}$, $z_i \in y_i s_i$. The latter condition is equivalent to the fact that, for each element i in $\{1,\ldots,n\}$, $y_i \in z_i s_i^*$. This means that $((z_1,\ldots,z_n),(y_1,\ldots,y_n)) \in (s_1^*,\ldots,s_n^*)\varsigma$.

(ii) For each element i in $\{1,\ldots,n\}$, we fix an element x_i in X_i. Then, by definition,

$$(x_1,\ldots,x_n) \in (y_1,\ldots,y_n)(p_1,\ldots,p_n)\varsigma \cap (z_1,\ldots,z_n)((q_1,\ldots,q_n)\varsigma)^*$$

if and only if, for each element i in $\{1,\ldots,n\}$, $x_i \in y_ip_i \cap z_iq_i^*$. Therefore, the sets

$$(y_1,\ldots,y_n)(p_1,\ldots,p_n)\varsigma \cap (z_1,\ldots,z_n)((q_1,\ldots,q_n)\varsigma)^*$$

and

$$y_1p_1 \cap z_1q_1^* \times \ldots \times y_np_n \cap z_nq_n^*$$

are equal. Thus, the claim follows from the fact that, for each element i in $\{1, \ldots, n\}$, $|y_i p_i \cap z_i q_i^*| = a_{p_i q_i s_i}$.

Theorem 7.2.3 *Let n be a positive integer. For each element i in $\{1, \ldots, n\}$, we fix a nonempty set X_i and a scheme S_i on X_i. Then the following hold.*

(i) *The set $(S_1 \times \ldots \times S_n)_\varsigma$ is a scheme on $X_1 \times \ldots \times X_n$.*

(ii) *Let us fix, for each element i in $\{1, \ldots, n\}$, elements p_i, q_i, and r_i in S_i. Then*

$$a_{(p_1,\ldots,p_n)_\varsigma (q_1,\ldots,q_n)_\varsigma (r_1,\ldots,r_n)_\varsigma} = a_{p_1 q_1 r_1} \cdots a_{p_n q_n r_n}.$$

PROOF. This follows from Lemma 7.2.1 and Lemma 7.2.2.

Let n be a positive integer. For each element i in $\{1, \ldots, n\}$, we fix a nonempty set X_i and a scheme S_i on X_i.

The scheme $(S_1 \times \ldots \times S_n)_\varsigma$ will be called the *direct product* of the schemes S_1, …, S_n.

We set $\tilde{X} := X_1 \times \ldots \times X_n$ and $\tilde{S} := (S_1 \times \ldots \times S_n)_\varsigma$.

For each element i in $\{1, \ldots, n\}$, we define $\tilde{S}_i$ to be the set of all elements $(s_1, \ldots, s_n)_\varsigma$ in $\tilde{S}$ such that, for each element j in $\{1, \ldots, n\} \setminus \{i\}$, $s_j = 1_{X_j}$.

We call S a *quasi-direct product* of the schemes S_1, …, S_n if there exists a morphism ϕ from $X \cup S$ to $\tilde{X} \cup \tilde{S}$ such that ϕ_X is bijective and, for any two elements i in $\{1, \ldots, n\}$ and $\tilde{s}$ in $\tilde{S}_i$, $|\tilde{s}\phi^{-1}| = 1$.

Note that the morphism ϕ in the definition of quasi-direct products is necessarily surjective; cf. Lemma 5.1.3(ii).

Note also that, if the morphism ϕ in the definition of quasi-direct products is injective, the second condition of the definition, namely that, for any two elements i in $\{1, \ldots, n\}$ and $\tilde{s}$ in $\tilde{S}_i$, $|\tilde{s}\phi^{-1}| = 1$, is necessarily satisfied. In particular, direct products are quasi-direct products.

The relationship between quasi-direct products of schemes and direct products of closed subsets of S (as defined in the previous section) is described in the following theorem which generalizes [11; Proposition 3.13].

Theorem 7.2.4 *Let n be a positive integer. For each element i in $\{1, \ldots, n\}$, we fix a nonempty set X_i and a scheme S_i on X_i. Then the following conditions are equivalent.*

(a) *The scheme S is a quasi-direct product of the schemes S_1, …, S_n.*

(b) *The scheme S possesses closed subsets T_1, …, T_n with $T_1 \times \ldots \times T_n = S$ such that, for any two elements i in $\{1, \ldots, n\}$ and x in X, $S_i \cong (T_i)_x$.*

PROOF. (a) ⇒ (b) We set $\tilde{X} := X_1 \times \ldots \times X_n$ and $\tilde{S} := (S_1 \times \ldots \times S_n)_\varsigma$.

For each element i in $\{1, \ldots, n\}$, we define $\tilde{S}_i$ to be the set of all elements $(s_1, \ldots, s_n)_\varsigma$ in $\tilde{S}$ such that, for each element j in $\{1, \ldots, n\} \setminus \{i\}$, $s_j = 1_{X_j}$.

Since S is assumed to be a quasi-direct product of the schemes $S_1, \ldots, S_n$, there exists a morphism ϕ from $X \cup S$ to $\tilde{X} \cup \tilde{S}$ such that ϕ_X is bijective and, for any two elements i in $\{1, \ldots, n\}$ and $\tilde{s}$ in $\tilde{S}_i$, $|\tilde{s}\phi^{-1}| = 1$.

For each element i in $\{1, \ldots, n\}$, we define

$$T_i := \tilde{S}_i \phi^{-1}.$$

Since $\tilde{S}_i$ is closed, T_i is closed; cf. Lemma 5.1.2(ii).

Since ϕ_X is bijective, we obtain from Lemma 5.1.3(i) that $T_1 \cdots T_n = S$.

Recall that, for each element i in $\{1, \ldots, n\}$, $\hat{T}_i$ is our notation for the closed subset generated by the union of the sets T_j with $j \in \{1, \ldots, n\} \setminus \{i\}$.

Let i be an element in $\{1, \ldots, n\}$. Then

$$(T_i \cap \hat{T}_i)\phi \subseteq T_i\phi \cap \hat{T}_i\phi \subseteq \tilde{S}_i \cap \hat{\tilde{S}}_i = \{1\phi\}.$$

Thus, as $\{1\} = (1\phi)\phi^{-1}$, $T_i \cap \hat{T}_i = \{1\}$.

Let i and j be elements in $\{1, \ldots, n\}$. We shall show that $T_i \subseteq N_S(T_j)$.

Let s be an element in T_i, and set $\tilde{s} := s\phi$. Then $\tilde{s} \in \tilde{S}_i$. From Lemma 7.2.2(ii) we obtain $\tilde{S}_j\tilde{s} = \tilde{s}\tilde{S}_j$. On the other hand, we have $\{s\} = \tilde{s}\phi^{-1}$. Thus, by Lemma 5.1.3(i),

$$T_j s = \tilde{S}_j\phi^{-1}\tilde{s}\phi^{-1} = (\tilde{S}_j\tilde{s})\phi^{-1} = (\tilde{s}\tilde{S}_j)\phi^{-1} = \tilde{s}\phi^{-1}\tilde{S}_j\phi^{-1} = sT_j.$$

Thus, as s has been chosen arbitrarily in T_i, we have shown that $T_i \subseteq N_S(T_j)$.

So far, we have proved that

$$T_1 \times \ldots \times T_n = S.$$

Let us now fix an element i in $\{1, \ldots, n\}$ and an element x in X. We set $\tilde{X}_i := x\phi\tilde{S}_i$. We have to show that $S_i \cong (T_i)_x$.

For each element w in xT_i, we set $w\psi := w\phi$. For each element s in T_i, we set $(s_{xT_i})\psi := s\phi$.

Since ϕ is a morphism, ψ is a morphism from $xT_i \cup (T_i)_{xT_i}$ to $\tilde{X}_i \cup \tilde{S}_i$.

Let p and q be elements in T_i with $(p_{xT_i})\psi = (q_{xT_i})\psi$. Then $p\phi = q\phi$. But, as $q\phi \in \tilde{S}_i$, $|(q\phi)\phi^{-1}| = 1$. Therefore, $p = q$. This proves that $\psi_{(T_i)_{xT_i}}$ injective. Thus, by Lemma 5.1.6(i), ψ is an injective homomorphism.

Let us now show that ψ is surjective. According to Lemma 5.1.3(ii), it suffices to show that ψ_{xT_i} is surjective.

In order to show that ψ_{xT_i} is surjective we fix an element $\tilde{w}$ in $\tilde{X}_i$. Then, by definition, $\tilde{w} \in x\phi\tilde{S}_i$. Therefore, there exists an element $\tilde{s}$ in $\tilde{S}_i$ such that $\tilde{w} \in x\phi\tilde{s}$.

Since ϕ is surjective, there exists an element w in X such that $w\phi = \tilde{w}$. Let us write s to denote the element in S which satisfies $w \in xs$.

Since ϕ is a morphism, we obtain from $w \in xs$ that $w\phi \in x\phi s\phi$. Thus, as $w\phi = \tilde{w}$, $\tilde{w} \in x\phi s\phi$. Thus, as $\tilde{w} \in x\phi\tilde{s}$, $s\phi = \tilde{s} \in \tilde{S}_i$. Thus, by definition, $s \in T_i$. Thus, as $w \in xs$, $w \in xT_i$. From $w \in xT_i$ we obtain $w\psi = w\phi$.

Since $\tilde{w}$ has been chosen arbitrarily in $\tilde{X}_i$, we have shown that ψ_{xT_i} is surjective.

It follows that ψ is an isomorphism from $xT_i \cup (T_i)_{xT_i}$ to $\tilde{X}_i \cup \tilde{S}_i$.

On the other hand, it is clear that $\tilde{S}_i \cong S_i$.

(b) $\Rightarrow$ (a) Let x be an element in X. We shall be done if we succeed in showing that S is a quasi-direct product of the schemes $(T_1)_x, \ldots, (T_n)_x$.

We define

$$\tilde{X} := xT_1 \times \ldots \times xT_n.$$

and

$$\tilde{S} := ((T_1)_{xT_1} \times \ldots \times (T_n)_{xT_n})_\varsigma.$$

Let y be an element in X, let i be an element in $\{1, \ldots, n\}$. From $T_i \times \hat{T}_i = S$ we obtain $y \in xT_i\hat{T}_i$ and $T_i \cap \hat{T}_i = \{1\}$. Thus, $|xT_i \cap y\hat{T}_i| = 1$. Let y_i be the element in $xT_i \cap y\hat{T}_i$. We set

$$y\phi := (y_1, \ldots, y_n).$$

Let s be an element in S. Then we find, for each element i in $\{1, \ldots, n\}$, a uniquely determined element t_i in T_i such that $s \in t_1 \cdots t_n$; cf. Theorem 7.1.1. We set

$$s\phi := ((t_1)_{xT_1}, \ldots, (t_n)_{xT_n})_\varsigma.$$

Then ϕ is an injective map from $X \cup S$ to $\tilde{X} \cup \tilde{S}$.

For each element i in $\{1, \ldots, n\}$, we define U_i to be the set of all elements $((s_1)_{xT_1}, \ldots, (s_n)_{xT_n})_\varsigma$ such that, for each element j in $\{1, \ldots, n\} \setminus \{i\}$, $(s_j)_{xT_j} = 1_{xT_j}$.

It follows from the definition of ϕ that, for any two elements i in $\{1, \ldots, n\}$ and f in U_i, $|f\phi^{-1}| = 1$. Therefore, we shall be done if we succeed in showing that ϕ is a morphism.

To this end, let y be an element in X, let s be an element in S, and let z be an element in ys.

Let $(y_1, \ldots, y_n) \in \tilde{X}$ be such that

$$y\phi = (y_1, \ldots, y_n),$$

and let $(z_1, \ldots, z_n) \in \tilde{X}$ be such that

$$z\phi = (z_1, \ldots, z_n).$$

Then, for each element i in $\{1, \ldots, n\}$, $y_i \in xT_i$ and $z_i \in xT_i$. Thus, for each element i in $\{1, \ldots, n\}$, there exists an element t_i in T_i such that $z_i \in y_i t_i$; cf. Lemma 2.1.4. It follows that

$$(y\phi, z\phi) = ((y_1, \ldots, y_n), (z_1, \ldots, z_n)) \in ((t_1)_{xT_1}, \ldots, (t_n)_{xT_n})\varsigma.$$

Now we shall be done if we succeed in showing that

$$s\phi = ((t_1)_{xT_1}, \ldots, (t_n)_{xT_n})\varsigma.$$

Let i be an element in $\{1, \ldots, n\}$. From the definition of y_i we obtain

$$y \in y_i \hat{T}_i \subseteq z_i t_i^* \hat{T}_i = z_i \hat{T}_i t_i^*.$$

Therefore, $z_i \hat{T}_i \cap yt_i$ is not empty. Let w be an element in $z_i \hat{T}_i \cap yt_i$.

From the definition of z_i we obtain $z \in z_i \hat{T}_i$. Thus, as $w \in z_i \hat{T}_i$, Lemma 2.1.4 yields

$$z \in w\hat{T}_i \subseteq yt_i\hat{T}_i.$$

It follows that $s \in t_i \hat{T}_i$.

Since i has been chosen arbitrarily in $\{1, \ldots, n\}$, we obtain $s \in t_1 \cdots t_n$. Thus, by definition, $s\phi = ((t_1)_{xT_1}, \ldots, (t_n)_{xT_n})\varsigma$.

The first part of the following theorem generalizes the first part of [11; Theorem 3.17]. Its second part is due to Pamela Ferguson and Alexandre Turull; it is the second part of [11; Theorem 3.17].

Theorem 7.2.5 *Assume S to have finite valency. Then the following hold.*

(i) *If $\{1\} \neq S$, S is a quasi-direct product of indecomposable schemes.*

(ii) *Let m and n be positive integers, and let T_1, ..., T_m and U_1, ..., U_n be indecomposable schemes. Assume that S is a quasi-direct product of T_1, ..., T_m as well as a quasi-direct product of U_1, ..., U_n.*

Assume that, for any two elements p and q in S, $qp = pq$. Then $m = n$ and there exists a permutation π of the set $\{1, \ldots, n\}$ such that, for each element i in $\{1, \ldots, n\}$, $T_i \cong U_{i\pi}$.

Proof. (i) By Theorem 7.1.3(ii), we find a positive integer n and closed subsets T_1, ..., T_n of S such that $T_1 \times \ldots \times T_n = S$ and, for each element i in $\{1, \ldots, n\}$, T_i is indecomposable.

Let x be an element in X. Then, by Theorem 7.2.4, S is a quasi-direct product of the schemes $(T_1)_x$, ..., $(T_n)_x$. For each element i in $\{1, \ldots, n\}$, T_i is indecomposable. Thus, for each such i, the scheme $(T_i)_x$ is indecomposable.

(ii) The scheme S is assumed to be a quasi-direct product of the schemes T_1, ..., T_m. Thus, S possesses closed subsets T'_1, ..., T'_m such that

$$T'_1 \times \ldots \times T'_m = S$$

and, for any two elements i in $\{1, \ldots, m\}$ and x in X,

$$T_i \cong (T'_i)_x;$$

see Theorem 7.2.4. Similarly, S possesses closed subsets $U'_1, \ldots, U'_n$ such that

$$U'_1 \times \ldots \times U'_n = S$$

and, for any two elements j in $\{1, \ldots, n\}$ and x in X,

$$U_j \cong (U'_j)_x.$$

For each element i in $\{1, \ldots, m\}$, T_i is assumed to be indecomposable. Thus, for each such i, the scheme $(T'_i)_x$ is indecomposable.

Similarly, we see that, for each element j in $\{1, \ldots, n\}$, $(U'_j)_x$ is indecomposable.

Now the claim follows from Theorem 7.1.5.

7.3 Semidirect Products

In this section, we define the semidirect product of two schemes which has been mentioned in the preface of these monograph. We follow an approach which, in this generality, was given first by Sejeong Bang, Mitsugu Hirasaka, and Sung-Yell Song in [3].

Let W be a finite set, and let A be a scheme on W.

Since W is assumed to be a finite set, A has finite valency. Thus, $O^\vartheta(A)$ is defined and, according to Theorem 4.1.3(i), $A//O^\vartheta(A)$ is a scheme on $W/O^\vartheta(A)$. Let us denote by π the natural homomorphism from $W \cup A$ to $W/O^\vartheta(A) \cup A//O^\vartheta(A)$.

According to Theorem 3.2.1(i), $O^\vartheta(A)$ is strongly normal in A. Thus, by Lemma 4.2.5(ii), $A//O^\vartheta(A)$ is thin. Thus, $(A//O^\vartheta(A))^\gamma$ is a group.

From Theorem 5.2.2(ii) we also know that $\mathrm{Stc}(S)$ is a group with respect to composition.

Let ζ be a group homomorphism from $(A//O^\vartheta(A))^\gamma$ to $\mathrm{Stc}(S)$, and let w be an element in W.

For any two elements s in S and a in A, we define $s_{\zeta,w}a$ to be the set of all pairs $((y, u), (z, v))$ with y, $z \in X$ and u, $v \in W$ satisfying $v \in ua$ and

$$z \in ys(d\pi\zeta),$$

where d denotes the uniquely determined element in A which satisfies

$$w \in ud.$$

For any two subsets R of S and B of A, we define $R_{\zeta,w}B$ to be the set of all elements $r_{\zeta,w}b$ where r is an element in R and b an element in B.

Lemma 7.3.1 *Let W be a set, let A be a scheme on W, let ζ be a group homomorphism from $(A//O^\vartheta(A))^\gamma$ to $\mathrm{Stc}(S)$, and let w be an element in W. Then we have the following.*

(i) *The set $S_{\zeta,w}A$ is a partition of $X \times W$.*

(ii) *We have $1_{X\times W} = 1_{\zeta,w}1$.*

PROOF. Let y, z be elements in X, and let u, v be elements in W.

Let d be the uniquely determined element in A with $w \in ud$, let π be the natural homomorphism from $W \cup A$ to $W/O^\vartheta(A) \cup A//O^\vartheta(A)$, and set $\sigma := d\pi\zeta$. Then $\sigma \in \mathrm{Stc}(S)$.

(i) Let s be the uniquely determined element in S with $z \in ys$, and let a be the uniquely determined element in A with $v \in ua$.

Since $\sigma \in \mathrm{Stc}(S)$, σ is bijective. Thus, by definition, we must have that $((y,u),(z,v)) \in (s\sigma^{-1})_{\zeta,w}a$.

Now let p, q be elements in S, and let b, c be elements in A such that $((y,u),(z,v)) \in p_{\zeta,w}b \cap q_{\zeta,w}c$. Then, by definition, $v \in ub \cap uc$ and $z \in yp\sigma \cap yq\sigma$.

From $v \in ub \cap uc$ we obtain $b = c$. From $z \in yp\sigma \cap yq\sigma$ we obtain $p\sigma = q\sigma$. Thus, as σ is bijective, $p = q$. It follows that $p_{\zeta,w}b = q_{\zeta,w}c$.

(ii) Since $\sigma \in \mathrm{Stc}(S)$, $1\sigma = 1$; cf. Lemma 5.2.3(ii). Thus, $((y,u),(z,v)) \in 1_{\zeta,w}1$ if and only if $(u,v) \in 1$ and $(y,z) \in 1$. This is the case if and only if $(y,u) = (z,v)$.

Lemma 7.3.2 *Let W be a set, let A be a scheme on W, let π denote the natural homomorphism from $W \cup A$ to $W/O^\vartheta(A) \cup A//O^\vartheta(A)$, let ζ be a group homomorphism from $(A//O^\vartheta(A))^\gamma$ to $\mathrm{Stc}(S)$, let w be an element in W, let s be an element in S, and let a be an element in A.*

(i) *We have $(s_{\zeta,w}a)^* = s^*(a\pi\zeta)_{\zeta,w}a^*$.*

(ii) *Let y, z be elements in X, and let u, v be elements in W such that $(z,v) \in (y,u)s_{\zeta,w}a$. Then, for any four elements p, q in S and b, c in A, we have*

$$|(y,u)p_{\zeta,w}b \cap (z,v)(q_{\zeta,w}c)^*| = a_{pq(b^*\pi\zeta)s}a_{bca}.$$

PROOF. (i) Let y, z be elements in X, and let u, v be elements in W such that $((y,u),(z,v)) \in s_{\zeta,w}a$. Then, by definition, $v \in ua$ and $z \in ys(d\pi\zeta)$, where d denotes the uniquely determined element in A which satisfies $w \in ud$.

From $z \in ys(d\pi\zeta)$ we obtain $y \in z(s(d\pi\zeta))^*$. On the other hand, as $d\pi\zeta \in \mathrm{Stc}(S)$, $s^*(d\pi\zeta) = (s(d\pi\zeta))^*$; cf. Lemma 5.2.3(iii). Thus,

$$y \in zs^*(d\pi\zeta).$$

Let us denote by d' the uniquely determined element in A which satisfies $w \in vd'$. From $w \in vd'$ and $v \in ua$ we obtain $w \in uad'$. Thus, as $w \in ud$, $d \in ad'$. Thus, by Lemma 5.1.1(ii), $d\pi \in a\pi d'\pi$, and then $d\pi\zeta = (a\pi\zeta)(d'\pi\zeta)$. Thus, as $y \in zs^*(d\pi\zeta)$, $y \in zs^*(a\pi\zeta)(d'\pi\zeta)$. Thus, as $u \in va^*$ and $w \in vd'$, $((z,v),(y,u)) \in s^*(a\pi\zeta)_{\zeta,w}a^*$.

(ii) Let x be an element in X, and let t be an element in W such that

$$(x,t) \in (y,u)p_{\zeta,w}b \cap (z,v)(q_{\zeta,w}c)^*.$$

From $(x,t) \in (y,u)p_{\zeta,w}b$ we obtain $t \in ub$ and $x \in yp(d\pi\zeta)$, where d denotes the uniquely determined element in A which satisfies $w \in ud$.

From $(z,v) \in (x,t)q_{\zeta,w}c$ we obtain $v \in tc$ and $z \in xq(d'\pi\zeta)$, where d' denotes the uniquely determined element in A which satisfies $w \in td'$.

Since $w \in ud$ and $u \in tb^*$, $w \in tb^*d$. Thus, as $w \in td'$, $d' \in b^*d$. Thus, by Lemma 5.1.1(ii), $d'\pi \in b^*\pi d\pi$, and then $d'\pi\zeta = (b^*\pi\zeta)(d\pi\zeta)$. Thus, as $z \in xq(d'\pi\zeta)$, $z \in xq(b^*\pi\zeta)(d\pi\zeta)$. Thus, as $x \in yp(d\pi\zeta)$,

$$x \in yp(d\pi\zeta) \cap z(q(b^*\pi\zeta)(d\pi\zeta))^*.$$

Thus, as $t \in ub \cap vc^*$,

$$(x,t) \in (yp(d\pi\zeta) \cap z(q(b^*\pi\zeta)(d\pi\zeta))^*) \times (ub \cap vc^*).$$

Thus, as (x,t) has been chosen arbitrarily in $(y,u)p_{\zeta,w}b \cap (z,v)(q_{\zeta,w}c)^*$,

$$(y,u)p_{\zeta,w}b \cap (z,v)(q_{\zeta,w}c)^* = (yp(d\pi\zeta) \cap z(q(b^*\pi\zeta)(d\pi\zeta))^*) \times (ub \cap vc^*).$$

On the other hand, we are assuming that $(z,v) \in (y,u)s_{\zeta,w}a$. Thus, by definition, $z \in ys(d\pi\zeta)$ and $v \in ua$. Thus,

$$|ye(d\pi\zeta) \cap z(q(b^*\pi\zeta)(d\pi\zeta))^*| = a_{p(d\pi\zeta)q(b^*\pi\zeta)(d\pi\zeta)s(d\pi\zeta)}$$

and

$$|ub \cap vc^*| = a_{bca}.$$

Finally recall that, as $d\pi\zeta \in \mathrm{Stc}(S)$,

$$a_{p(d\pi\zeta)q(b^*\pi\zeta)(d\pi\zeta)s(d\pi\zeta)} = a_{pq(b^*\pi\zeta)s}.$$

Thus,

$$|(y,u)p_{\zeta,w}b \cap (z,v)(q_{\zeta,w}c)^*| = a_{pq(b^*\pi\zeta)g}a_{bca}.$$

This finishes the proof of Lemma 7.3.2.

Theorem 7.3.3 *Let W be a set, let A be a scheme on W, and let ζ be a group homomorphism from $(A//O^{\vartheta}(A))^{\gamma}$ to* Stc(S).

(i) *For each element w in W, $S_{\zeta,w}A$ is a scheme on $X \times W$.*

(ii) *For any two elements u and v in W, $S_{\zeta,u}A = S_{\zeta,v}A$.*

PROOF. (i) This follows from Lemma 7.3.1 together with Lemma 7.3.2.

(ii) Let p and q be elements in S, let v and w be elements in W, and let b and c be elements in A such that $p_{\zeta,v}b \cap q_{\zeta,w}c$ is not empty. We shall show that $p_{\zeta,v}b = q_{\zeta,w}c$.

Assuming that $p_{\zeta,v}b \cap q_{\zeta,w}c$ is not empty we find elements y and z in X and t and u in W such that

$$((y,t),(z,u)) \in p_{\zeta,v}b \cap q_{\zeta,w}c.$$

Let d be the uniquely defined element in A satisfying $v \in td$, and let e be the uniquely defined element in A satisfying $w \in te$.

Let us denote by π the natural homomorphism from $W \cup A$ to $W/O^{\vartheta}(A) \cup A//O^{\vartheta}(A)$.

From $((y,t),(z,u)) \in p_{\zeta,v}b$ and $v \in td$ we obtain $z \in yp(d\pi\zeta)$. Similarly, as $((y,t),(z,u)) \in q_{\zeta,w}b$ and $w \in te$, $z \in yq(e\pi\zeta)$. Thus, as S is a scheme,

$$p(d\pi\zeta) = q(e\pi\zeta).$$

Note also that $u \in tb \cap tc$. Thus, as A is assumed to be a scheme, $b = c$.

Let y' and z' be elements in X, and let t' and u' be elements in W such that

$$((y',t'),(z',u')) \in p_{\zeta,v}b.$$

We shall be done if we succeed in showing that $((y',t'),(z',u')) \in q_{\zeta,w}c$.

From $((y',t'),(z',u')) \in p_{\zeta,v}b$ we obtain $u' \in t'b$ and $z' \in y'p(d\pi\zeta)$, where $v \in t'd$. From $z' \in y'p(d\pi\zeta)$ and $p(d\pi\zeta) = q(e\pi\zeta)$ we obtain

$$z' \in y'q(e\pi\zeta).$$

Let e' be the uniquely defined element in A satisfying $w \in t'e'$, and let r be the element in A satisfying $t' \in tr$. From $w \in t'e'$ and $t' \in tr$ we obtain $w \in tre$. Thus, as $w \in te$, $e \in re'$.

From $t' \in vd^*$ and $v \in td$ we obtain $t' \in tdd^*$. Thus, as $t' \in tr$, $r \in dd^*$. Now recall that, by Theorem 3.2.1(ii), $dd^* \subseteq O^{\vartheta}(A)$, so that $r \in O^{\vartheta}(A)$.

From $e \in re'$ and $r \in O^{\vartheta}(A)$ we obtain $e \in O^{\vartheta}(A)e'$. It follows that $e\pi = e'\pi$. Thus, as $z' \in y'q(e\pi\zeta)$,

$$z' \in y'q(e'\pi\zeta).$$

Thus, as $w \in t'e'$, $u' \in t'b$, and $b = c$, $((y', t'), (z', u')) \in q_{\varsigma,w}c$.

Let W, A, and ς be as before, and let w be an element in W. Theorem 7.3.3(ii) allows us to write $S_\varsigma A$ instead of $S_{\varsigma,w}A$. Similarly, for any two elements s in S and a in A, we may write $s_\varsigma a$ instead of $s_{\varsigma,w}a$.

We call $S_\varsigma A$ the *semidirect product* of S and A *with respect to* ς. The scheme S is called the *kernel* of the semidirect product $S_\varsigma A$, A its *complement.*

Occasionally, we do not need to specify the group homomorphism ς which defines the semidirect product $S_\varsigma A$. In this case, we shall just write SA instead of $S_\varsigma A$.

Corollary 7.3.4 *Let A be a scheme, let π denote the natural homomorphism from $W \cup A$ to $W/O^\vartheta(A) \cup A//O^\vartheta(A)$, and let ς be a group homomorphism from $(A//O^\vartheta(A))^\gamma$ to* $\mathrm{Stc}(S)$. *Then, for any six elements p, q, r in S and a, b, c in A, the following hold.*

(i) *We have* $a_{(p_\varsigma b)(q_\varsigma c)(r_\varsigma a)} = a_{pq(b^*\pi\varsigma)r}a_{bca}$.

(ii) *We have* $r_\varsigma a \in (p_\varsigma b)(q_\varsigma c)$ *if and only if* $r \in pq(b^*\pi\varsigma)$ *and* $a \in bc$.

PROOF. (i) This follows from Theorem 7.3.3(i) together with Lemma 7.3.2(ii).

(ii) This follows from (i).

Theorem 7.3.5 *Let A be a scheme, and let ς be a group homomorphism from $(A//O^\vartheta(A))^\gamma$ to* $\mathrm{Stc}(S)$.

(i) *The sets $S_\varsigma\{1\}$ and $\{1\}_\varsigma A$ are closed subsets of $S_\varsigma A$.*

(ii) *For any two elements s in S and a in A, $\{s_\varsigma a\} = (s_\varsigma 1)(1_\varsigma a)$.*

(iii) *We have $\{1_\varsigma 1\} = S_\varsigma\{1\} \cap \{1\}_\varsigma A$ and $(S_\varsigma\{1\})(\{1\}_\varsigma A) = S_\varsigma A$.*

(iv) *The closed subset $S_\varsigma\{1\}$ is normal in $S_\varsigma A$.*

PROOF. (i) From Lemma 7.3.2(i) we obtain $(s_\varsigma 1)^* = (s^*)_\varsigma 1 \in S_\varsigma 1$ for each element s in S. From Corollary 7.3.4(ii) we obtain $(p_\varsigma 1)(q_\varsigma 1) = \{s_\varsigma 1 \mid s \in pq\}$ for any two elements p and q in S. Thus, $S_\varsigma\{1\}$ is a closed subset of $S_\varsigma A$.

From Lemma 5.2.3(ii) we obtain $1(a\pi\varsigma) = 1$ for each element a in A. Thus, for each element a in A, $(1_\varsigma a)^* = 1_\varsigma a^* \in 1_\varsigma A$; cf. Lemma 7.3.2(i). For any two elements b and c in A, $(1_\varsigma b)(1_\varsigma c) = \{1_\varsigma a \mid a \in bc\}$; cf. Corollary 7.3.4(ii). Thus, we have shown that $\{1\}_\varsigma A$ is a closed subset of $S_\varsigma A$.

(ii) This follows immediately from Corollary 7.3.4(ii).

(iii) The first claim is obvious, the second claim follows from (i).

(iv) According to Lemma 2.5.1(ii), we just have to show that $\{1\}_\varsigma A$ normalizes $S_\varsigma\{1\}$. In order to do so we fix elements s in S and a in A. We have to show that $(s_\varsigma 1)(1_\varsigma a) \subseteq (1_\varsigma a)(S_\varsigma\{1\})$.

From (ii) we know that $\{s_\zeta a\} = (s_\zeta 1)(1_\zeta a)$. From Corollary 7.3.4(ii) we also obtain

$$s_\zeta a \in (1_\zeta a)(s(a\pi\zeta)_\zeta 1) \subseteq (1_\zeta a)(S_\zeta 1).$$

(Note that $a^*\pi\zeta = (a\pi\zeta)^{-1}$.) Thus, $(s_\zeta 1)(1_\zeta a) \subseteq (1_\zeta a)(S_\zeta\{1\})$.

Note that, according to Lemma 2.5.1(ii), the last condition in Theorem 7.3.5 is equivalent to $\{1\}_\zeta A \subseteq N_{S_\zeta A}(S_\zeta\{1\})$.

7.4 A Characterization of Semidirect Products

It is the purpose of this (short) section to show that the four conditions given in Theorem 7.3.5 are sufficient to identify a scheme as a semidirect product.

Throughout this section, the letters T and U will stand for closed subsets of S satisfying $\{1\} = T \cap U$ and $U \subseteq N_S(T)$. We shall always assume that, for any two elements t in T and u in U, $1 = |tu|$.

The latter hypothesis implies that, for any two elements t in T and u in U, u^*tu contains exactly one element which is in T; cf. Lemma 2.6.4(ii). In the following, we shall denote this element by $t\sigma_u$.

Lemma 7.4.1 *We have the following.*

(i) *For any two elements t in T and u in U, $tu = ut\sigma_u$.*

(ii) *For any three elements p, q, and r in T with $r \in pq$, we have $\sigma_p\sigma_q = \sigma_r$.*

(iii) *For any two elements t in T and u in $O^\vartheta(U)$, $t\sigma_u = t$.*

(iv) *For each element u in U, $\sigma_u \in \mathrm{Stc}(T)$.*

PROOF. (i) Let t be an element in T, and let u be an element in U. Then, by definition, $t\sigma_u \in u^*tu$. Thus, there exists an element s in tu such that $t\sigma_u = u^*s$. From $s \in tu$ and $1 = |tu|$ we obtain $\{s\} = tu$. From $t\sigma_u = u^*s$ we obtain $s \in ut\sigma_u$; cf. Lemma 1.3.3(ii). Thus, as $1 = |ut\sigma_u|$, $\{s\} = ut\sigma_u$. It follows that $tu = ut\sigma_u$.

(ii) Let t be an element in T, and let p, q be elements in U. Then, by (i), $tp = pt\sigma_p$ and $t\sigma_p q = qt\sigma_p\sigma_q$. Thus, $tpq = pt\sigma_p q = pqt\sigma_p\sigma_q$. Thus, for each element r in pq,

$$t\sigma_r \in r^*tr \cap T \subseteq q^*p^*tpq \cap T = q^*p^*pqt\sigma_p\sigma_q \cap T.$$

Thus, there exists an element u in q^*p^*pq such that $t\sigma_r \in ut\sigma_p\sigma_q \cap T$.

From $t\sigma_r \in ut\sigma_p\sigma_q$ we obtain $u \in T$, from $u \in q^*p^*pq$ we obtain $u \in U$. Thus, $u \in T \cap U$. However, we are assuming that $\{1\} = T \cap U$. Thus, $u = 1$. Thus, as $t\sigma_r \in ut\sigma_p\sigma_q$, $t\sigma_r = t\sigma_p\sigma_q$. Thus, as t has been chosen arbitrarily in T, we have shown that $\sigma_p\sigma_q = \sigma_r$.

(iii) Let u be an element in $O^{\vartheta}(U)$. Then, by Theorem 3.2.1(ii), there exist elements $u_1, \ldots, u_n$ in U such that $u \in u_1^* u_1 \cdots u_n^* u_n$. Thus, the claim follows from (ii).

(iv) Let u be an element in U. Then, by (ii), $\sigma_{u^*}\sigma_u = \sigma_1$. Thus, σ_u is bijective. In order to show that $\sigma_u \in \mathrm{Stc}(T)$, we now fix three elements p, q, and r in T. We shall see that $a_{p\sigma_u q\sigma_u r\sigma_u} = a_{pqr}$.

Let y and z be elements in X such that $z \in yr$. Since $r\sigma_u \in u^*ru$, $r \in ur\sigma_u u^*$. Thus, as $z \in yr$, $z \in yur\sigma_u u^*$. Thus, there exists an element v in yu such that $z \in vr\sigma_u u^*$. Since $z \in vr\sigma_u u^*$, there exists an element w in $vr\sigma_u$ such that $z \in wu^*$.

Let x be an element in $yp \cap zq^*$. Since $x \in yp$, $y \in xp^*$. Thus, as $v \in yu$, $v \in xp^*u$. On the other hand, we know from (i) that $p^*u = up^*\sigma_u$. Thus, $v \in xup^*\sigma_u$. Thus, there exists an element x' in xu such that $v \in x'p^*\sigma_u$. From $p^*\sigma_u = (p\sigma_u)^*$ and $v \in x'p^*\sigma_u$ we obtain $x' \in vp\sigma_u$.

Let x'' be an element in $vp\sigma_u \cap xu$. Then $x'' \in x'(u^*u \cap T) = \{x'\}$. Thus, we have a bijective map from $yp \cap zq^*$ to $vq\sigma_u \cap w(q\sigma_u)^*$. It follows that $a_{p\sigma_u q\sigma_u r\sigma_u} = a_{pqr}$.

Thus, as p, q, and r have been chosen arbitrarily in T, we have shown that $\sigma_u \in \mathrm{Stc}(T)$.

Theorem 7.4.2 *For any two elements y in X and z in yTU, $T_y \cong T_z$.*

PROOF. We are assuming that $z \in yTU$. Thus, there exist elements t in T and u in U such that $z \in ytu$. Since $z \in ytu$, there exists an element s in ut such that $z \in ys$. Since $s \in tu$ and $1 = |tu|$, $\{s\} = tu$. Thus, as $u^*tu \cap T$ is assumed to be not empty, $u^*s \cap T$ is not empty. Thus, there exists an element t' in T such that $t' \in u^*s$. Since $t' \in u^*s$, $s \in ut'$; cf. Lemma 1.3.3(ii). Thus, as $z \in ys$, $z \in yuT$. Thus, there exists an element z' in yu such that $z \in z'T$. Thus, we may assume that $z' = z$. Thus, $z \in yu$.

For each element x in yT, we define $x\phi$ to be the uniquely determined element in $xu \cap zT$. For each element t in T, we define $t\phi$ to be the uniquely defined element in $u^*tu \cap T$.

Then ϕ is an isomorphism from $yT \cup T_{yT}$ to $zt \cup T_{zT}$.

The following theorem is the converse of Theorem 7.3.5.

Theorem 7.4.3 *For each element x in X, there exists a group homomorphism ς from $(U_{xU})//O^{\vartheta}(U_{xU})$ to $\mathrm{Stc}(T_{xT})$ such that $(TU)_x \cong (T_x)_\varsigma(U_x)$.*

PROOF. Let us write π to denote the natural homomorphism from $xU \cup U_{xU}$ to

$$xU/O^{\vartheta}(U_{xU})_{xU} \cup U_{xU}//O^{\vartheta}(U_{xU}).$$

Then, for any two elements t in T and u in U,

$$t_{xT}(u_{xU}\pi\zeta) = (t\sigma_u)_{xT};$$

cf. Lemma 7.4.1(ii), (iii).

Let w be an element in xTU. Since we are assuming that $\{1\} = T \cap U$, there exists a uniquely determined element w_T in xT such that $w \in w_TU$. From $w \in xTU$ we also obtain $w \in xUT$. Thus, there exists a uniquely determined element w_U in xU such that $w \in w_UT$. We set

$$w\chi := (w_T, w_U).$$

Let s be an element in TU. Then there exist uniquely determined elements t in T and u in U such that $\{s\} = tu$; cf. Lemma 2.1.2. We define

$$s\phi := (t_{xT})_{\zeta,x}(u_{xU})$$

and claim that ϕ is an isomorphism with respect to χ.

Let us first show that χ is a bijective map from X to $xT \times xU$.

Let u and v be elements in X such that $u\chi = v\chi$. Then, by definition, $u_T = v_T$ and $u_U = v_U$. Since $u \in u_TU$ and $v \in v_TU$, $u_T = v_T$ yields $v \in uU$. Similarly $u_U = v_U$ yields $v \in uT$. Thus, $v \in u(T \cap U)$. Thus, as we are assuming that $\{1\} = T \cap U$, $u = v$. This shows that χ is injective.

Let y be an element in xT, and let z be an element in xU. From $y \in xT$ we obtain $x \in yT$. Thus, as $z \in xU$, $z \in yTU$. Thus, as $TU \subseteq UT$, $z \in yUT$. Thus, there exists an element w in yU such that $z \in wT$. Since $y \in xT$ and $w \in yU$, $y = w_T$. Similarly, as $z \in xU$ and $w \in zT$, $z = w_U$. Thus, $w\chi = (y, z)$.

Thus, we have shown that χ is a bijective map from X to $xT \times xU$.

That ϕ is a bijective map from TU_{xTU} to $(T_{xT})_\zeta(U_{xU})$ follows immediately from the definition of ϕ.

Let us now show that ϕ is a homomorphism with respect to χ. In order to do this, we fix elements y and z in X, and we call s the uniquely determined element in S which satisfies $z \in ys$. We have to show that $z\chi \in y\chi s\phi$.

Let t and u be the uniquely determined elements in T and U such that $\{s\} = tu$. We have to show that

$$(z_T, z_U) \in (y_T, y_U)(t_{xT})_{\zeta,x}(u_{xU}).$$

Thus, we have to show that $z_U \in y_Uu$ and that $z_T \in y_Tt_{xT}(d_{xU}\pi\zeta)$, where d denotes the uniquely determined element in U which satisfies $x \in y_Ud$.

Since $z \in ys \subseteq ytu$, there exists an element w in yt such that $z \in wu$. From $w \in yt$ and $y \in y_UT$ we obtain $w \in y_UT$. Thus, as $z \in wu$, $z \in y_UTu$. Thus, $z \in y_UuT$. Thus, there exists an element z' in y_Uu such that $z \in z'T$. From $z' \in y_Uu$ and $y_U \in xU$ we obtain $z' \in xU$. Thus, as $z' \in zT$, $z' = z_U$. It follows that $z_U \in y_Uu$.

By definition, $x \in y_U d$ and $y_U \in yT$. Thus, $x \in yTd$. But as $d \in U$, $Td \subseteq dT$. Thus, $x \in ydT$. Thus, there exists an element y' in yd such that $x \in y'T$. From $y_T \in yU \cap xT$ we, therefore, obtain $y' = y_T$. Thus, as $y' \in yd$, $y_T \in yd$.

Similarly, we obtain $z_T \in wd$.

From $y \in y_T d^*$, $w \in yt$, and $z_T \in wd$ we obtain $z_T \in y_T(d^*td \cap T)$. Thus, as $t\sigma_d$ is our notation for the only element in $d^*td \cap T$, $z_T \in y_T t\sigma_d$. Thus, as $z_T \in xT$ and $t\sigma_d \in T$, $z_T \in y_T(t\sigma_d)_{xT}$. Thus, by the above, $z_T \in y_T t_{xT}(d_{xU}\pi\zeta)$.

8

From Thin Schemes to Modules

In this chapter, we shall develop some of the fundamental aspects of the representation theoretic part of scheme theory. Representations of (finite) schemes reflect the arithmetic structure of schemes. They are useful in cases where the structure constants underly extreme constraints.

The central notion in representation theory of finite schemes is the one of an associative ring. Rings give rise to modules.

The concept of a ring goes back to Richard Dedekind; cf. [7]. The first abstract definition of a ring was given by Adolf Fraenkel; cf. [12]. The use of modules in the theory of finite groups is due to Emmy Noether [32; III]. The idea to generalize the relationship between modules and finite groups to a fruitful relationship between modules and schemes of finite valency had been recognized first by Donald Higman; cf. [20].

Since rings as well as modules are built from commutative groups and groups are identified with thin schemes (via the group correspondence) one may view the notion of a ring and the one of a module as part of 'thin scheme theory'.

The first section of this chapter provides general observations on modules over associative rings with 1. The collection includes the Homomorphism Theorem and the Isomorphism Theorem for modules over associative rings with 1.

In the second section, we shall look at commutative associative rings with 1. We prove that the set of all elements of a commutative associative ring D with 1 which are integral over a unitary subring of D forms a ring.

In Section 8.3, we focus on completely reducible modules over associative rings with 1. Section 8.4 deals with irreducible modules over these rings. In Section 8.5, we combine the results obtained in these two sections to obtain the famous (and complete) description of semisimple associative rings with 1 which was first given by Joseph Wedderburn and Emil Artin.

The interest in semisimple rings is based on Theorem 9.1.5(ii), a result in which we shall see that each scheme of finite valency gives rise to semisimple rings, the so-called scheme rings.

In Section 8.6, we shall provide a few basic facts on characters of semisimple rings.

In the last section of this chapter, in Section 8.7, we present identities about roots of unity in integral domains. The results will be useful in Section 12.4 where we shall investigate Coxeter sets of cardinality 2.

8.1 Rings and Modules

Let M be an additively written group (with neutral element 0).

For each element m in M with $m + m = m$, one has

$$m = m + 0 = m + (m + (-m)) = (m + m) + (-m) = m + (-m) = 0.$$

The group M is called *commutative* if, for any two elements k and l in M, $k + l = l + k$.

Let us now assume M to be commutative, and let D be a further additively written commutative group.

A map from $M \times D$ to M is called an *operation of D on M*.

Let us fix an operation of D on M. For any two elements m in M and d in D, we shall write md to denote the image of (m, d) under this operation.

The (commutative) group M is called a *group over D* or a *D-group* if, for any three elements k, l in M and d in D,

$$(k + l)d = kd + ld$$

and, for any three elements m in M and b, c in D,

$$m(b + c) = mb + mc.$$

Let us assume M to be a D-group.

For each element m in M, we have $m0 + m0 = m(0 + 0) = m0$. Thus, the above observation tells us that, for each element m in M, $m0 = 0$. Similarly, one obtains $0d = 0$ for each element d in D.

Let m be an element in M and d an element in D. Then we have

$$md + (-m)d = (m + (-m))d = 0d = 0,$$

so that $(-m)d = -(md)$. Similarly, one obtains $m(-d) = -(md)$.

The D-group M is called *unital* if there exists an element d in $D \setminus \{0\}$ such that, for each element m in M,

$$md = m.$$

An (additively written) group is called a *ring* if it is a group over itself.

Let us assume D to be a ring.

Assuming D to be a ring we have an operation of D on D or, what is the same, an operation on D in the sense of Section 5.5. This operation on D is called the *multiplication* of D, and one stays with the convention that bc denotes the image of (b, c) under the multiplication on D whenever b and c are elements in D.

A subgroup C of the additive group of D is called a *subring* if, with respect to the restriction of the multiplication on D to $C \times C$, C is a C-group.

We define $Z(D)$ to be the set of all elements z in D such that, for each element d in D, $dz = zd$.

Note that $0 \in Z(D)$. Moreover, for any two elements d in D and z in $Z(D)$, we have

$$d(-z) = -(dz) = -(zd) = (-z)d.$$

Finally, for any three elements d in D and x, y in $Z(D)$, one has

$$d(x + y) = dx + dy = xd + yd = (x + y)d.$$

Thus, $Z(D)$ is a subgroup of the additive group D. It is called the *center* of D.

The ring D is called *commutative* if $Z(D) = D$.

Since $Z(D)$ is a subgroup of the additive group D, the additive group D is a $Z(D)$-group with respect to the restriction of the multiplication of D to $D \times Z(D)$.

Let x and y be elements in $Z(D) \setminus \{0\}$, and let us assume that, for each element d in D, $dx = d$ and $dy = d$. Then $x = xy = yx = y$. This shows that, if the $Z(D)$-group D is unital, then there exists at most one element z in $Z(D) \setminus \{0\}$ such that, for each element d in D, $dz = d$. In this case, we shall denote this unique element by 1. (Thus, by definition, $0 \neq 1$.)

The ring D is called a *ring with* 1 if the $Z(D)$-group D is unital.

The D-group M is called a *module over* D or a D-*module* if, for any three elements m in M and b, c in D,

$$m(bc) = (mb)c.$$

Assume M to be a D-module. Then, for each subring C of D, M is a C-module.

A ring is called *associative* if it is a module over itself.

If D is an associative ring with 1 and M a D-module, we always assume M to be unital.

For the remainder of this section, we shall assume D to be an associative ring with 1.

Assuming D to be associative we obtain

$$d(xy) = (dx)y = (xd)y = x(dy) = x(yd) = (xy)d.$$

for any three elements d in D and x, y in $Z(D)$. Thus, $Z(D)$ is a subring of the ring D.

Let d be an element in D. An element c in D is called a *multiplicative inverse* of d if $dc = 1 = cd$.

Let d be an element in D, and let b and c be multiplicative inverses of d. Then

$$b = b1 = b(dc) = (bd)c = 1c = c.$$

Thus, each element in D has at most one multiplicative inverse.

An element d in D is called a *unit* if it has at least one multiplicative inverse. The multiplicative inverse of a unit d of D is denoted by d^{-1}.

The set of all units of D is denoted by $U(D)$.

Recall that, for each element d in D, $0d = 0 = d0$. Thus, as $0 \neq 1$, $U(D) \subseteq D \setminus \{0\}$. The ring D is called a *division ring* if $U(D) = D \setminus \{0\}$.

A commutative division ring is called a *field*.

A commutative subring C of D is called a *subfield* if, for each element c in $C \setminus \{0\}$, $c \in U(D)$ and $c^{-1} \in C$.

A (unital) module over a field C is called a *vector space over* C.

Let us now assume M to be a (unital) D-module.

Since M is assumed to be unital, D possesses an element e such that, for each element m in M, $me = m$. It follows that, for each element m in M,

$$m1 = (me)1 = m(e1) = me = m.$$

Let L be a subset of M, and let C be a subset of D. If one of the two sets C or L is empty, we set $LC := \{0\}$. If both C and L are not empty, we define LC to be the set of all finite sums of products lc with $l \in L$ and $c \in C$. If there exists an element l in L with $\{l\} = L$, we write lC instead of LC. Similarly, if there exists an element c in C with $\{c\} = C$, we write Lc instead of LC.

For any two nonempty subsets C of D and L of M, the set LC is called the *complex product* of L and C. We also speak of the *complex multiplication between* M and D.

A subgroup L of the additive group M is called a *submodule* of M if $LD \subseteq L$.

Note that a subgroup L of M is a submodule of M if, for any two elements l in L and d in D, $ld \in L$.

Note also that $\{0\}$ and M are submodules of M. Moreover, for any two submodules K and L of M, $K \cap L$ is a submodule of M.

A submodule L of the D-module D is called an *ideal* of D if $DL \subseteq L$.

It is obvious that $\{0\}$ and D are ideals of D. If, conversely, these two ideals are the only ideals of D and $\{0\} \neq D$, then the ring D is called *simple*.

Let K and L be nonempty subsets of M. We define $K + L$ to be the set of all sums $k + l$ with $k \in K$ and $l \in L$. If there exists an element k in K with $\{k\} = K$, we write $k + L$ instead of $K + L$. Similarly, if there exists an element l in L with $\{l\} = L$, we write $K + l$ instead of $K + L$.

For any two nonempty subsets K and L of M, the set $K + L$ is called the *complex sum of* K and L. The associated operation on the set of all nonempty subsets of M will be referred to as the *complex addition in* M.

It is easy to see that the complex sum of two submodules of M is a submodule of M.

Let us now fix a submodule L of M.

We define M/L to be the set of all sets $L + m$ with $m \in M$.

It is easy to see that the restriction of the complex addition to M/L is an operation on M/L. Moreover, one verifies easily that, with respect to this addition, M/L is a commutative group.

For any two elements m in M and d in D, we define

$$(L + m)d := L + md.$$[1]

It is easy to see that, for any three elements h, k in M and d in D,

$$((L + h) + (L + k))d = (L + h)d + (L + k)d.$$

Moreover, for any three elements m in M and b, c in D, one has

$$(L + m)(b + c) = (L + m)b + (L + m)c.$$

Thus, M/L is a D-group.

Note also that, for any three elements m in M and b, c in D,

$$(L + m)(bc) = ((L + m)b)c.$$

Thus, M/L is a D-module.

Finally, for each element m in M, we have

$$(L + m)1 = L + m,$$

so that the D-module M/L is unital.

In fact, all four of these equations follow from the corresponding equation for the D-group M.

The unital D-module M/L is called the *factor module* of M *over* L.

---

[1] This is not the complex multiplication between M and D.

Let $\tilde{M}$ be a D-module, and let ϕ be a map from M to $\tilde{M}$.

The map ϕ is called a *homomorphism* if, for any two elements k and l in M,

$$(k+l)\phi = k\phi + l\phi$$

and, for any two elements m in M and d in D,

$$(md)\phi = (m\phi)d.$$

Sometimes, in order to emphasize the underlying ring, one says *D-module homomorphism* or *D-homomorphism* instead of homomorphism.

We define $\ker(\phi)$ to be the set of all elements m in M satisfying $m\phi = 0$.

It is easy to see that $\ker(\phi)$ is a submodule of M and that $M\phi$ is a submodule of $\tilde{M}$. (By $M\phi$ we mean, as is common, the set of all elements $m\phi$ with $m \in M$.)

The D-modules M and $\tilde{M}$ are called *isomorphic*, if there exists a bijective homomorphism from M to $\tilde{M}$. Sometimes, we shall write $M \cong \tilde{M}$ in order to indicate that M and $\tilde{M}$ are isomorphic.

The following theorem is called the *Homomorphism Theorem* for modules over associative rings with 1.

Theorem 8.1.1 *Let ϕ be a homomorphism from M. Then* $\ker(\phi)$ *is a submodule of M, and $M\phi$ is an D-module with* $M/\ker(\phi) \cong M\phi$.

PROOF. It is obvious that $\ker(\phi)$ is a submodule of M.

For any two elements k and l in M, we have $k\phi + l\phi = (k+l)\phi \in M\phi$. Also, for any two elements m in M and d in D, we have $(m\phi)d = (md)\phi \in M\phi$. Therefore, $M\phi$ is an D-module.

Let us abbreviate $L := \ker(\phi)$.

For any two elements h and k in M, we have $h\phi = k\phi$ if and only if $(h-k)\phi = 0$, and this is equivalent to $h - k \in L$. Note also that $h - k \in L$ is equivalent to $L + h = L + k$. This gives us a bijective map ι from M/L to $M\phi$ which, for each element m in M, maps $L + m$ to $m\phi$.

For any two elements h and k in M, we have

$$\begin{aligned}((L+h)+(L+k))\iota &= (L+(h+k))\iota = (h+k)\phi \\ &= h\phi + k\phi = (L+h)\iota + (L+k)\iota.\end{aligned}$$

For any two elements m in M and d in D, we have

$$((L+m)d)\iota = (L+md)\iota = (md)\phi = (m\phi)d = ((L+m)\iota)d.$$

Thus, ι is a D-homomorphism.

The following theorem is the *Isomorphism Theorem* for modules over associative rings with 1.

Theorem 8.1.2 *For any two submodules K and L of M, we have $K/K\cap L \cong (K+L)/L$.*

PROOF. For each element k in K, we define $k\phi := L+k$. Then ϕ is a surjective map from K to $(K+L)/L$.

For any two elements h and k in M, we have

$$(h+k)\phi = L+(h+k) = (L+h)+(L+k) = h\phi + k\phi.$$

For any two elements k in K and d in D, we have

$$(kd)\phi = L+kd = (L+k)d = (k\phi)d.$$

Therefore, ϕ is a homomorphism.

Note that $\ker(\phi) = K\cap L$. Thus, the claim follows from Theorem 8.1.1.

A group homomorphism from the additive group M to itself is called a (*group*) *endomorphism* of M. Let us write $\mathrm{End}(M)$ to denote the set of all endomorphisms of the additive group M.

Let c and d be elements in $\mathrm{End}(M)$.

For each element m in M, we define

$$m(c+d) := mc + md.$$

It is straightforward to see that, as c, $d \in \mathrm{End}(M)$, $c+d \in \mathrm{End}(M)$. This shows that *componentwise addition* is an operation on $\mathrm{End}(M)$.

Note also that the composition of elements in $\mathrm{End}(M)$ is an operation on $\mathrm{End}(M)$.

It is easy to see that, with respect to componentwise addition and composition (as multiplication), $\mathrm{End}(M)$ is a ring. Note also that M is a module over $\mathrm{End}(M)$.

A D-homomorphism from M to M is called a *D-endomorphism* of M. We define

$$\mathrm{End}_D(M)$$

to be the set of all D-endomorphisms of M.

Note that $\mathrm{End}_D(M)$ is a subring of $\mathrm{End}(M)$. In particular, M is a module over $\mathrm{End}_D(M)$.

For each element d in D, we define

$$d_M : M \to M,\ m \mapsto md.$$

Note that, for each element d in D, $d_M \in \mathrm{End}(M)$.
We set $D_M := \{d_M \mid d \in D\}$.

Lemma 8.1.3 *Set $E := \mathrm{End}_D(M)$. Then $D_M \subseteq \mathrm{End}_E(M)$.*

PROOF. For any three elements m in M, d in D, and e in E, we have

$$(md_M)e = (md)e = (me)d = (me)d_M.$$

Thus, $d_M \in \mathrm{End}_E(M)$.

For each element d in D, we define $B_M(d)$ to be the kernel of d_M. For each subset C of D, we define $B_M(C)$ to be the intersection of the sets $B_M(c)$ with $c \in C$.

Lemma 8.1.4 *Set $E := \mathrm{End}_D(M)$. Then, for each subset C of D, $B_M(C)$ is a submodule of the E-module M.*

PROOF. Considering that intersections of submodules are submodules, this follows from Theorem 8.1.1 together with Lemma 8.1.3.

Let m be an element in M. We define

$$m_D : D \rightarrow M, \; d \mapsto md.$$

Then m_D is a D-homomorphism from the D-module D to M.
For each element m in M, we define $A_D(m)$ to be the kernel of m_D. For each subset L of M, we define $A_D(L)$ to be the intersection of the sets $A_D(l)$ with $l \in L$.

Lemma 8.1.5 *For each subset L of M, $A_D(L)$ is a submodule of the D-module D.*

PROOF. Considering that intersections of submodules are submodules, this follows from Theorem 8.1.1.

Lemma 8.1.6 *We have the following.*

(i) *For each subset C of D, $CD \subseteq A_D(B_M(C))$.*

(ii) *Set $E := \mathrm{End}_D(M)$. Then, for each subset L of M, $LE \subseteq B_M(A_D(L))$.*

PROOF. (i) Let c be an element in C. Then, for each element m in $B_M(C)$, $mc = 0$. Thus, by definition, $c \in A_D(B_M(C))$.
Since c has been chosen arbitrarily in C, we have shown that $C \subseteq A_D(B_M(C))$. Thus, the claim follows from Lemma 8.1.5.

(ii) Let l be an element in L. Then, for each element d in $A_D(L)$, $ld = 0$. Thus, by definition, $l \in B_M(A_D(L))$.

Since l has been chosen arbitrarily in L, we have shown that $L \subseteq B_M(A_D(L))$. Thus, the claim follows from Lemma 8.1.4.

A subring C of D is called *unitary* if $1 \in C$.

Note that the intersection of any two unitary subrings of D is a unitary subring of D. In particular, D possesses a uniquely determined smallest unitary subring. Let us (for the moment) denote this unitary subring by $\breve{D}$.

If $|\breve{D}|$ is finite, we call $|\breve{D}|$ the *characteristic* of D. Otherwise, we say that D has characteristic 0.

8.2 Integrality in Associative Rings with 1

Throughout this section, the letter D stands for an associative ring with 1.

Let C be a unitary subring of D.

Let A be a subset of D such that, for any two elements c in C and a in A, $ca = ac$. Then we shall denote by $C[A]$ the smallest subring of D containing C and A as subsets. If there exists an element a in A with $\{a\} = A$, we write $C[a]$ instead of $C[A]$.

Let us now assume D to be commutative.

An element d in D is called *integral over* C if there exist elements $c_0, \ldots, c_{n-1}$ in C such that

$$d^n + c_{n-1}d^{n-1} + \ldots + c_1 d + c_0 = 0.$$

We define $I_D(C)$ to be the set of all elements in D which are integral over C.

The ring D is called *finite over* C if there exists a finite subset A of D such that $C[A] = D$.

A D-module M is called *finitely generated* if it possesses a finite subset L with $LD = M$. Occasionally, we shall refer to the elements in L as *generators* of M.

Lemma 8.2.1 *Let C be a unitary subring of D such that D is finite over C and integral over C. Then the C-module D is finitely generated.*

Proof. Since D is assumed to be finite over C, D possesses a finite subset A such that $C[A] = D$. Since D is assumed to be integral over C, each element of A is integral over C. Thus, for each element a in A, there exists a positive integer n_a together with elements $c_{a,0}, \ldots, c_{a,n_a-1}$ in C such that

$$a^{n_a} + c_{a,n_a-1}a^{n_a-1} + \ldots + c_{a,1}a + c_{a,0} = 0.$$

Let a be an element in A. We set

$$B_a := \{1, a, \ldots, a^{n_a - 1}\}.$$

Then $a^{n_a} \in B_a C$. Thus, for each element d in $B_a C$, $ad \in B_a C$, and that means that $aB_a C \subseteq B_a C$.

We define B to be the set of all products

$$\prod_{a \in A} b_a$$

such that, for each element a in A, $b_a \in B_a$. Then, as A is finite, so is B.

Recall that, for each element a in A, $aB_a C \subseteq B_a C$. Thus, for each element a in A, $aBC \subseteq BC$. Thus, as $C[A] = D$, $DBC \subseteq BC$. Thus, as $1 \in BC$, $BC = D$.

Lemma 8.2.2 *Let B be a finite subset of D. For any two elements a and b in B, let d_{ab} be an element in D such that, for each element b in B,*

$$\sum_{a \in B} a d_{ab} = 0.$$

Then, for each element b in B, $b\det(d_{ab}) = 0$.

Proof. Set $n := |B|$. Then there exist elements $b_1, \ldots, b_n$ in B such that $\{b_1, \ldots, b_n\} = B$.

Set $A := (b_1, \ldots, b_n)$. For any two elements i and j in $\{1, \ldots, n\}$, we set $d_{ij} := d_{b_i b_j}$. We set $M := (d_{ij})$. Then $AM = 0$.

For any two elements i and j in $\{1, \ldots, n\}$, we set

$$n_{ij} := (-1)^{i+j} \det(M_{kl})_{k \neq i, l \neq j}.$$

We set $N := (n_{ij})$. Then $MN = \det(M)I$.

From $AM = 0$ and $MN = \det(M)I$ we obtain $A\det(M)I = 0$. Thus, for each element b in B, $b\det(M) = 0$.

Lemma 8.2.3 *Let B be a finite subset of D, and let C be a unitary subring of D such that BC is a subring of D. Then $BC \subseteq I_D(C)$.*

Proof. Let d be an element in BC. We have to show that $d \in I_D(C)$.

Let b be an element in B. Then, as $d \in BC$ and BC is assumed to be a subring of D, $bd \in BC$. Thus, there exists, for each element a in B, an element c_{ab} in C such that

$$bd = \sum_{a \in B} a c_{ab}.$$

For any two elements a and b in B such that $a \neq b$, we set $d_{ab} := -c_{ab}$. For each element b in B, we set $d_{bb} := d - c_{bb}$. Then, for each element b in B,

$$\sum_{a \in B} a d_{ab} = 0.$$

Thus, for each element b in B, $b\det(d_{ab}) = 0$; cf. Lemma 8.2.2.

Recall that D is assumed to be a ring with 1, and BD is assumed to be a unitary subring of D. Thus, $1 \in BC$. Thus, as, for each element b in B, $b\det(d_{ab}) = 0$, $1\det(d_{ab}) = 0$. It follows that $\det(d_{ab}) = 0$. However, $\det(d_{ab})$ is a monic polynomial in d with coefficients in C. In particular, d is integral over C.

The following result is due to Richard Dedekind; cf. [7; §160].

Theorem 8.2.4 *For each unitary subring C of D, $I_D(C)$ is a subring of D.*

Proof. We set $I := I_D(C)$ and pick an element d in I^2. We have to show that $d \in I$.

Since $d \in I^2$, there exists a finite subset A of I such that $d \in A^2$. Since A is a finite subset of I, $C[A]$ possesses a finite subset B such that $C[A] = BC$; cf. Lemma 8.2.1.

Since $A \subseteq BC$ and BC is a subring of D, $d \in A^2 \subseteq (BC)^2 \subseteq BC$. On the other hand, as B is finite and BC a subring of D, Lemma 8.2.3 yields that $BC \subseteq I$. Thus, $d \in I$.

Lemma 8.2.5 *Assume D to be a field of characteristic 0. Let us denote by Q the smallest subfield of D and by Z the smallest unitary subring of D. Then $Z = I_Q(Z)$.*

Proof. Let q be an element in $I_Q(Z)$. We have to show that $q \in Z$.

Since $q \in Q$, there exist elements d in Z and c in $Z \setminus \{0\}$ such that $qc = d$. Since D is assumed to have characteristic 0, we may assume that d and c are coprime in Z.

We are assuming that $q \in I_Q(Z)$. Thus, there exist elements $z_0, \ldots, z_{n-1}$ in Z such that

$$q^n + z_{n-1}q^{n-1} + \ldots + z_1 q + z_0 = 0.$$

Multiplying this equation by c^n we obtain

$$d^n = -(z_{n-1}d^{n-1} + z_{n-2}d^{n-2}c + \ldots + z_1 dc^{n-2} + r_0 c^{n-1})c.$$

Thus, c divides d. Thus, as that d and c are assumed to be coprime in Z, $c^{-1} \in Z$. It follows that $q = dc^{-1} \in Z$.

Assume D to be a field, let C be a subfield of D, and let d be an element in D. Instead of saying that d is integral over C one also calls d *algebraic* over C. If $C = I_D(C)$, one says that C is *algebraically closed* in D.

One says that D is *algebraically closed* if, for each field E which contains D as a subfield, D is algebraically closed in E.

Lemma 8.2.6 *Assume D to be a division ring, and let C be a subfield of D with $C \subseteq Z(D)$. Assume C to be algebraically closed and D to be a finitely generated C-module. Then $C = D$.*

PROOF. Let d be an element in D. Since we are assuming that $C \subseteq Z(D)$, $C[d]$ is a field.

We are also assuming that D is finitely generated as a C-module, so that $C[d]$ is finitely generated as a C-module. Thus, as we are assuming C to be algebraically closed, we must have $C = C[d]$, and that means that $d \in C$.

Since d has been chosen arbitrarily in D, we have shown that $C = D$.

8.3 Completely Reducibility

Throughout this section, the letter D stands for an associative ring with 1, the letter M for a D-module. We shall focus on the structure of M.

A submodule L of the D-module M different from M is called *maximal* if L and M are the only submodules of M containing L (as a subset).

The D-module M is called *irreducible* if $\{0\}$ is a maximal submodule of M.

Note that $\{0\}$ is not an irreducible D-module.

Let K and L be submodules of M. If $\{0\} = K \cap L$, we write $K \oplus L$ instead of $K + L$.

Lemma 8.3.1 *Let K be a submodule of M, and let L be an irreducible submodule of M such that $L \not\subseteq K$. Then K is a maximal submodule of M if and only if $K \oplus L = M$.*

PROOF. Since L is assumed to be irreducible and $L \not\subseteq K$, $\{0\} = K \cap L$.

Since $L \not\subseteq K$, $K \neq K + L$. Thus, assuming that K is maximal we obtain $K + L = M$.

Let us now assume that $K + L = M$, and let H be a submodule of M such that $K \subseteq H$. Then, referring to the group correspondence we obtain from Lemma 2.2.1 that

$$H = H \cap (K + L) = K + (H \cap L).$$

However, L is assumed to be irreducible. Thus, $\{0\} = H \cap L$ or $H \cap L = L$. In the first case, we obtain $K = H$, in the second one, we obtain $H = M$.

The module M is called *completely reducible* if, for each submodule L of M, there exists a submodule K of M such that $K \oplus L = M$.

Note that $\{0\}$ is completely reducible.

For the remainder of this section, we shall look at the case where M is completely reducible.

Lemma 8.3.2 *Assume that M is completely reducible. Then so is each submodule of M.*

PROOF. Let L be a submodule of M, and let K be a submodule of L.

Since M is assumed to be completely reducible, there exists a submodule H of M such that $H \oplus K = M$.

From $H + K = M$ and $K \subseteq L$ we obtain $(L \cap H) + K = L$; cf. Lemma 2.2.1(ii) (together with the group correspondence). From $\{0\} = H \cap K$, we obtain $\{0\} = (L \cap H) \cap K$. Thus, $(L \cap H) \oplus K = L$.

Lemma 8.3.3 *Assume that M is completely reducible and that $\{0\} \neq M$. Then M contains an irreducible submodule.*

PROOF. Let m be an element in $M \setminus \{0\}$. Then, by Zorn's Lemma, there exists a maximal element L in the set of all submodules of M not containing m. Since M is assumed to be completely reducible, M possesses a submodule K such that $K \oplus L = M$.

Since $m \notin L$, $L \neq M$. Thus, as $K + L = M$, $\{0\} \neq K$.

We claim that K is irreducible, and in order to prove this we assume, by way of contradiction, that K is not irreducible.

Assuming K to be not irreducible we obtain from $\{0\} \neq K$ a submodule J of K with $\{0\} \neq J \neq K$.

Since M is assumed to be completely reducible and K is a submodule of M, K must be completely reducible; cf. Lemma 8.3.2. Thus, as J is a submodule of K, K possesses a submodule I such that $I \oplus J = K$

The maximal choice of L yields $m \in I \oplus L$ and $m \in J \oplus L$. However, as $\{0\} = K \cap L$ and $\{0\} = I \cap J$,

$$L = (I \oplus L) \cap (J \oplus L).$$

Thus, $m \in L$, contrary to the choice of L.

This contradiction shows that K is irreducible.

Let $\mathcal{H}$ be a set of submodules of the D-module M.

We say that M is the *sum* of the elements in $\mathcal{H}$ if each element of M is sum of finitely many elements in the union of the elements in $\mathcal{H}$. (The sum of zero elements in M is 0.)

Assume that M is equal to the sum of the elements in $\mathcal{H}$. We say that M is the *direct* sum of the elements in $\mathcal{H}$ if, for each element H in $\mathcal{H}$, $\{0\} = H \cap \hat{H}$, where $\hat{H}$ denotes the sum of the submodules in $\mathcal{H}$ different from H.

Proposition 8.3.4 *The following statements are equivalent.*

(a) *The module M is sum of irreducible submodules of M.*

(b) *The module M is completely reducible.*

(c) *The module M is a direct sum of irreducible submodules of M.*

PROOF. (a) $\Rightarrow$ (b) In order to do this we assume that there exists a set $\mathcal{H}$ of irreducible submodules of M such that M is the sum of the elements in $\mathcal{H}$.

Let L be a submodule of M, and let us define $\mathcal{K}$ to be the set of all submodules K of M with $\{0\} = K \cap L$. By Zorn's Lemma, $\mathcal{K}$ has a maximal element. We pick such a maximal element and call it K.

Assume that $K + L \neq M$. Then there exists an element H in $\mathcal{H}$ such that $H \not\subseteq K + L$. As H is irreducible, $H \not\subseteq K + L$ yields $\{0\} = H \cap (K + L)$.

From $\{0\} = H \cap (K + L)$ we obtain $(H + K) \cap L \subseteq K$. However, we have $\{0\} = K \cap L$. Thus, $\{0\} = (H + K) \cap L$, contrary to the (maximal) choice of K.

(b) $\Rightarrow$ (c) We define L to be a maximal direct sum of irreducible submodules of M. According to Zorn's Lemma, we find such a submodule. Then, by (b), there exists a submodule K of M such that $K \oplus L = M$.

Let us assume that $\{0\} \neq K$. Since M is assumed to be completely reducible and K is a submodule of M, K is completely reducible; cf. Lemma 8.3.2. Thus, by Lemma 8.3.3, K possesses an irreducible submodule H. It follows that $H \oplus L$ is a direct sum of irreducible submodules and $L \neq H \oplus L$, contrary to the choice of L.

Thus, we have shown that $\{0\} = K$, so that $L = M$.

(c) $\Rightarrow$ (a) This is obvious.

Lemma 8.3.5 *Let $\mathcal{H}$ be a set of irreducible submodules of M such that M is the sum of the elements in $\mathcal{H}$, and let L be an irreducible submodule of M. Then there exists an element H in $\mathcal{H}$ such that $H \cong L$.*

PROOF. Since L is assumed to be a submodule of M, M possesses a submodule K such that $K \oplus L = M$; cf. Proposition 8.3.4. Since L is assumed to be irreducible, K is a maximal submodule of M; cf. Lemma 8.3.1. In particular, $K \neq M$. Thus, there exists an element H in $\mathcal{H}$ such that $H \not\subseteq K$.

Since K is a maximal submodule of M and H an irreducible submodule of M with $H \not\subseteq K$, $H \oplus K = M$; cf. Lemma 8.3.1.

From $H \oplus K = M$ and $K \oplus L = M$ we obtain $H \cong M/K \cong L$; cf. Theorem 8.1.2.

We call M *homogeneous* if M is a sum of irreducible submodules of M which are isomorphic in pairs.

Note that, as a consequence of Lemma 8.3.3 and Lemma 8.3.5, each completely reducible D-module has maximal homogeneous submodules.

Theorem 8.3.6 *Assume that M is completely reducible, and let $\mathcal{H}$ denote the set of all maximal homogeneous submodules of M. Then the following hold.*

(i) *The module M is the direct sum of the elements in $\mathcal{H}$.*

(ii) *For each irreducible submodule L of M, there exists exactly one element H in $\mathcal{H}$ such that $L \subseteq H$.*

(iii) *Let K and L be irreducible submodules of M. Then $K \cong L$ if and only if there exists an element H in $\mathcal{H}$ such that $K \subseteq H$ and $L \subseteq H$.*

PROOF. (i) We are assuming M to be completely reducible. Thus, by Proposition 8.3.4, M is sum of irreducible submodules of M. In particular, M is sum of the elements in $\mathcal{H}$.

Let us fix an element H in $\mathcal{H}$, and define (as we did earlier) $\hat{H}$ to be the sum of the submodules in $\mathcal{H}$ different from H. Then, by Lemma 8.3.5, $\{0\} = H \cap \hat{H}$. Thus, as H has been chosen arbitrarily in $\mathcal{H}$, M is direct sum of the elements in $\mathcal{H}$.

(ii) Let L be an irreducible submodule of M, and let H be the sum of all submodules of M isomorphic to L. Then $L \subseteq H \in \mathcal{H}$. That L is not subset of two different elements of $\mathcal{H}$ follows from Lemma 8.3.5.

(iii) This follows from (ii).

Assume that M is completely reducible and that $\{0\} \neq M$.

From Lemma 8.3.3 we know that M contains an irreducible submodule L. Since M is assumed to be completely reducible, there exists a submodule K of M such that $K \oplus L = M$. According to Lemma 8.3.1, K is a maximal submodule of M, so that M has at least one maximal submodule.

We define $J(M)$ to be the intersection of all maximal submodules of M and call this submodule the *Jacobson radical* of M.

The module M is called *artinian* if each set of submodules of M has minimal elements.

Theorem 8.3.7 *Assume that $\{0\} \neq M$. Then we have the following.*

(i) *If M is completely reducible, $\{0\} = J(M)$.*

(ii) *If M is artinian and $\{0\} = J(M)$, M is completely reducible.*

PROOF. (i) We set $J := J(M)$. Then J is a submodule of M. Thus, as M is assumed to be completely reducible, there exists a submodule L of M such that $L \oplus J = M$.

Let us assume, by way of contradiction, that $\{0\} \neq J$. Then $L \neq M$, so that M possesses a maximal submodule K with $L \subseteq K$. Since K is a maximal submodule of M, $J \subseteq K$. Thus, $M = L \oplus J \subseteq K$, contradiction.

(ii) Let K be minimal among the submodules of M with completely reducible factor module. We shall be done if we succeed in showing that K is contained in each maximal submodule of M.

Let us assume, by way of contradiction, that M possesses a maximal submodule L with $K \not\subseteq L$. Then $K + L = M$. From this we conclude that

$$(K/K \cap L) \oplus (L/K \cap L) = M/K \cap L.$$

With the help of Theorem 8.1.2 we obtain from $K + L = M$ also that

$$K/K \cap L \cong M/L$$

and that

$$L/K \cap L \cong M/K.$$

Now the choice of K yields $K \cap L = K$. This means that $K \subseteq L$, contrary to the choice of L.

Theorem 8.3.6 tells us that the structure of a completely reducible D-module depends on the structure of its irreducible submodules. This is the reason why we shall now have a closer look at irreducible D-modules.

8.4 Irreducible Modules over Associative Rings with 1

Throughout this section, the letter D stands for an associative ring with 1, the letter M for an irreducible D-module.

Recall that, for each element m in M, $A_D(m)$ is our notation for the set of all elements d in D such that $md = 0$.

Recall that, for each element m in M, $A_D(m)$ is a submodule of the D-module D; cf. Lemma 8.1.5.

Lemma 8.4.1 *For each element m in $M \setminus \{0\}$, $D/A_D(m) \cong M$.*

PROOF. Let m be an element in $M \setminus \{0\}$. Since $1m_D = m$ and $0 \neq m$, $\{0\} \neq Dm_D$. Thus, as Dm_D is a submodule of M and M is assumed to be irreducible, we must have that $Dm_D = M$. Thus, looking at the definition of $A_D(m)$ the claim follows from Theorem 8.1.1.

Lemma 8.4.2 *We have* $\{0\} = MJ(D)$.

PROOF. We are assuming that M is irreducible. Thus, for each element m in $M \setminus \{0\}$, $A_D(m)$ is a maximal submodule of the D-module D. Thus, by definition, $J(D) \subseteq A_D(M)$, and that means that $\{0\} = MJ(D)$.

Recall that $\mathrm{End}_D(M)$ is our notation for the ring of all D-endomorphisms of M. (We saw that $\mathrm{End}_D(M)$ is a ring with respect to componentwise addition and composition as multiplication.)

The following lemma is often referred to as *Schur's Lemma.*

Lemma 8.4.3 *The ring* $\mathrm{End}_D(M)$ *is a division ring.*

PROOF. Let e be an element in $\mathrm{End}_D(M)$ such that $0 \neq e$. We have to show that e possesses a multiplicative inverse in $\mathrm{End}_D(M)$.

From $0 \neq e$ we obtain $\ker(e) \neq M$. Thus, as $\ker(e)$ is a submodule of M and M is assumed to be irreducible, we conclude that $\{0\} = \ker(e)$. It follows that e is injective.

From $0 \neq e$ we also obtain $\{0\} \neq Me$. Thus, as Me is a submodule of M and M is assumed to be irreducible, we conclude that $Me = M$, and that means that e is surjective.

Let us now see that $e^{-1} \in \mathrm{End}_D(M)$. In order to see this, we fix elements m in M and d in D. Then

$$((md)e^{-1})e = (md)(e^{-1}e) = (m(e^{-1}e))d = ((me^{-1})e)d = ((me^{-1})d)e.$$

Thus, as e is injective, $(md)e^{-1} = (me^{-1})d$. Thus, as m has been chosen arbitrarily in M and d arbitrarily in D, we have shown that $e^{-1} \in \mathrm{End}_D(M)$.

For the remainder of this section, we set

$$E := \mathrm{End}_D(M).$$

According to Lemma 8.4.3, the ring E is a division ring.

Lemma 8.4.4 *For each finite subset* L *of* M, *we have* $LE = B_M(A_D(L))$.

PROOF.[2] From Lemma 8.1.6(ii) we know that $LE \subseteq B_M(A_D(L))$. Thus, we just have to show that $B_M(A_D(L)) \subseteq LE$. We set $A := A_D(L)$.

If L is empty, $A = D$. From $A = D$ (together with $\{0\} \neq M$ and $1 \in D$) we obtain $\{0\} = B_M(A)$. It follows that $B_M(A) \subseteq LE$.

Let us now assume that L is not empty. In this case, we fix an element l in L, and we set $L' := L \setminus \{l\}$ and $A' := A_D(L')$. Then, by induction,

[2] Our proof follows the one given for [28; Theorem 13.15].

$$B_M(A') \subseteq L'E$$

If $A = A'$, $B_M(A) = B_M(A')$. Thus, as $L'E \subseteq LE$, $B_M(A) \subseteq LE$.

Let us now assume that $A \neq A'$. Then $A' \not\subseteq A_D(\{l\})$, so that $\{0\} \neq lA'$. Thus, as lA' is a submodule of M and M is assumed to be irreducible, $lA' = M$.

Let m be an element in $B_M(A)$. We shall be done if we succeed in showing that $m \in LE$.

Let b and c be elements in A' such that $lb = lc$. Then $l(b - c) = 0$. Thus, $b - c \in A' \cap A_D(l) = A$. Thus, as we are assuming that $\{0\} = mA$, we conclude that $m(b - c) = 0$. It follows that $mb = mc$.

For each element a in A', we define

$$(la)e := ma.$$

Since $lA' = M$, we just saw that e is a map from M to M. It is obvious that e is a group endomorphism of M. However, for any two elements a in A' and d in D, we have

$$((la)d)e = (l(ad))e = m(ad) = (ma)d = ((la)e)d.$$

Thus, $e \in E$.

Note, finally, that, for each element a in A',

$$(m - le)a = ma - (le)a = ma - (la)e = 0.$$

This means that $m - le \in B_M(A')$. Thus, as $B_M(A') \subseteq L'E$, $m - le \in L'E$. Thus, as $e \in E$,

$$m \in L'E + lE = LE.$$

This proves the lemma.

The following theorem, due to Nathan Jacobson, is often referred to as *Jacobson Density Theorem.*

Theorem 8.4.5 *Let L be a finite subset of M linearly independent with respect to E. Then, for each element e in $\mathrm{End}_E(M)$, there exists an element d in D such that, for each element l in L, $ld = le$.*

PROOF. We may assume that L is not empty. We fix an element l in L, and we set $L' := L \setminus \{l\}$ and $A' := A_D(L')$.

By induction, there exists an element d' in D such that, for each element m in L', $md' = me$.

Since $l \notin L'$, $l \notin L'E$. Thus, by Lemma 8.4.4, $l \notin B_M(A')$. Thus, as M is assumed to be irreducible, $lA' = M$. Thus, there exists an element a in A' such that $la = le - ld'$.

Set $d := a + d'$. Then $ld = le$ and, for each element m in L',

$$md = m(a + d') = ma + md' = me.$$

This proves the theorem.

The ring D is called *artinian* if the D-module D is artinian.

Lemma 8.4.6 *If D is artinian, M is a finitely generated E-module.*

PROOF. Define $\mathcal{L}$ to be the set of all finite subsets of M, and define $\mathcal{A}$ to be the set of all sets $A_D(L)$ with $L \in \mathcal{L}$.

Since D is assumed to be artinian, $\mathcal{L}$ possesses an element L such that $A_D(L)$ is minimal in $\mathcal{A}$. We set $A := A_D(L)$.

Let m be an element in M.

If $\{0\} \neq mA$, $A \not\subseteq A_D(L \cup \{m\})$. Thus, as $A_D(L \cup \{m\}) \subseteq A$, this contradicts the (minimal) choice of A.

Thus, we have that $\{0\} = mA$, and that means that $m \in B_M(A)$. Thus, by Lemma 8.4.4, $m \in LE$.

Since m has been chosen arbitrarily in M, we have shown that $M = LE$.

Recall that D_M is our notation for the set of all elements d_M with $d \in D$.

In Lemma 8.1.3 we saw that $D_M = \mathrm{End}_E(M)$.

The following lemma is often referred to as *Double Centralizer*.

Lemma 8.4.7 *Assume D to be artinian. Then $D_M = \mathrm{End}_E(M)$.*

PROOF. According to Lemma 8.1.3, it is enough to show that $\mathrm{End}_E(M) \subseteq D_M$. In order to show this, we pick an element in $\mathrm{End}_E(M)$ and call it e. We have to show that $e \in D_M$.

According to Lemma 8.4.6, there exists a finite set L such that $M = LE$. In particular, we may assume L to be linearly independent with respect to E. Thus, according to Theorem 8.4.5, there exists an element d in D such that, for each element l in L, $ld = le$.

For each element m in M, we define $md_M = md$. Then $d_M \in \mathrm{End}_E(M)$, and d_M coincides with e on L. It follows that $e = d_M$. In particular, $e \in D_M$.

Let $\tilde{D}$ be a ring, and let ϕ be a map from D to $\tilde{D}$.

The map ϕ is called a *ring homomorphism* or simply a *homomorphism* if, for any two elements b and c in D,

$$(b + c)\phi = b\phi + c\phi$$

and

$$(bc)\phi = b\phi c\phi.$$

A ring $\tilde{D}$ is called *isomorphic* to D if there exists a bijective ring homomorphism from D to $\tilde{D}$. We shall write $D \cong \tilde{D}$ in order to indicate that D and $\tilde{D}$ are isomorphic rings.

Theorem 8.4.8 *Assume D to be artinian and simple. Then, $D \cong \mathrm{End}_E(M)$.*

PROOF. The set $A_D(M)$ is an ideal of D. Moreover, $1 \notin A_D(M)$. Thus, as we are assuming D to be simple, $\{0\} = A_D(M)$. It follows that $D \cong D_M$. Thus, our claim follows from Lemma 8.4.7.

8.5 Semisimple Associative Rings with 1

An associative ring D with 1 is called *semisimple* if the D-module D is completely reducible.

In this section, we shall combine our results about completely reducible modules obtained in Section 8.3 with those about irreducible modules (and artinian simple rings) obtained in Section 8.4 in order to give a complete description of semisimple rings.

For the remainder of this section, we shall now fix an associative ring with 1 and call it D.

Lemma 8.5.1 *Let $\mathcal{L}$ be a set of submodules of the D-module D such that D is the sum of the elements in $\mathcal{L}$. Then there exists a finite subset $\mathcal{K}$ of $\mathcal{L}$ such that D is the sum of the elements in $\mathcal{K}$.*

PROOF. We are assuming that D is a ring with 1. Since D is the sum of the elements in $\mathcal{L}$, there exists a finite subset $\mathcal{K}$ of $\mathcal{L}$ and, for each element K in $\mathcal{K}$, an element d_K in K such that

$$1 = \sum_{K \in \mathcal{K}} d_K.$$

Thus, for each element d in D,

$$d = 1d = (\sum_{K \in \mathcal{K}} d_K)d = \sum_{K \in \mathcal{K}} d_K d.$$

For each element K in $\mathcal{K}$, we have $d_K d \in K$. Thus, as d has been chosen arbitrarily, we have shown that D is the sum of the elements in $\mathcal{K}$.

Recall that a subgroup L of a D-module M is called a submodule of M if $LD \subseteq L$. Recall that a submodule L of the D-module D is called an ideal of D if $DL \subseteq L$. It is obvious that $\{0\}$ and D are ideals of D.

Let C be an ideal of D such that $\{0\} \neq C$. We call C a *minimal* ideal of D if $\{0\}$ and C are the only ideals of D contained (as subsets) in C.

Proposition 8.5.2 *Assume D to be semisimple. Then each maximal homogeneous submodule of the D-module D is a minimal ideal of D.*

PROOF. Let H be a maximal homogeneous submodule of the D-module D, let d be an element in D, and let K be an irreducible submodule of H. Since K is a submodule of the D-module D, so is dK.

Let K be an element in $\mathcal{K}$. Then $dK = Kd_D$ is a submodule of D. Therefore, as K is irreducible, we have either $dK \cong K$ or $\{0\} = dK$; cf. Theorem 8.1.1. Thus, $dK \subseteq H$.

Since d has been chosen arbitrarily in D, we conclude that $DK \subseteq H$.

Since K has been chosen arbitrarily among the irreducible submodules of H, we conclude that $DH \subseteq H$. (Recall that H is homogeneous. In particular, H is the sum of its irreducible submodules.) This means that H is an ideal of D.

Since H is an ideal of D different from $\{0\}$, H contains a minimal ideal C of D. We shall prove that $C = H$, and in order to see that we assume, by way of contradiction, that $C \neq H$. From $C \neq H$ we obtain an irreducible submodule K of H such that $\{0\} = K \cap C$. It follows that $\{0\} = KC$.

Let L be an irreducible submodule of C. Then, as H is homogeneous, there exists a bijective D-homomorphism ϕ from K to L. Thus, $LC = (K\phi)C = (KC)\phi$. Thus, as $\{0\} = KC$, $\{0\} = LC$.

Since L has been chosen arbitrarily, we have shown that $\{0\} = C^2$.

Since the D-module D is assumed to be semisimple, D possesses a submodule K such that $K \oplus C = D$. Then $\{0\} = KC$. Thus, as D is assumed to be a ring with 1 and $\{0\} = C^2$,

$$C = DC = (K + C)C = \{0\},$$

contrary to the choice of C as minimal ideal of D.

This contradiction shows that $C = H$, so that H is a minimal ideal of D.

The following two theorems are the main theorems about semisimple associative rings with 1. They are due to Emil Artin; cf. [2]. Less general versions have been given earlier by Joseph Wedderburn; cf. [39; Theorem 10] and [39; Theorem 17].

Theorem 8.5.3 *Assume D to be semisimple, and let $\mathcal{H}$ denote the set of all maximal homogeneous submodules of the D-module D. Then we have the following.*

(i) *The set $\mathcal{H}$ is finite.*

(ii) *The D-module D is the direct sum of the elements in $\mathcal{H}$.*

(iii) *For any two elements K and L in $\mathcal{H}$, we have $\{0\} = KL$.*

(iv) *Each element in $\mathcal{H}$ is a direct sum of finitely many of its (pairwise isomorphic) irreducible submodules.*

PROOF. (i) The D-module D is sum of the elements in $\mathcal{H}$. Thus, by Lemma 8.5.1, $\mathcal{H}$ is a finite set.

(ii) This is an application of Theorem 8.3.6(i) to the D-module D.

(iii) Let K and L be elements in $\mathcal{H}$. According to Proposition 8.5.2, K is an ideal of D. Thus, $KL \subseteq K$. Similarly, as L is an ideal of D, $KL \subseteq L$. However, we know from (ii) that $\{0\} = K \cap L$.

(iv) Let H be an element in $\mathcal{H}$. Then H is a submodule of the D-module D.

Since D is assumed to be semisimple, the D-module D is completely reducible. Thus, by Lemma 8.3.2, H is completely reducible. Thus, by Proposition 8.3.4, H is a direct sum of irreducible submodules. Now, considering that H is a ring with 1, the claim follows from Lemma 8.5.1.

Theorem 8.5.4 *Assume D to be semisimple, let H be a maximal homogeneous submodule of the D-module D, let M be an irreducible submodule of the D-module D contained in H, and set $E := \mathrm{End}_D(M)$. Then we have the following.*

(i) *We have $H \cong \mathrm{End}_E(M)$.*

(ii) *Let C be an algebraically closed subfield of $Z(E)$ such that E is a finitely generated vector space over C. Then $C = E$ and $H \cong \mathrm{End}_C(M)$.*

PROOF. (i) Let C be an ideal of H. For each element K in $\mathcal{H} \setminus \{H\}$, we have $\{0\} = HK$. Thus, C is an ideal of D. Thus, by Proposition 8.5.2, $\{0\} = C$ or $C = H$.

This proves that H is a simple ring. That H is artinian follows from Theorem 8.5.3(iv). Thus, the claim follows from Theorem 8.4.8.

(ii) Applying Lemma 8.2.6 to E in place of D, we obtain $C = E$. Thus, the claim follows from (i).

Corollary 8.5.5 *Assume D to be semisimple, and let C be an algebraically closed subfield of $Z(D)$. The ring D is commutative if and only if, for each irreducible D-module M of D, $\dim_C(M) = 1$.*

PROOF. This follows from Theorem 8.5.3(iii) together with Theorem 8.5.4(ii).

Theorem 8.5.3 together with Theorem 8.5.4 gives a complete picture about semisimple rings. The following theorem gives a complete picture about the collection of all modules over semisimple rings.

Theorem 8.5.6 *If D is semisimple, we have the following.*

(i) *Each D-module is completely reducible.*

(ii) *Each irreducible D-module is isomorphic to a submodule of the D-module D.*

PROOF. (i) Let M be a D-module, and let m be an element in M.

Since D is assumed to be a ring with 1, there exists a finite set $\mathcal{K}$ of irreducible submodules of the D-module D such that 1 is element of the sum C of the elements of $\mathcal{K}$. It follows that $m = m1 \in mC$.

Note that mC is the sum of the submodules mK with $K \in \mathcal{K}$. We shall now see that, for each element K in $\mathcal{K}$, mK is contained in an irreducible submodule of M.

Let K be an element in $\mathcal{K}$. Then $mK = Km_D$ is a submodule of M. Thus, as K is irreducible, we must have $mK \cong K$ or $mK = \{0\}$; cf. Theorem 8.1.1.

We have seen that m is element of a sum of irreducible submodules of M. Thus, as m has been chosen arbitrarily, we have shown that M is sum of irreducible submodules of M. Thus, the claim follows from Proposition 8.3.4.

(ii) Let M be an irreducible D-module, and let m be an element in $M \setminus \{0\}$. Then, by Lemma 8.4.1,

$$D/A_D(m) \cong M.$$

Since D is assumed to be semisimple, there exists a submodule K of the D-module D such that $K \oplus A_D(m) = D$. Thus, by Theorem 8.1.2,

$$D/A_D(m) \cong K.$$

It follows that $M \cong K$.

8.6 Characters of Associative Rings with 1

Characters of associative rings D with 1 arise from D-modules M when D contains a subfield C in its center such that M is a finitely generated vector space over C. In this section, we shall look at this situation.

Throughout this section, the letter D stands for an associative ring with 1. The letter C stands for a subfield of D with $C \subseteq Z(D)$.

Let M be a D-module. Then, as $C \subseteq D$, M is a vector space over C. We assume that M is a finitely generated vector space over C.

Recall that, for each element d in D, d_M is our notation for the map from M to M which maps each element m in M to md.

We saw earlier that, for each element d in D, $d_M \in \mathrm{End}(M)$. Since $C \subseteq Z(D)$, we now also have that, for any three elements m in M, c in C, and d in D,

$$(mc)d_M = (mc)d = m(cd) = m(dc) = (md)c = (md_M)c.$$

Thus, for each element d in D, $d_M \in \mathrm{End}_C(M)$.

For each element d in D, we shall denote by $\mathrm{tr}(d_M)$ the trace of the C-endomorphism d_M, and we set

$$\chi_M(d) := \mathrm{tr}(d_M).$$

Then χ_M is a homomorphism from D to C. This map is called the C-*character* of D *afforded by* M. If there is no danger of ambiguity, we shall simply speak about characters instead of C-characters.

The integer $\chi_M(\sigma_1)$ ($= \dim_C(M)$) is called the *degree* of χ_M. Character of degree 1 are called *linear.*

Note that the sum of two characters is a character.

Theorem 8.6.1 *Let M and $\tilde{M}$ be D-modules finitely generated as vector spaces over C. Then, if $M \cong \tilde{M}$, $\chi_M = \chi_{\tilde{M}}$.*

PROOF. Let d be an element in D, and let B be a basis of M. Then, for any two elements k and l in B, there exists an element c_{kl} in C such that, for each element k in B,

$$kd = \sum_{l \in H} l c_{kl}.$$

In particular,

$$\chi_M(d) = \sum_{b \in B} c_{bb}.$$

On the other hand, since M and $\tilde{M}$ are assumed to be isomorphic, there exists a bijective D-homomorphism ϕ from M to $\tilde{M}$. In particular, $B\phi$ is a basis of $\tilde{M}$, and, for each element k in B,

$$(k\phi)d = (kd)\phi = (\sum_{l \in H} l c_{kl})\phi = \sum_{l \in H} (l c_{kl})\phi = \sum_{l \in H} (l\phi) c_{kl}.$$

It follows that

$$\chi_{\tilde{M}}(d) = \sum_{b \in B} c_{bb}.$$

Thus, as d has been chosen arbitrarily in D, we have shown that $\chi_M = \chi_{\tilde{M}}$.

A character of D afforded by an irreducible D-module is called *irreducible.*

Let us denote by $\mathrm{Irr}(D)$ the set of all irreducible characters of D.

Looking at Theorem 8.5.3(i), (iv) and Theorem 8.5.6(i) one has a complete picture about the set of all irreducible modules over a semisimple ring. Thus, if all these irreducible modules are finitely generated vector spaces over C, one can expect a clear picture about the set of all irreducible characters of a semisimple ring, too.

For the remainder of this section, we shall now assume D to be semisimple and to be finitely generated as a vector space over C.

Assuming D to be semisimple we obtain from Theorem 8.5.6(i) and Proposition 8.3.4 that each character of D is the sum of irreducible characters of D. Let us, therefore, look at the set of all irreducible characters of D.

As earlier (in Section 8.3 and Section 8.5), we shall denote by $\mathcal{H}$ the set of the maximal homogeneous submodules of the D-module D.

For each element H in $\mathcal{H}$, D possesses an irreducible character ψ_H such that, for each irreducible submodule K of H, $\chi_K = \psi_H$; cf. Lemma 8.3.5 together with Theorem 8.6.1.

Theorem 8.6.2 *The following hold.*

(i) *We have* $\{\psi_H \mid H \in \mathcal{H}\} = \mathrm{Irr}(D)$.

(ii) *For any two different elements K and L in $\mathcal{H}$, we have* $\psi_K(L) = \{0\}$.

PROOF. (i) Let χ be an element in $\mathrm{Irr}(D)$. Then, by definition, there exists an irreducible D-module M such that $\chi = \chi_M$.

By Theorem 8.5.6(ii), there exists an irreducible submodule K of the D-module D such that $M \cong K$. From $M \cong K$ we obtain $\chi_M = \chi_K$; cf. Theorem 8.6.1.

By Theorem 8.3.6(ii) there exists exactly one element H in $\mathcal{H}$ such that $K \subseteq H$. Since K is a submodule of H, $\chi_K = \psi_H$.

It follows that $\chi = \psi_H$.

(ii) Let K and L be elements in $\mathcal{H}$. Then, by Theorem 8.5.3(iii), $\{0\} = KL$.

Let l be an element in L, and let J be an irreducible submodule of K.

Since $J \subseteq K$, $\chi_J = \psi_K$. From $J \subseteq K$ we also obtain $Jl_J = Jl = 0$. Thus, $\mathrm{tr}(l_J) = 0$, so that, by definition, $\chi_J(l) = 0$. Thus, as $\chi_J = \psi_K$, $\psi_K(l) = 0$.

For each element H in $\mathcal{H}$, there exists a uniquely determined element 1_H in H such that

$$1 = \sum_{H \in \mathcal{H}} 1_H;$$

see Theorem 8.5.3(ii). Note that 1_H is the multiplicative neutral element of the ring H.

Lemma 8.6.3 *For any two elements H in $\mathcal{H}$ and d in D, we have* $\psi_H(d) = \psi_H(1_H d)$.

PROOF. Using Theorem 8.6.2(ii) for the last equation we obtain

$$\psi_H(d) = \psi_H(1d) = \psi_H(\sum_{K \in \mathcal{H}} 1_K d) = \sum_{K \in \mathcal{H}} \psi_H(1_K d) = \psi_H(1_H d);$$

note that, for each element K in $\mathcal{K}$, $1_K d \in K$.

Theorem 8.6.2(i) says that the map

$$\Psi : \mathcal{H} \to \mathrm{Irr}(D), \ H \mapsto \psi_H$$

is surjective. In the first part of the following theorem, we shall see that Ψ is bijective if the characteristic of C does not divide any of the degrees of the irreducible characters of D.

Theorem 8.6.4 *Assume that, for each irreducible character χ of D, the characteristic of C does not divide $\chi(\sigma_1)$. Then we have the following.*

(i) *The set* $\mathrm{Irr}(D)$ *is a linearly independent subset of* $\mathrm{Hom}_C(D, C)$.

(ii) *We have* $|\mathcal{H}| = |\mathrm{Irr}(D)|$.

(iii) *If C is algebraically closed,*

$$\chi_D = \sum_{\chi \in \mathrm{Irr}(D)} \chi(\sigma_1)\chi.$$

Proof. (i) From Lemma 8.6.3 we know that, for each element H in $\mathcal{H}$, $\psi_H(1) = \psi_H(1_H)$. From Theorem 8.6.2(ii) we know that, for any two different elements K and L in $\mathcal{H}$, $\psi_K(1_L) = 0$. Thus, the claims are consequences of Theorem 8.6.2(i).

(ii) This follows from (i) together with Theorem 8.6.2(i).

(iii) Assume C to be algebraically closed, and let H be an element in $\mathcal{H}$. Since $1 \in C$, $1_H C$ is a subfield of H, so that we may apply Theorem 8.5.4(ii).

Let M be an irreducible H-module. Then, by Theorem 8.5.4(ii), $H \cong \mathrm{End}_C(M)$. Thus, H is the direct sum of $\chi(\sigma_1)$ copies of M, so that the desired equation follows from (i).

Corollary 8.6.5 *Assume that, for each irreducible character χ of D, the characteristic of C does not divide $\chi(\sigma_1)$. Then we have*

$$\dim_C(D) = \sum_{\chi \in \mathrm{Irr}(D)} \chi(\sigma_1)^2.$$

Proof. This follows immediately from Theorem 8.6.4(iii).

8.7 Roots of Unity in Integral Domains

A commutative associative ring D with 1 is called an *integral domain* if the product of any two elements in $D \setminus \{0\}$ is in $D \setminus \{0\}$.

Throughout this section, the letter D stands for an integral domain.

Lemma 8.7.1 *Let d be an element in $D \setminus \{1\}$, and let n be a positive integer such that $d^n = 1$. Then we have*

$$\sum_{i=0}^{n-1} d^i = 0.$$

PROOF. We have

$$\sum_{i=0}^{n-1} d^i(d-1) = d^n - 1.$$

Thus, our claim follows from the hypotheses that $d - 1 \neq 0$ and $d^n - 1 = 0$.

Lemma 8.7.2 *Let d be an element in $D \setminus \{1\}$, and let n be a positive integer such that $d^{2n+1} = 1$. Then we have*

$$\sum_{i=0}^{n} (d^i + d^{-i})^2 = 2n + 3$$

and

$$\sum_{i=0}^{n} (d^{i+1} - d^{-(i+1)} + d^i - d^{-i})^2 = -(2n+1)(d + d^{-1} + 2).$$

PROOF. We are assuming that $d^{2n+1} = 1$. Thus, by Lemma 8.7.1,

$$\sum_{i=0}^{2n} d^i = 0.$$

Since we are assuming that $d^{2n+1} = 1$, this implies

$$\sum_{i=0}^{n} (d^{2i} + d^{-2i}) = \sum_{i=0}^{n} (d^{2i} + d^{2(n-i)+1}) = \sum_{i=0}^{n} (d^{2i} + d^{2i+1}) = 1.$$

It follows that

$$\sum_{i=0}^{n} (d^i + d^{-i})^2 = \sum_{i=0}^{n} (d^{2i} + 2 + d^{-2i}) = 2(n+1) + 1,$$

and that is our first claim.

From

$$\sum_{i=0}^{n}(d^{2i} + d^{-2i}) = 1$$

we also obtain

$$\sum_{i=0}^{n}(d^{2i+1} + d^{-(2i+1)}) = 1$$

and

$$\sum_{i=0}^{n}(d^{2(i+1)} + d^{-2(i+1)}) = d + d^{-1} - 1.$$

Thus, as $(d^{i+1} - d^{-(i+1)} + d^i - d^{-i})^2$ is equal to

$$(d^{2(i+1)} + d^{-2(i+1)}) + (d^{2i} + d^{-2i}) + 2(d^{2i+1} + d^{-(2i+1)}) - 2(d + d^{-1} + 2),$$

we conclude that

$$\sum_{i=0}^{n}(d^{i+1} - d^{-(i+1)} + d^i - d^{-i})^2 = -(2n+1)(d + d^{-1} + 2),$$

and that is our second claim.

Lemma 8.7.3 *Let d be an element in $D \setminus \{-1, 1\}$, and let n be a positive integer such that $d^{2n} = 1$. Then we have*

$$\sum_{i=0}^{n-1}(d^{i+1} - d^{-(i+1)})^2 + (d^i - d^{-i})^2 = -4n$$

and

$$\sum_{i=0}^{n-1}(d^{i+1} - d^{-(i+1)})(d^i - d^{-i}) = -n(d + d^{-1}).$$

PROOF. We are assuming that $d^{2n} = 1$. Thus, we obtain from Lemma 8.7.1 that

$$\sum_{i=0}^{n-1} d^{2i}(d+1) = \sum_{i=0}^{2n-1} d^i = 0.$$

Thus, as $d + 1 \neq 0$ and D is assumed to be an integral domain,

$$\sum_{i=0}^{n-1} d^{2i} = 0.$$

From this we obtain

$$\sum_{i=0}^{n-1}(d^i - d^{-i})^2 = \sum_{i=0}^{n-1} d^{2i} - 2 + d^{-2i} = -2n.$$

Multiplying

$$\sum_{i=0}^{n-1} d^{2i} = 0$$

by d^2 we obtain

$$\sum_{i=0}^{n-1} d^{2(i+1)} = 0.$$

Thus,

$$\sum_{i=0}^{n-1} (d^{i+1} - d^{-(i+1)})^2 = \sum_{i=0}^{n-1} d^{2(i+1)} - 2 + d^{-2(i+1)} = -2n,$$

so that

$$\sum_{i=0}^{n-1} (d^{i+1} - d^{-(i+1)})^2 + (d^i - d^{-i})^2 = -4n,$$

and that is our first claim.

Multiplying

$$\sum_{i=0}^{n-1} d^{2i} = 0$$

by d we obtain

$$\sum_{i=0}^{n-1} d^{2i+1} = 0.$$

Therefore, we also obtain

$$\sum_{i=0}^{n-1} (d^{i+1} - d^{-(i+1)})(d^i - d^{-i}) = \sum_{i=0}^{n-1} (d^{2i+1} - d - d^{-1} + d^{-(2i+1)}) = -n(d + d^{-1}),$$

and that is our second claim.

Lemma 8.7.4 *Assume that D has characteristic 0, and let us denote by Z the smallest unitary subring of D. Let d be an element in D for which there exists a positive integer n with $d^n = 1$. Then, if $d + d^{-1} \in Z$, $d^4 = 1$ or $d^6 = 1$.*

PROOF. For each integer z, we have

$$(d^z + d^{-z})(d + d^{-1}) = (d^{z+1} + d^{-(z+1)}) + (d^{z-1} + d^{-(z-1)}).$$

Thus, as we are assuming that $d + d^{-1} \in Z$, induction yields that, for each integer z, $d^z + d^{-z} \in Z$. (Note that $d^0 + d^{-0} \in Z$.)

We are assuming that there exists a positive integer n with $d^n = 1$. Thus, the set of all sums $d^z + d^{-z}$ with integral exponent z is finite. Let m be an integer such that $d^m + d^{-m}$ is maximal (with respect to the natural ordering of Z).

Since $d^n = 1$, $d^{-n} = 1$. Thus, $d^n + d^{-n} = 2$. Thus, $2 \leq d^m + d^{-m}$.

From

$$(d+d^{-1})^2(d^m+d^{-m}) = (d^{m+2}+d^{-(m+2)})+(d^{m-2}+d^{-(m-2)})+2(d^m+d^{-m}),$$

we also obtain

$$(d+d^{-1})^2(d^m+d^{-m}) \leq 4(d^m+d^{-m}).$$

Thus, as $d^m + d^{-m}$ is positive, we obtain $(d+d^{-1})^2 \leq 4$.

Assume that $(d+d^{-1})^2 = 4$. Then $d^2 - 2 + d^{-2} = 0$. It follows that $(d^2-1)^2 = d^4 - 2d^2 + 1 = 0$. Thus, as D is assumed to be an integral domain, $d^2 - 1 = 0$, and that means that $d^2 = 1$.

Assume that $(d+d^{-1})^2 = 1$. Then $d^2 + 1 + d^{-2} = 0$. It follows that $d^6 - 1 = (d^4 + d^2 + 1)(d^2 - 1) = 0$, whence $d^6 = 1$.

Assume that $d + d^{-1} = 0$. Then $d^2 + 1 = 0$. Thus, $d^4 - 1 = (d^2+1)(d^2-1) = 0$, whence $d^4 = 1$.

9

Scheme Rings

In this chapter, S is assumed to have finite valency. With the help of S we shall define, for each field C, an associative ring with 1. We shall denote this ring by CS and call it the scheme ring of S over C.

In the first section of this chapter, we shall prove that CS is semisimple if the characteristic of the field C does not divide any of the integers $|s^*|$ with $s \in S$. This enables us to refer to some of the results about semisimple rings which we obtained in Section 8.5. We shall also see that $C \subseteq Z(CS)$, so that we may speak about characters of CS and refer to results about characters which we obtained in Section 8.6. The results of the first section include the orthogonality relations for fields the characteristic of which does not divide any of the integers $|s^*|$ with $s \in S$.

In the second section, we derive the Schur relations in the case where the characteristic of the underlying base field does not divide any of the integers $|s^*|$ with $s \in S$ and is algebraically closed.

In Section 9.3, we assume the underlying base field C to be the field of the complex numbers. We establish a relationship between closed subsets of S and characters of CS.

In Section 9.4, we obtain a few general results about closed subsets.

In Section 9.5, we shall apply the orthogonality relations in order to investigate the case where $|S| \leq 5$.

In the last of the six sections of this chapter, we shall briefly look at scheme rings of S under the assumption that S is generated by a constrained set of involutions.

9.1 Basic Facts

In this section, the letter C stands for a field.

Let us fix, for each element x in X, an element c_x in C. We write

$$\sum_{x \in X} c_x x$$

to denote the map from X to C which maps each element x in X to c_x.

The set CX of all maps from X to C is a vector space over C with respect to componentwise addition and componentwise multiplication with elements of C.

Each element x in X can be identified with the map from X to C which maps x to 1 and each element different from x to 0. Thus, X can be viewed as a subset of CX.

Recall that S is assumed to have finite valency. Thus, X is finite, so that X is a basis of the vector space CX.

Let s be an element in S. Then, as X is a basis of CX, there exists a uniquely defined element σ_s in $\mathrm{End}_C(CX)$ such that, for each element x in X,

$$x\sigma_s = \sum_{y \in xs} y.$$

For each nonempty subset R of S, we define CR to be the set of all finite sums of products $c\sigma_r$ with $c \in C$ and $r \in R$.

Note that, for each nonempty subset R of S, CR is a vector space over C with respect to componentwise addition and componentwise multiplication with elements of C. The set $\{\sigma_r \mid r \in R\}$ is a basis of CR.

Lemma 9.1.1 *The following statements hold.*

(i) *For any two elements p and q in S, we have*

$$\sigma_p\sigma_q = \sum_{s \in S} a_{pqs}\sigma_s.$$

(ii) *Let n be an integer with $2 \leq n$, and let $r_1, \ldots, r_n$ be elements in S. Then*

$$\sigma_{r_1} \cdots \sigma_{r_n} = \sum_{s \in S} a_{r_1 \ldots r_n s}\sigma_s.$$

PROOF. (i) For each element x in X, we have

$$x(\sigma_p\sigma_q) = (x\sigma_p)\sigma_q = (\sum_{y \in xp} y)\sigma_q = \sum_{y \in xp} y\sigma_q = \sum_{y \in xp}\sum_{z \in yq} z$$

and

$$x(\sum_{s \in S} a_{pqs}\sigma_s) = \sum_{s \in S} a_{pqs}x\sigma_s = \sum_{s \in S} a_{pqs}\sum_{z \in xs} z = \sum_{s \in S}\sum_{z \in xs} a_{pqs}z.$$

This proves the desired equation.

(ii) Assume that $3 \leq n$. Referring to (i) induction yields

$$\begin{aligned}\sigma_{r_1} \cdots \sigma_{r_n} &= (\sum_{q \in S} a_{r_1 \ldots r_{n-1} q} \sigma_q) \sigma_{r_n} = \sum_{q \in S} a_{r_1 \ldots r_{n-1} q} (\sigma_q \sigma_{r_n}) \\ &= \sum_{q \in S} a_{r_1 \ldots r_{n-1} q} \sum_{s \in S} a_{q r_n s} \sigma_s = \sum_{s \in S} (\sum_{q \in S} a_{r_1 \ldots r_{n-1} q} a_{q r_n s}) \sigma_s \\ &= \sum_{s \in S} a_{r_1 \ldots r_n s} \sigma_s,\end{aligned}$$

and that proves the claim.

From Lemma 9.1.1(i) we obtain that CS is a subring of $\mathrm{End}_C(CX)$.

The associative ring CS is a ring with 1. We call CS the *scheme ring* of S *over* C. The field C is called the *base field* of CS.

Since CS is a subring of $\mathrm{End}_C(CX)$, CX is a CS-module. We call CX the *standard module* of CS.

Since $1 \in S$, $\sigma_1 \in CS$. Thus, the elements of C can be identified with the multiples of σ_1. Thus, C can be viewed as a subfield of $Z(CS)$, the center of CS.

The fact that $C \subseteq Z(CS)$ enables us to define characters for each CS-module which is finitely generated over C.

Recall that the standard module of CS is finitely generated over C. The character of CS afforded by the standard module is called the *standard character* of CS. We shall denote the standard character of CS by χ_{CX}.

In the following lemma, we collect information about the standard character.

Lemma 9.1.2 *The following statements hold.*

(i) *We have* $\chi_{CX}(\sigma_1) = n_S$.

(ii) *For each element* s *in* $S \setminus \{1\}$, *we have* $\chi_{CX}(\sigma_s) = 0$.

(iii) *For any two elements* p *and* q *in* S, $\chi_{CX}(\sigma_{p^*}\sigma_q) = \delta_{pq}|p^*|$.

(iv) *For each element* s *in* S, *let* c_s *be an element in* C. *Set*

$$\sigma := \sum_{s \in S} c_s \sigma_s.$$

Then, for each element s *in* S, $\chi_{CX}(\sigma_{s^*}\sigma) = c_s|s^*|$.

PROOF. (i) This equation follows immediately from the definition of χ_{CX}.

(ii) These equations follow immediately from the definition of χ_{CX}.

(iii) Let p and q be elements in S. Referring to Lemma 9.1.1(i) we obtain from (i) and (ii) that

$$\chi_{CX}(\sigma_{p^*}\sigma_q) = \chi_{CX}(\sum_{s\in S} a_{p^*qs}\sigma_s) = \sum_{s\in S} a_{p^*qs}\chi_{CX}(\sigma_s) = a_{p^*q1}n_S.$$

Thus, the claim follows from the definition of n_{s^*} together with Lemma 1.1.2(ii) and Lemma 1.1.3(i).

(iv) This follows from (iii) because CS is distributive.

The standard module possesses an irreducible submodule which induces a character of CS all values of which can be computed explicitly. In order to introduce this module we (temporarily) set

$$j := \sum_{x\in X} x.$$

Note that, for each element s in S,

$$j\sigma_s = \sum_{x\in X}\sum_{y\in xs} y = n_{s^*}\sum_{x\in X} x = n_{s^*}j.$$

Thus, Cj is a submodule of the CS-module CX.[1]

We call Cj the *principal module* of CS. The character afforded by the principal module is called the *principal character* of CS. Instead of χ_{Cj} we shall write 1_{CS}.

The above equation tells us that, for each element s in S,

$$1_{CS}(\sigma_s) = n_{s^*}.$$

Having introduced the standard module and the principal module of CS we shall now introduce a third module of CS, the CS-module CS. This module is defined by the multiplication of CS. The CS-module CS is usually called the *regular module* of CS.

Let M be a CS-module. For each element ϕ in $\mathrm{End}_C(M)$, we define

$$\phi\zeta_M := \sum_{s\in S}\frac{1}{n_s}\sigma_{s^*}\phi\sigma_s.$$

It is obvious that ζ_M is a C-module endomorphism of $\mathrm{End}_C(M)$ to $\mathrm{End}_C(M)$.

The following lemma says that, in fact, the image of ζ_M is a subset of $\mathrm{End}_{CS}(M)$.

Lemma 9.1.3 *Let M be a CS-module, and let ϕ be an element in* $\mathrm{End}_C(M)$. *Then $\phi\zeta_M \in$* $\mathrm{End}_{CS}(M)$.

Proof. For any two elements s in S and ϕ in $\mathrm{End}_C(M)$, we have

[1] Recall that Cj denotes the set of all elements cj with $c \in C$.

$$\sigma_s(\phi\zeta_M) = \sigma_s \sum_{p\in S} \frac{1}{n_p}\sigma_{p^*}\phi\sigma_p = \sum_{p\in S}\frac{1}{n_p}\sigma_s\sigma_{p^*}\phi\sigma_p$$
$$= \sum_{p\in S}\frac{1}{n_p}\sum_{q\in S} a_{sp^*q}\sigma_q\phi\sigma_p = \sum_{q\in S}\sum_{p\in S}\frac{1}{n_p}a_{ps^*q^*}\sigma_q\phi\sigma_p$$
$$= \sum_{q\in S}\sum_{p\in S}\frac{1}{n_{q^*}}a_{q^*sp}\sigma_q\phi\sigma_p = \sum_{q\in S}\frac{1}{n_{q^*}}\sigma_q\phi\sum_{p\in S}a_{q^*sp}\sigma_p$$
$$= \sum_{q\in S}\frac{1}{n_{q^*}}\sigma_q\phi\sigma_{q^*}\sigma_s = (\phi\zeta_M)\sigma_s.$$

(The third and the seventh equations follow from Lemma 9.1.1(i), the fourth from Lemma 1.1.1(ii), and the fifth from Lemma 1.1.3(ii).)

The second part of the following theorem gives the transformation matrix of ζ_{CS} with respect to the basis $\{\sigma_s \mid s \in S\}$ of CS. In Lemma 1.1.6, this matrix has been proven to be 'stochastic' in the sense that each of its column sums is equal to n_S.

Theorem 9.1.4 *Writing ζ instead of ζ_{CS} we have the following.*

(i) *We have $(CS)\zeta \subseteq Z(CS)$.*

(ii) *For each element p in S, we have*

$$\sigma_p\zeta = \sum_{q\in S}(\sum_{s\in S}\frac{a_{s^*psq}}{n_s})\sigma_q.$$

(iii) *We have $\sigma_1\zeta \in CO^\vartheta(S)$.*

(iv) *There exists an element γ in $\mathrm{End}_C(CS)$ such that $\gamma\zeta = \sigma_1$.*

PROOF. (i) This is an application of Lemma 9.1.3 to the CS-module CS.

(ii) Let p be an element in S. Then, by Lemma 9.1.1(ii),

$$\sigma_p\zeta = \sum_{s\in S}\frac{1}{n_s}\sigma_{s^*}\sigma_p\sigma_s = \sum_{s\in S}\frac{1}{n_s}\sum_{q\in S}a_{s^*psq}\sigma_q = \sum_{q\in S}(\sum_{s\in S}\frac{a_{s^*psq}}{n_s})\sigma_q,$$

and that proves the claim.

(iii) From the second equation of Lemma 1.1.1(i) we obtain $a_{s^*1sq} = a_{s^*sq}$ for any two elements q and s in S. Thus, by (ii),

$$\sigma_1\zeta = \sum_{q\in S}(\sum_{s\in S}\frac{a_{s^*sq}}{n_s})\sigma_q.$$

For each element q in S, we define

$$c_q := \sum_{s\in S}\frac{a_{s^*sq}}{n_s}.$$

Then

$$\sigma_1\zeta = \sum_{q\in S} c_q\sigma_q.$$

Let q be an element in S such that $c_q \neq 0$. Then there exists an element s in S such that $1 \leq a_{s^*sq}$. This means that $q \in s^*s$. Thus, by Theorem 3.2.1(ii), $q \in O^\vartheta(S)$.

Since $q \in S$ has been chosen arbitrarily with $0 \neq c_q$, we conclude that $1\zeta \in CO^\vartheta(S)$.

(iv) Let us denote by γ the uniquely determined element of $\mathrm{End}_C(CS)$ which, for each element s in S, satisfies

$$\sigma_s\gamma = \delta_{1s}\sigma_s.$$

Then, for each element q in S, we have

$$\begin{aligned}\sigma_q(\gamma\zeta) &= \sigma_q\sum_{s\in S}\frac{1}{n_s}\sigma_{s^*}\gamma\sigma_s = \sum_{s\in S}\frac{1}{n_s}\sigma_q\sigma_{s^*}\gamma\sigma_s \\ &= \sum_{s\in S}\frac{1}{n_s}\sum_{p\in S}a_{qs^*p}\sigma_p\gamma\sigma_s = \sum_{s\in S}\frac{a_{qs^*1}}{n_s}\sigma_s = \sigma_q.\end{aligned}$$

(The last equation follows from Lemma 1.1.3(i).)

Let us now look at the Jacobson radical $J(CS)$ of the scheme ring CS.

The third part of the following theorem was first proved by Heinrich Maschke; cf. [30].

Theorem 9.1.5 *We have the following.*

(i) *Let us denote by R the set of all elements s in S such that the characteristic of C divides $|s^*|$. Then, $J(CS) \subseteq CR$.*

(ii) *Assume that, for each element s in S, the characteristic of C does not divide $|s^*|$. Then CS is semisimple.*

(iii) *Assume S to be thin. If the characteristic of C does not divide the valency of S, CS is semisimple.*

Proof. (i) Let σ be an element in $J(CS)$. Then $\sigma \in CS$. Thus, there exists, for each element s in S, an element c_s in C such that

$$\sigma = \sum_{s\in S} c_s\sigma_s.$$

Let s be an element in S such that $c_s \neq 0$. We shall be done if we succeed in showing that the characteristic of C divides $|s^*|$.

Let

$$\{0\} = V_0 \subseteq V_1 \subseteq \ldots \subseteq V_n = CX$$

be a series of submodules of CX such that, for each element i in $\{1, \ldots, n\}$, V_i/V_{i-1} is irreducible. Then, for each element i in $\{1, \ldots, n\}$, $V_i \sigma_{s^*} \sigma \subseteq V_{i-1}$; cf. Lemma 8.4.2. Therefore, $\chi_{CX}(\sigma_{s^*}\sigma) = 0$.

On the other hand, by Lemma 9.1.2(iv), $\chi_{CX}(\sigma_{s^*}\sigma) = c_s|s^*|$.

From $\chi_{CX}(\sigma_{s^*}\sigma) = 0$ and $\chi_{CX}(\sigma_{s^*}\sigma) = c_s|s^*|$ we obtain $c_s|s^*| = 0$. Thus, as $c_s \neq 0$, $|s^*| = 0$. This means that the characteristic of C divides $|s^*|$.

(ii) This follows from (i) and Theorem 8.3.7(ii).

(iii) Referring to Lemma 1.5.1 this follows from (ii).

For the remainder of this section, we assume that, for each element s in S, the characteristic of C does not divide $|s^*|$. According to Theorem 9.1.5(ii), this implies that CS is semisimple so that we may refer to results obtained in Section 8.5 and Section 8.6.

Let χ be an irreducible character of CS. Since CS is semisimple, we obtain from Theorem 8.6.2(i) together with Theorem 8.6.4(ii) that there exists exactly one maximal homogeneous submodule H_χ of the CS-module CS such that $\chi = \psi_{H_\chi}$. Set

$$\epsilon_\chi := 1_{H_\chi}.$$

Then

$$1 = \sum_{\chi \in \mathrm{Irr}(CS)} \epsilon_\chi.$$

From Theorem 8.5.6(i) we know that there exists, for each irreducible character χ of CS, a non-negative integer m_χ such that

$$\chi_{CX} = \sum_{\chi \in \mathrm{Irr}(CS)} m_\chi \chi.$$

For each irreducible character χ of CS, the element m_χ is called the *multiplicity* of χ.

Lemma 9.1.6 *Let χ be an irreducible character of CS. Then the following hold.*

(i) *For each element σ in H_χ, we have*

$$\sigma = \frac{m_\chi}{n_S} \sum_{s \in S} \frac{\chi(\sigma_{s^*}\sigma)}{n_{s^*}} \sigma_s.$$

(ii) *We have*

$$\epsilon_\chi = \frac{m_\chi}{n_S} \sum_{s \in S} \frac{\chi(\sigma_{s^*})}{n_{s^*}} \sigma_s.$$

(iii) *The characteristic of C does not divide m_χ.*

PROOF. (i) Let σ be an element in H_χ. Then $\sigma \in CS$. Thus, there exists, for each element s in S, an element c_s in C such that

$$\sigma = \sum_{s \in S} c_s \sigma_s.$$

Let s be an element in S. Then, as $\sigma \in H_\chi$, $\sigma_{s^*}\sigma \in H_\chi$; cf. Proposition 8.5.2. Therefore, by Theorem 8.6.2(ii), $\chi_{CX}(\sigma_{s^*}\sigma) = m_\chi \chi(\sigma_{s^*}\sigma)$.
On the other hand, by Lemma 9.1.2(iv), $\chi_{CX}(\sigma_{s^*}\sigma) = c_s|s^*|$. Therefore,

$$c_s = \frac{m_\chi \chi(\sigma_{s^*}\sigma)}{|s^*|}.$$

Thus, the claim follows from Lemma 1.1.2(ii).

(ii) This follows from (i) and Lemma 8.6.3.

(iii) This follows from (ii), because $\epsilon_\chi \neq 0$.

The equations in the following theorem are called the *orthogonality relations* for schemes with finite valency.

Theorem 9.1.7 *For any two irreducible characters ϕ and ψ of CS, the following hold.*

(i) *For each element s in S, we have*

$$\frac{1}{n_S} \sum_{p \in S} \sum_{q \in S} \frac{a_{s^*pq}}{n_{p^*}} \phi(\sigma_{p^*})\psi(\sigma_q) = \delta_{\phi\psi} \frac{\phi(\sigma_{s^*})}{m_\phi}.$$

(ii) *We have*

$$\frac{1}{n_S} \sum_{s \in S} \frac{1}{n_{s^*}} \phi(\sigma_{s^*})\psi(\sigma_s) = \delta_{\phi\psi} \frac{\phi(\sigma_1)}{m_\phi}.$$

PROOF. (i) From Lemma 9.1.6(ii) and Lemma 9.1.1(i) we obtain

$$\begin{aligned}
\epsilon_\phi \epsilon_\psi &= \frac{m_\phi}{n_S} \Big(\sum_{p \in S} \frac{\phi(\sigma_{p^*})}{n_{p^*}} \sigma_p\Big) \frac{m_\psi}{n_S} \Big(\sum_{q \in S} \frac{\psi(\sigma_{q^*})}{n_{q^*}} \sigma_q\Big) \\
&= \frac{m_\phi m_\psi}{n_S^2} \sum_{p \in S} \sum_{q \in S} \frac{\phi(\sigma_{p^*})\psi(\sigma_{q^*})}{n_{p^*} n_{q^*}} \sigma_p \sigma_q \\
&= \frac{m_\phi m_\psi}{n_S^2} \sum_{s \in S} \Big(\sum_{p \in S} \sum_{q \in S} \frac{\phi(\sigma_{p^*})\psi(\sigma_{q^*})}{n_{p^*} n_{q^*}} a_{pqs}\Big) \sigma_s.
\end{aligned}$$

On the other hand, as $\epsilon_\phi \in H_\phi$ and $\epsilon_\psi \in H_\psi$, $\epsilon_\phi \epsilon_\psi = \delta_{\phi\psi}\epsilon_\phi$. Referring once again to the equation

$$\epsilon_\phi = \frac{m_\phi}{n_S}\sum_{s\in S}\frac{\phi(\sigma_{s^*})}{n_{s^*}}\sigma_s$$

we, therefore, obtain

$$\frac{m_\phi m_\psi}{n_S}\sum_{p\in S}\sum_{q\in S}\frac{\phi(\sigma_{p^*})\psi(\sigma_{q^*})a_{pqs}}{n_{p^*}n_{q^*}} = \delta_{\phi\psi}\frac{m_\phi \phi(\sigma_{s^*})}{n_{s^*}}$$

for each element s in S. Recall that, by Lemma 9.1.6(iii), $1 \leq m_\phi$ and $1 \leq m_\psi$. Therefore we conclude that, for each element s in S,

$$\frac{1}{n_S}\sum_{p\in S}\sum_{q\in S}\frac{a_{pq^*s}n_{s^*}}{n_{p^*}n_q}\phi(\sigma_{p^*})\psi(\sigma_q) = \delta_{\phi\psi}\frac{\phi(\sigma_{s^*})}{m_\phi}.$$

Finally, recall that, by Lemma 1.1.1(ii) and Lemma 1.1.3(ii),

$$a_{pq^*s}n_{s^*} = a_{qp^*s^*}n_{s^*} = a_{s^*pq}n_q.$$

(ii) This follows from (i) by setting $s^* = 1$; use the first equation of Lemma 1.1.1(i).

Lemma 9.1.8 *The following statements hold.*

(i) *For each non-principal irreducible character χ of CS, we have*

$$\sum_{s\in S}\chi(\sigma_s) = 0.$$

(ii) *We have $m_{1_{CS}} = 1$.*

PROOF. (i) Set $\psi = 1_{CS}$ in Theorem 9.1.7(ii).

(ii) Set $\phi = 1_{CS} = \psi$ in Theorem 9.1.7(ii).

Let M be a CS-module. Recall from Lemma 9.1.3 that, for each element ϕ in $\mathrm{End}_C(M)$, $\phi\zeta_M \in \mathrm{End}_{CS}(M)$. Here is an analog of Theorem 9.1.4(iv).

Theorem 9.1.9 *Let M be a CS-module. Then there exists an element ϕ in $\mathrm{End}_C(M)$ such that $\phi\zeta_M$ is the identity on M.*

PROOF. We are assuming that, for each element s in S, the characteristic of C does not divide $|s^*|$. Thus, by Theorem 9.1.5(ii), CS is semisimple.

Let M be an irreducible CS-module. Then, by Theorem 8.5.6(i), we may assume that M is a submodule of the CS-module CS. Since CS is semisimple, CS possesses a submodule L such that $L \oplus M = CS$; cf. Proposition 8.3.4.

For any two elements l in L and m in M, we define $(l+m)\pi := m$. Then, as M is a submodule of CS, we obtain $\pi\sigma_s = \sigma_s\pi$ for each element s in S.

From Theorem 9.1.4(iv) we know that there exists an element γ in $\mathrm{End}_C(CS)$ such that $\gamma\zeta_{CS} = \sigma_1$. Set $\phi := \pi\gamma\pi$. Then $\phi \in \mathrm{End}_C(M)$ and

$$\begin{aligned}\phi\zeta_M &= \sum_{s\in S}\frac{1}{n_s}\sigma_{s^*}\phi\sigma_s = \sum_{s\in S}\frac{1}{n_s}\sigma_{s^*}\pi\gamma\pi\sigma_s \\ &= \pi(\sum_{s\in S}\frac{1}{n_s}\sigma_{s^*}\gamma\sigma_s)\pi = \pi\gamma\zeta_{CS}\pi = \pi\sigma_1\pi = \pi\sigma_1.\end{aligned}$$

It follows that, for each element m in M, $m\phi\zeta_M = m$.

9.2 Algebraically Closed Base Fields

In this section, the letter C stands for a field. We assume that, for each element s in S, the characteristic of C does not divide $|s^*|$, so that, according to Theorem 9.1.5(ii), CS is semisimple. In addition, we assume C to be algebraically closed.

Recall that, for each element ϕ in $\mathrm{End}_C(CS)$,

$$\phi\zeta_{CS} := \sum_{s\in S}\frac{1}{n_s}\sigma_{s^*}\phi\sigma_s.$$

The following result is due to Donald Higman; cf. [20; (7.1)].

Theorem 9.2.1 *Writing ζ instead of ζ_{CS}, we have $(CS)\zeta = Z(CS)$.*

PROOF. From Theorem 9.1.4(i) we know that $(CS)\zeta \subseteq Z(CS)$. Thus, we just have to prove that $Z(CS) \subseteq (CS)\zeta$.

Referring to Theorem 8.5.3(ii) and Theorem 8.5.4(ii) we shall be done if we succeed in showing that $\epsilon_\chi \in (CS)\zeta$.

Let K be an irreducible submodule of the CS-module CS such that $K \subseteq H_\chi$. Then, by Theorem 8.5.4(ii), $H_\chi \cong \mathrm{End}_C(K)$. Thus, by Theorem 9.1.9, there exists an element σ in H_χ such that, for each element k in K, $k(\sigma\zeta) = k$.

Since H_χ is an ideal of CS, $\sigma\zeta \in H_\chi$. Thus, $\sigma\zeta = \epsilon_\chi$.

Corollary 9.2.2 *The matrix*

$$(\sum_{s\in S}\frac{a_{s^*psq}}{n_s})_{pq}$$

has rank $|\mathrm{Irr}(CS)|$.

PROOF. From Theorem 8.6.4(ii) (together with Theorem 8.5.3(ii) and Theorem 8.5.4(ii)) we conclude that

$$\dim_C(Z(CS)) = |\mathrm{Irr}(CS)|.$$

Thus, the claim follows from Theorem 9.1.4(ii) together with Theorem 9.2.1.

We shall now see that the assumption that C is algebraically closed gives rise to a generalization of the orthogonality relations.

Let χ be an irreducible character of CS, and let M be an irreducible CS-module such that $\chi_M = \chi$.

Fix a basis B_χ of M. Then, for any three elements t, u in B_χ and σ in CS, there exists an element $c^\chi_{tu}(\sigma)$ in C such that, for each element t in B_χ,

$$t\sigma = \sum_{u \in B_\chi} c^\chi_{tu}(\sigma)u.$$

Let v and w be elements in B_χ. Then, by Theorem 8.5.4(ii), there exists

$$\epsilon^\chi_{vw} \in H_\chi$$

such that, for any two elements t and u in B_χ,

$$c^\chi_{tu}(\epsilon^\chi_{vw}) = \delta_{tw}\delta_{uv}.$$

In other words, we have $w\epsilon^\chi_{vw} = v$ and, for each element s in $B_\chi \setminus \{w\}$, we have $s\epsilon^\chi_{vw} = 0$.

Lemma 9.2.3 *Let χ be an irreducible character of CS, and let t and u be elements in B_χ. Then the following hold.*

(i) *For each element σ in CS, we have $\chi(\sigma\epsilon^\chi_{tu}) = c^\chi_{tu}(\sigma)$.*

(ii) *We have*

$$\epsilon^\chi_{tu} = \frac{m_\chi}{n_S}\sum_{s \in S}\frac{c^\chi_{tu}(\sigma_{s^*})}{n_{s^*}}\sigma_s.$$

PROOF. (i) For each element v in B_χ,

$$v\sigma\epsilon^\chi_{tu} = \sum_{w \in B_\chi} c^\chi_{vw}(\sigma)w\epsilon^\chi_{tu} = \sum_{w \in B_\chi} c^\chi_{vw}(\sigma)\delta_{wu}t = c^\chi_{vu}(\sigma)t.$$

Therefore,

$$\chi(\sigma\epsilon^\chi_{tu}) = c^\chi_{tu}(\sigma).$$

(ii) By definition, $\epsilon^\chi_{tu} \in H_\chi$. Thus, the desired equality follows from (i) together with Lemma 9.1.6(i).

In the case where S is thin, the equations of the following theorem are known as *Schur relations.*

Theorem 9.2.4 *Let ϕ and ψ be irreducible characters of CS, let t and u be elements in B_ϕ, and let v and w be elements in B_ψ. Then the following hold.*

(i) *For each element s in S, we have*

$$\frac{1}{n_S}\sum_{p\in S}\sum_{q\in S}\frac{a_{s^*pq}}{n_{p^*}}c^\phi_{tu}(\sigma_{p^*})c^\psi_{vw}(\sigma_q) = \delta_{\phi\psi}\delta_{tw}\frac{c^\phi_{vu}(\sigma_{s^*})}{m_\phi}.$$

(ii) *We have*

$$\frac{1}{n_S}\sum_{s\in S}\frac{1}{n_{s^*}}c^\phi_{tu}(\sigma_{s^*})c^\psi_{vw}(\sigma_s) = \delta_{\phi\psi}\delta_{tw}\delta_{uv}\frac{1}{m_\phi}.$$

PROOF. (i) From Lemma 9.2.3(ii) and Lemma 9.1.1(i) we obtain

$$\begin{aligned}\epsilon^\phi_{tu}\epsilon^\psi_{vw} &= \frac{m_\phi}{n_S}(\sum_{p\in S}\frac{c^\phi_{tu}(\sigma_{p^*})}{n_{p^*}}\sigma_p)\frac{m_\psi}{n_S}(\sum_{q\in S}\frac{c^\psi_{vw}(\sigma_{q^*})}{n_{q^*}}\sigma_q)\\ &= \frac{m_\phi m_\psi}{n_S^2}\sum_{p\in S}\sum_{q\in S}\frac{c^\phi_{tu}(\sigma_{p^*})c^\psi_{vw}(\sigma_{q^*})}{n_{p^*}n_{q^*}}\sigma_p\sigma_q\\ &= \frac{m_\phi m_\psi}{n_S^2}\sum_{s\in S}(\sum_{p\in S}\sum_{q\in S}\frac{c^\phi_{tu}(\sigma_{p^*})c^\psi_{vw}(\sigma_{q^*})}{n_{p^*}n_{q^*}}a_{pqs})\sigma_s.\end{aligned}$$

But, by definition, $\epsilon^\phi_{tu}\in H_\phi$ and $\epsilon^\psi_{vw}\in H_\psi$. Thus, $\epsilon^\phi_{tu}\epsilon^\psi_{vw} = \delta_{\phi\psi}\delta_{tw}\epsilon^\phi_{vu}$. Referring once again to the equation

$$\epsilon^\phi_{vu} = \frac{m_\phi}{n_S}\sum_{s\in S}\frac{c^\phi_{vu}(\sigma_{s^*})}{n_{s^*}}\sigma_s$$

we, therefore, obtain

$$\frac{m_\phi m_\psi}{n_S}\sum_{p\in S}\sum_{q\in S}\frac{c^\phi_{tu}(\sigma_{p^*})c^\psi_{vw}(\sigma_{q^*})a_{pqs}}{n_{p^*}n_{q^*}} = \delta_{\phi\psi}\delta_{tw}\frac{m_\phi c^\phi_{vu}(\sigma_{s^*})}{n_{s^*}}$$

for each element s in S. Recall that, by Lemma 9.1.6(iii), $m_\phi \neq 0 \neq m_\psi$. Therefore, for each element s in S, we must have

$$\frac{1}{n_S}\sum_{p\in S}\sum_{q\in S}\frac{a_{pq^*s}n_{s^*}}{n_{p^*}n_q}c^\phi_{tu}(\sigma_{p^*})c^\psi_{vw}(\sigma_q) = \delta_{\phi\psi}\delta_{tw}\frac{c^\phi_{vu}(\sigma_{s^*})}{m_\phi}.$$

Finally, recall that, by Lemma 1.1.1(ii) and Lemma 1.1.3(ii),

$$a_{pq^*s}n_{s^*} = a_{qp^*s^*}n_{s^*} = a_{s^*pq}n_q.$$

(ii) This equation follows from (i) by setting $s^* = 1$; use the first equation of Lemma 1.1.1(i).

Lemma 9.2.5 *Let χ be an irreducible character of CS, and let s be an element in S. Then $\chi(\sigma_s)$ is integral over the smallest unitary subring of C.*

PROOF. Let us denote by Z the smallest unitary subring of C. The C-module CS is finitely generated. Thus, σ_s is zero of a monic polynomial over Z. Thus, each characteristic root of σ_s is integral over Z. Thus, as $\chi(\sigma_s)$ is a sum of characteristic roots of σ_s, the claim follows from Theorem 8.2.4.

9.3 Scheme Rings over the Field of Complex Numbers

In this section, the letter C stands for the field of complex numbers. The scheme ring over the field of complex numbers provides us with an additional tool to work with, the norm function of C.

Lemma 9.3.1 *Let s be an element in S, and let c be a characteristic root of σ_s. Then $|c| \leq n_s$.*

PROOF. Since c is a characteristic root of σ_s, there exists an element m in CX such that $m \neq 0$ and $m\sigma_s = cm$.

Since $m \in CX$, there exists, for each element x in X, an element c_x in C such that

$$m = \sum_{x \in X} c_x x.$$

Thus, as

$$(\sum_{x \in X} c_x x)\sigma_s = \sum_{x \in X} c_x x\sigma_s = \sum_{x \in X} c_x \sum_{y \in xs} y = \sum_{x \in X}(\sum_{y \in xs^*} c_y)x$$

and

$$c(\sum_{x \in X} c_x x) = \sum_{x \in X} cc_x x,$$

we obtain from $m\sigma_s = cm$ that

$$cc_x = \sum_{y \in xs^*} c_y$$

for each element x in X.

Let us now pick an element z in X such that, for each of the finitely many elements x in X, $|c_x| \leq |c_z|$. Then, by Lemma 1.1.2(i),

$$|c||c_z| = |cc_z| = |\sum_{y \in zs^*} c_y| \leq \sum_{y \in zs^*} |c_y| \leq n_{s^*}|c_z|.$$

But $c_z \neq 0$. Therefore, $|c| \leq n_{s^*}$.

Recall, finally, that we are assuming S to have finite valency. Thus, $n_{s^*} = n_s$. Thus, as $|c| \leq n_{s^*}$, $|c| \leq n_s$.

Corollary 9.3.2 *Let s be an element in S, and let χ be a character of CS. Then the following hold.*

(i) *We have $|\chi(\sigma_s)| \leq n_s\chi(\sigma_1)$.*

(ii) *Let M be an irreducible CS-module such that $\chi_M = \chi$. Then $\chi(\sigma_s) = n_s\chi(\sigma_1)$ if and only if, for each element m in M, $m\sigma_s = n_s m$.*

Proof. (i) Since C is assumed to be the field of complex numbers, C is algebraically closed. Thus, $\chi(\sigma_s)$ is the sum of $\chi(\sigma_1)$ characteristic roots of σ_s. Thus, the claim follows from Lemma 9.3.1.

(ii) Assume that, for each element m in M, $m\sigma_s = n_s m$. Then we obviously have that $\chi(\sigma_s) = n_s\chi(\sigma_1)$.

Let us now assume that $\chi(\sigma_s) = n_s\chi(\sigma_1)$. We set $e := \chi(\sigma_1)$, and we define $c_1, \ldots, c_e$ to be the characteristic roots of σ_s. Then, referring to Lemma 9.3.1 we obtain

$$|\chi(\sigma_s)| = |c_1 + \ldots + c_e| \leq |c_1| + \ldots + |c_e| \leq n_s e = n_s\chi(\sigma_1) = \chi(\sigma_s) = |\chi(\sigma_s)|.$$

This proves that, for each element i in $\{1, \ldots, e\}$, $c_i = n_s$.

The following two results are due to Akihide Hanaki; cf. [17; Theorem 3.4] and [17; Theorem 3.5].

Lemma 9.3.3 *Let T be a normal closed subset of S. Then there exists a ring homomorphism ϕ from CS to $C(S//T)$ such that, for each element s in S, $\phi(\sigma_s) = a_{sTs}\sigma_{s^T}$.*

Proof. Let p and q be elements in S. Referring to Lemma 9.1.1(i) and to Theorem 4.1.3(ii) we obtain

$$\begin{aligned}
\phi(\sigma_p)\phi(\sigma_q) &= a_{pTp}\sigma_{p^T}a_{qTq}\sigma_{q^T} \\
&= a_{pTp}a_{qTq} \sum_{s^T \in S//T} a_{p^T q^T s^T}\sigma_{s^T} \\
&= a_{pTp}a_{qTq} \sum_{s^T \in S//T} \sum_{u \in pT} \sum_{v \in qT} a_{uvs}(n_T)^{-1}\sigma_{s^T} \\
&= \sum_{s^T \in S//T} (\sum_{u \in pT} \sum_{v \in qT} a_{pTp}a_{qTq}a_{uvs}(n_T)^{-1})\sigma_{s^T}
\end{aligned}$$

and

$$\phi(\sigma_p\sigma_q) = \phi(\sum_{s\in S} a_{pqs}\sigma_s) = \sum_{s\in S} a_{pqs}\phi(\sigma_s)$$
$$= \sum_{s\in S} a_{pqs}a_{sTs}\sigma_{s^T} = \sum_{s^T\in S//T} (\sum_{w\in sT} a_{pqw}a_{wTw})\sigma_{s^T}.$$

Thus, the claim follows from Lemma 2.5.4(iii).

Lemma 9.3.4 *Let T be a normal closed subset of S, and let M be a $C(S//T)$-module. Then M is a CS-module such that, for any two elements m in M and s in S, $m\sigma_s = a_{sTs}m\sigma_{s^T}$.*

Proof. Let m be an element in M, and let p and q be elements in S. We have to show that $m(\sigma_p\sigma_q) = (m\sigma_p)\sigma_q$.
Referring to Lemma 9.1.1(i) we obtain

$$m(\sigma_p\sigma_q) = m(\sum_{s\in S} a_{pqs}\sigma_s) = \sum_{s\in S} a_{pqs}m\sigma_s$$
$$= \sum_{s\in S} a_{pqs}a_{sTs}m\sigma_{s^T} = m(\sum_{s\in S} a_{pqs}a_{sTs}\sigma_{s^T}).$$

Thus, as

$$(m\sigma_p)\sigma_q = (a_{pTp}m\sigma_{p^T})\sigma_q = a_{qTq}a_{pTp}m\sigma_{p^T}\sigma_{q^T} = m(a_{qTq}a_{pTp}\sigma_{p^T}\sigma_{q^T}),$$

the desired equation follows from Lemma 9.3.3.

Let T be a normal closed subset of S. The following corollary shows how to obtain characters of CS from characters of $C(S//T)$.

Corollary 9.3.5 *Let T be a normal closed subset of S, let χ be a character of $C(S//T)$. For each element s in S, we define*

$$\bar{\chi}(\sigma_s) := a_{sTs}\chi(\sigma_{s^T}).$$

Then $\bar{\chi}$ is a character of CS.

Proof. This follows immediately from Lemma 9.3.4.

For each character χ of CS, we define

$$T(\chi) := \{s \in S \mid \chi(\sigma_s) = n_s\chi(\sigma_1)\}.$$

The following two results are due to Akihide Hanaki; cf. [18; Theorem 3.2] and [18; Theorem 3.5].

Lemma 9.3.6 *For each character χ of CS, $T(\chi)$ is a closed subset of S.*

PROOF. Let p and q be elements in $T(\chi)$. Then, by definition,

$$\chi(\sigma_p) = n_p\chi(\sigma_1)$$

and

$$\chi(\sigma_q) = n_q\chi(\sigma_1).$$

According to Lemma 2.1.6, we shall be done if we succeed in showing that $pq \subseteq T(\chi)$.

From Lemma 9.1.1(i), Corollary 9.3.2(ii), and Lemma 1.1.3(iv) we obtain

$$\sum_{s\in S} a_{pqs}\chi(\sigma_s) = \chi(\sum_{s\in S} a_{pqs}\sigma_s) = \chi(\sigma_p\sigma_q) = n_pn_q\chi(\sigma_1) = \sum_{s\in S} a_{pqs}n_s\chi(\sigma_1).$$

Now recall from Corollary 9.3.2(i) that, for each element s in S, $|\chi(\sigma_s)| \leq n_s\chi(\sigma_1)$. Moreover, for each element s in S, $0 \leq a_{pqs}$. Thus, for each element s in S with $1 \leq a_{pqs}$, we must have $\chi(\sigma_s) = n_s\chi(\sigma_1)$. This means that, for each element s in pq, $\chi(\sigma_s) = n_s\chi(\sigma_1)$. Thus, for each element s in pq, $s \in T(\chi)$.

Lemma 9.3.7 *For each normal closed subset T of S, there exists a character χ of S such that $T = T(\chi)$.*

PROOF. Let T be a normal closed subset of S. For any two elements s in S and χ in $\mathrm{Irr}(C(S/\!/T))$, we define

$$\bar{\chi}(\sigma_s) := a_{sTs}\chi(\sigma_{s^T}).$$

Then, by Corollary 9.3.5, $\bar{\chi}$ is a character of CS.

We define

$$\zeta := \sum_{\chi\in\mathrm{Irr}(C(S/\!/T))} m_\chi\bar{\chi},$$

and we claim that $T = T(\zeta)$.

For each element t in T, we have $a_{tTt} = n_t$. Thus, for each element t in T,

$$\begin{aligned}\zeta(\sigma_t) &= \sum_{\chi\in\mathrm{Irr}(C(S/\!/T))} m_\chi\bar{\chi}(\sigma_t) = \sum_{\chi\in\mathrm{Irr}(C(S/\!/T))} m_\chi a_{tTt}\chi(\sigma_{t^T}) \\ &= n_t \sum_{\chi\in\mathrm{Irr}(C(S/\!/T))} m_\chi\chi(\sigma_{1^T}) = n_t\chi_{C(X/T)}(\sigma_{1^T}) = n_tn_{X/T}.\end{aligned}$$

(The last equation follows from Lemma 9.1.2(i).) In particular, $\zeta(\sigma_1) = n_{X/T}$. Thus, for each element t in T, $\zeta(\sigma_t) = n_t\zeta(\sigma_1)$, and that means that $T \subseteq T(\zeta)$.

In order to show that $T(\zeta) \subseteq T$ we fix an element s in $T(\zeta)$. We shall be done if we succeed in showing that $s \in T$.

Since $s \in T(\zeta)$,

$$\zeta(\sigma_s) = n_s \zeta(\sigma_1) = n_s \sum_{\chi \in \mathrm{Irr}(C(S//T))} m_\chi \bar{\chi}(\sigma_1) \neq 0.$$

On the other hand,

$$\begin{aligned}\zeta(\sigma_s) &= \sum_{\chi \in \mathrm{Irr}(C(S//T))} m_\chi \bar{\chi}(\sigma_s) \\ &= a_{sTs} \sum_{\chi \in \mathrm{Irr}(C(S//T))} m_\chi \chi(\sigma_{s^T}) = a_{sTs} \chi_{C(X/T)}(\sigma_{s^T}).\end{aligned}$$

Thus, $\chi_{C(X/T)}(\sigma_{s^T}) \neq 0$.

From $\chi_{C(X/T)}(\sigma_{s^T}) \neq 0$ we obtain $1^T = s^T$; cf. Lemma 9.1.2(ii). Thus, by Lemma 4.1.1, $s \in T$.

Since s in $T(\zeta)$ has been chosen arbitrarily, we have shown that $T(\zeta) \subseteq T$.

9.4 Closed Subsets

In this section, the letter C stands for a field. We shall look at the relationship between representations of CS and closed subsets of S.

For each subset R of S, we denote by $C[R]$ the smallest unitary subring of CS which contains C and $\{\sigma_r \mid r \in R\}$.

Theorem 9.4.1 *Let T be a closed subset of S, and let R be a subset of T such that $C[R] = CT$. Then $\langle R \rangle = T$.*

Proof. Let t be an element in T. Then $\sigma_t \in CT$. Thus, as we are assuming that $C[R] = CT$, there exist elements $r_1, \ldots, r_m$ in R and positive integers $e_1, \ldots, e_m$ such that

$$\sigma_{r_1}^{e_1} \cdots \sigma_{r_m}^{e_m} \notin C(T \setminus \{t\}).$$

Thus, setting

$$n := \sum_{i=1}^{m} e_i$$

we obtain from Lemma 9.1.1(ii) that $t \in R^n$.

Since t has been chosen arbitrarily in T, the claim follows from Lemma 2.3.2.

Let R be a nonempty subset of S. We define

$$\sigma_R := \sum_{r \in R} \sigma_r.$$

(Recall that n_R is our notation for the sum of the integers n_r with $r \in R$.)

Lemma 9.4.2 *A nonempty subset R of S is closed if and only if $(\sigma_R)^2 = n_R\sigma_R$.*

PROOF. Let R be a nonempty subset of S. Then, by Lemma 9.1.1(i),

$$(\sigma_R)^2 = (\sum_{p\in R}\sigma_p)(\sum_{q\in R}\sigma_q) = \sum_{p\in R}\sum_{q\in R}\sum_{s\in R} a_{pqs}\sigma_s = \sum_{s\in R}(\sum_{p\in R}\sum_{q\in R} a_{pqs})\sigma_s.$$

From Lemma 1.4.3 we know that, for each element s in S,

$$\sum_{p\in R}\sum_{q\in R} a_{pqs} = n_R$$

if and only if $R^*s \subseteq R$. Thus, $(\sigma_R)^2 = n_R\sigma_R$ if and only if R is closed.

Lemma 9.4.3 *For each closed subset T of S, the following hold.*

(i) *The closed subset T is normal in S if and only if $\sigma_T \in Z(CS)$.*

(ii) *Let M be a CS-module, and assume that the characteristic of C does not divide n_T. Then $M = \ker(\sigma_T) \oplus \{m \in M \mid m\sigma_T = n_T m\}$.*

PROOF. (i) Let us assume that T is normal in S. Then, for each element s in S, $Ts = sT$; cf. Lemma 2.5.2(ii).
On the other hand, we have

$$\sigma_T\sigma_s = \sum_{t\in T}\sigma_t\sigma_s = \sum_{t\in T}\sum_{r\in Ts} a_{tsr}\sigma_r = \sum_{r\in Ts}\sum_{t\in T} a_{tsr}\sigma_r = \sum_{r\in Ts} a_{Tsr}\sigma_r$$

and, similarly,

$$\sigma_s\sigma_T = \sum_{r\in sT} a_{sTr}\sigma_r.$$

Thus, our claim follows from Lemma 2.5.4(i).
(ii) Let m be an element in M. Set $t := n_T m - m\sigma_T$ and $u := m\sigma_T$. Then $n_T m = t + u$. On the other hand, with the help of Lemma 9.4.2, one obtains

$$t\sigma_T = (n_T m - m\sigma_T)\sigma_T = n_T m\sigma_T - m\sigma_T^2 = m(n_T\sigma_T - \sigma_T^2) = 0$$

and

$$u\sigma_T = m\sigma_T^2 = n_T m\sigma_T = n_T u.$$

Now the claim follows from the hypothesis that the characteristic of C does not divide n_T.

Corollary 9.4.4 *For each involution l of S, the following hold.*

(i) *We have $\sigma_l^2 = n_l\sigma_1 + (n_l - 1)\sigma_l$.*

(ii) *For each CS-module M, we have*

$$M = \{m \in M \mid m\sigma_l = -m\} \oplus \{m \in M \mid m\sigma_l = n_l m\}.$$

PROOF. Since l is assumed to be an involution of S, $\{1, l\}$ is closed. Thus, (i) follows from Lemma 9.4.2, and (ii) follows from Lemma 9.4.3(ii).

Lemma 9.4.5 *Let L be a set of two different involutions. Assume that C is algebraically closed and that $C[L] = CS$. Then, for each irreducible CS-module M, $\dim_C(M) \leq 2$.*

PROOF. Let M be an irreducible CS-module, and let us denote by h and k the elements in L.

Since C is assumed to be algebraically closed, there exists an element m in M such that $m \neq 0$ and $Cm\sigma_h\sigma_k = Cm$. Let us define $U := Cm + Cm\sigma_h$. Then $\{0\} \neq U$.

From Corollary 9.4.4(i) we obtain

$$(Cm\sigma_h)\sigma_h = Cm\sigma_h^2 \subseteq Cm + Cm\sigma_h = U,$$

whence $U\sigma_h \subseteq U$. Similarly, Corollary 9.4.4(i) yields $U\sigma_k \subseteq U$.

Recall that we are assuming that $C[L] = CS$. Therefore, U is a submodule of M. But $\{0\} \neq U$ and M is assumed to be irreducible. Therefore, $U = M$. It follows that $\dim_C(M) \leq 2$.

Proposition 9.4.6 *Assume that C has characteristic* 0. *Let s be an element in S, and let T be a closed subset of S.*

(i) *The vector space $C(sT)$ is a CT-module.*

(ii) *For each element t in T, we have*

$$\chi_{C(sT)}(\sigma_t) = \sum_{r \in sT} a_{rtr}.$$

(iii) *The character 1_{CT} has multiplicity* 1 *in $\chi_{C(sT)}$.*

PROOF. Let p be an element in sT, and let t be an element in T. Then, by Lemma 9.1.1(i),

$$\sigma_p\sigma_t = \sum_{q \in S} a_{ptq}\sigma_q.$$

(i) Let q be an element in S such that $1 \leq a_{ptq}$. Then, by definition, $q \in pt \subseteq pT$. But, as $p \in sT$, $pT = sT$; cf. Lemma 2.1.4. Therefore, $q \in sT$.

Since q has been chosen arbitrarily in S satisfying $1 \leq a_{ptq}$, the above equality yields that $\sigma_p\sigma_t \in C(sT)$.

(ii) This claim follows from the above equality together with the fact that $\{\sigma_q \mid q \in sT\}$ is a basis of $C(sT)$.

(iii) From Lemma 2.1.1 we know that, for each element r in sT, $a_{rTr} = a_{rTs}$. Thus,

$$\sum_{t\in T}\sum_{r\in sT} a_{rtr} = \sum_{r\in sT}\sum_{t\in T} a_{rtr} = \sum_{r\in sT}\sum_{t\in T} a_{rts} = \sum_{t\in T}\sum_{r\in sT} a_{rts}.$$

Thus, referring to (ii) and to the second equation of Lemma 1.1.3(iii) we obtain

$$\sum_{t\in T}\chi_{C(sT)}(\sigma_t) = \sum_{t\in T}\sum_{r\in sT} a_{rtr} = \sum_{t\in T}\sum_{r\in sT} a_{rts} = \sum_{t\in T}\sum_{r\in S} a_{rts} = \sum_{t\in T} n_{t^*} = n_T.$$

Now recall that, for each element χ in $\mathrm{Irr}(CT)\setminus\{1_{CT}\}$,

$$\sum_{t\in T}\chi(\sigma_t) = 0;$$

cf. Lemma 9.1.8(i). Thus, the claim follows from

$$\sum_{t\in T} 1_{CT}(\sigma_t) = n_T$$

together with the hypothesis that C has characteristic 0.

For each element s in S, we define

$$C_s := \{\sigma \in CS \mid \sigma\sigma_s = n_s\sigma\}.$$

For each nonempty subset R of S, we define C_R to be the intersection of the sets C_r such that $r \in R$.

Recall that, for each closed subset T of S, S/T is our notation for the set of all left cosets of T in S.

Theorem 9.4.7 *Assume that C has characteristic 0. Then, for each closed subset T of S, we have $|S/T| = \dim_C(C_T)$.*

PROOF. Let R be a left transversal of S. Then, by Lemma 2.1.4, $\{rT \mid r \in R\}$ is a partition of S. In particular, CS is a direct sum of the submodules $C(rT)$ with $r \in R$.

Let σ be an element in C_T. Since $\sigma \in CS$, there exists, for each element r in R, an element $\sigma^{(r)}$ in $C(rT)$ such that

$$\sigma = \sum_{r\in R}\sigma^{(r)}.$$

Let t be an element in T. Then, as $\sigma \in C_T$,

$$\sum_{r\in R}\sigma^{(r)}\sigma_t = \sigma\sigma_t = n_t\sigma = n_t\sum_{r\in R}\sigma^{(r)} = \sum_{r\in R} n_t\sigma^{(r)}.$$

But, for each element r in R, $n_t\sigma^{(r)} \in C(rT)$ and, by Proposition 9.4.6(i), $\sigma^{(r)}\sigma_t \in C(rT)$. Thus, as CS is a direct sum of the submodules $C(rT)$ with $r \in R$, we conclude that, for each element r in R, $\sigma^{(r)}\sigma_t = n_t\sigma^{(r)}$. Thus, for each element r in R, $\sigma^{(r)} \in C_t$.

Since $t \in T$ has been chosen arbitrarily, we obtain $\sigma^{(r)} \in C_T$ for each element r in R. It follows that

$$C_T = \sum_{r\in R}(C(rT) \cap C_T).$$

Thus, the claim follows from Proposition 9.4.6(iii).

Let us now see what we can say about characters of semidirect products of S and specific thin schemes.

Let A be a subgroup of $\mathrm{Aut}(S)$. For each element a in A, we define $a\zeta := a_S$. (Recall that, for each element a in $\mathrm{Aut}(S)$, a_S is our notation for $a \cap (S \times S)$.) From Lemma 5.2.2(iii) we know that ζ is a group homomorphism from A to $\mathrm{Stc}(S)$. Thus, we may consider the semidirect product $S_\zeta A$ of S and A with respect to ζ; cf. Section 7.3.

In the following theorem, we shall see that the standard module CX of CS is a module also for the scheme ring $C(S_\zeta A)$ of $S_\zeta A$ over C. We shall refer to this fact in the proof of Proposition 12.6.3.

Theorem 9.4.8 *Let A be a subgroup of $\mathrm{Aut}(S)$. For each element a in A, we define $a\zeta := a_S$. For any three elements x in X, s in S, and a in A, we set*

$$x\sigma_{s_\zeta a} := \sum_{y\in xasa} y.$$

Then we have the following.

(i) *For any two elements x in X and s in S, $x\sigma_{s_\zeta 1} = x\sigma_s$.*

(ii) *For any two elements x in X and a in A, $x\sigma_{1_\zeta a} = xa$.*

(iii) *For any two subsets $\{c_x \mid x \in X\}$ and $\{c_{(s,a)} \mid (s,a) \in S \times A\}$ of C, we define*

$$(\sum_{x\in X} c_x x)(\sum_{(s,a)\in S\times A} c_{(s,a)}\sigma_{s_\zeta a}) := \sum_{x\in X}\sum_{(s,a)\in S\times A} c_x c_{(s,a)} x\sigma_{s_\zeta a}.$$

Then CX is an $C(S_\zeta A)$-module.

(iv) *Let ϕ denote the character of $C(S_\zeta A)$ afforded by CX. Let s be an element in S, and let a be an element in A. Then*

$$\phi(\sigma_{s_\zeta a}) = |\{x \in X \mid x \in xas\}|.$$

PROOF. (i) This follows immediately from the definition of $x\sigma_{s_\zeta 1}$.

(ii) This follows immediately from the definition of $x\sigma_{1_\varsigma a}$.

(iii) Let p and q be elements in S, and let b and c be elements in A. We shall prove first that, for each element x in X,

$$x(\sigma_{p_\varsigma b}\sigma_{q_\varsigma c}) = (x\sigma_{p_\varsigma b})\sigma_{q_\varsigma c}.$$

First of all, with the help of Lemma 9.1.1(i) and Corollary 7.3.4(i), we obtain

$$\sigma_{p_\varsigma b}\sigma_{q_\varsigma c} = \sum_{(s,a)\in S\times A} a_{(p_\varsigma b)(q_\varsigma c)(s_\varsigma a)}\sigma_{s_\varsigma a} = \sum_{s\in S} a_{pqb^*s}\sigma_{s_\varsigma bc}.$$

Therefore, for each element x in X,

$$\begin{aligned} x(\sigma_{p_\varsigma b}\sigma_{q_\varsigma c}) &= x\sum_{s\in S} a_{pqb^*s}\sigma_{s_\varsigma bc} = \sum_{s\in S} a_{pqb^*s}x\sigma_{s_\varsigma bc} \\ &= \sum_{s\in S} a_{pqb^*s} \sum_{y\in(xbc)(sbc)} y = \sum_{s\in S} a_{(pbc)(qc)(sbc)} \sum_{y\in xbcsbc} y \\ &= \sum_{s\in S}\sum_{y\in(xbc)(sbc)} a_{(pbc)(qc)(sbc)}y = \sum_{s\in S}\sum_{y\in(xbc)s} a_{(pbc)(qc)s}y. \end{aligned}$$

On the other hand, for each element x in X,

$$\begin{aligned} (x\sigma_{p_\varsigma b})\sigma_{q_\varsigma c} &= \sum_{y\in(xb)(pb)} y\sigma_{q_\varsigma c} = \sum_{y\in(xb)(pb)}\sum_{z\in(yc)(qc)} z \\ &= \sum_{yc\in(xbc)(pbc)}\sum_{z\in(yc)(qc)} z = \sum_{s\in S}\sum_{z\in(xbc)s} a_{(pbc)(qc)s}z. \end{aligned}$$

Thus, we have proved the desired equation.

Now looking at the definition of an $C(S_\varsigma A)$-module, we see that the above equation suffices in order to show that CX is an $C(S_\varsigma A)$-module.

(iv) This follows immediately from the definition of $x\sigma_{s_\varsigma a}$, where $x \in X$, $s \in S$, and $a \in A$.

Let a be an element in A, and set

$$\mathrm{Fix}_X(a) := \{x \in X \mid xa = x\}.$$

Then, by Theorem 9.4.8, $\phi(\sigma_{1_\varsigma a}) = |\mathrm{Fix}_X(a)|$.

9.5 Schemes with at most Five Elements

In this section, we shall see how some of the representation theoretic results of the previous sections can be applied successfully if $|S|$ is small.

We start with two results due to Donald Higman; cf. [20; (4.2)].

Lemma 9.5.1 *Let C be an algebraically closed field of characteristic* 0. *Then, if* $|\mathrm{Irr}(CS)| = 2$, $|S| = 2$.

PROOF. We are assuming that $|\mathrm{Irr}(CS)| = 2$. Thus, there exists an irreducible character χ of CS such that $\{1_{CS}, \chi\} = \mathrm{Irr}(CS)$.

Let s be an element in $S \setminus \{1\}$. Then, as $1_{CS}(\sigma_s) = n_{s^*}$, we obtain from Lemma 9.1.8(ii) that

$$n_{s^*} + m_\chi \chi(\sigma_s) = m_{1_{CS}} 1_{CS}(\sigma_s) + m_\chi \chi(\sigma_s) = \chi_{CX}(\sigma_s) = 0.$$

It follows that

$$\chi(\sigma_s) = \frac{-n_{s^*}}{m_\chi}.$$

Let us write Q to denote the smallest subfield of C, and let Z denote the smallest unitary subring of C. Then $\chi(\sigma_s) \in Q$. On the other hand, by Lemma 9.2.5, $\chi(\sigma_s)$ is integral over Z. Thus, as C is assumed to have characteristic 0, $\chi(\sigma_s) \in Z$. It follows that

$$\chi(\sigma_s) \leq -1.$$

Now recall that, by Corollary 8.6.5, $|S| = 1 + \chi(\sigma_1)^2$. Thus,

$$\sum_{s \in S} \chi(\sigma_s) = \chi(\sigma_1) + \sum_{s \in S \setminus \{1\}} \chi(\sigma_s) \leq \chi(\sigma_1) - (|S| - 1) = \chi(\sigma_1) - \chi(\sigma_1)^2.$$

Thus, by Lemma 9.1.8(i), $\chi(\sigma_1) = \chi(\sigma_1)^2$. It follows that $\chi(\sigma_1) = 1$. Thus, as $|S| = 1 + \chi(\sigma_1)^2$, $|S| = 2$.

Theorem 9.5.2 *If* $|S| \leq 5$, *S is commutative.*

PROOF. This is an immediate consequence of Lemma 9.5.1 and Corollary 8.6.5.

The second part of the following theorem is due to Paul Erdös, Alfred Rényi, and Vera Sós; cf. [9; Theorem 6]. Its third part is a result of Alan Hoffman and Robert Singleton; cf. [27].

Theorem 9.5.3 *Assume that $S \setminus \{1\}$ contains symmetric elements p and q such that $p \neq q$ and $\{1, p, q\} = S$.*

(i) *The rational number*

$$\frac{(a_{ppp} - a_{ppq})(n_S - 1) + 2n_p}{\sqrt{(a_{ppp} - a_{ppq})^2 + 4(n_p - a_{ppq})}}$$

is an integer.

(ii) *If* $a_{ppp} = 1$, *then* $a_{ppq} \neq 1$.

(iii) *If* $a_{ppp} = 0$ *and* $a_{ppq} = 1$, $(n_p, n_S) \in \{(2,5), (3,10), (7,50), (57,3250)\}$.

PROOF. (i) Let C be an algebraically closed field of characteristic 0, and let χ be a non-principal irreducible character of CS. Then, as $|S| = 3$, χ is linear; cf. Corollary 8.6.5. Thus, by Lemma 9.1.1(i),

$$\begin{aligned}\chi(\sigma_p)^2 &= \chi(\sigma_p^2)\\ &= \chi(n_p\sigma_1 + a_{ppp}\sigma_p + a_{ppq}\sigma_q)\\ &= n_p + a_{ppp}\chi(\sigma_p) + a_{ppq}\chi(\sigma_q).\end{aligned}$$

But, by Lemma 9.1.8(i),

$$\chi(\sigma_q) = -\chi(\sigma_p) - 1.$$

(Recall that, by hypothesis, $\{1, p, q\} = S$.) Therefore,

$$\chi(\sigma_p)^2 = n_p + a_{ppp}\chi(\sigma_p) + a_{ppq}(-\chi(\sigma_p) - 1),$$

so that

$$\chi(\sigma_p)^2 - (a_{ppp} - a_{ppq})\chi(\sigma_p) - (n_p - a_{ppq}) = 0.$$

From Corollary 8.6.5 we know that there exist elements ϕ and ψ in $\mathrm{Irr}(CS) \setminus \{1_{CS}\}$ such that $\phi \neq \psi$ and $\{1_{CS}, \phi, \psi\} = \mathrm{Irr}(CS)$. Without loss of generality, we may assume that

$$\phi(\sigma_p) = \frac{1}{2}(a_{ppp} - a_{ppq} + \sqrt{(a_{ppp} - a_{ppq})^2 + 4(n_p - a_{ppq})})$$

and

$$\psi(\sigma_p) = \frac{1}{2}(a_{ppp} - a_{ppq} - \sqrt{(a_{ppp} - a_{ppq})^2 + 4(n_p - a_{ppq})}).$$

From Lemma 9.1.8(ii) we know that $m_{1_{CS}} = 1$. Therefore,

$$1 + m_\phi + m_\psi = \sum_{\chi \in \mathrm{Irr}(CS)} m_\chi \chi(\sigma_1) = \chi_{CX}(\sigma_1) = n_S.$$

From $1_{CS}(\sigma_p) = n_{p^*}$ and $p^* = p$ we obtain

$$n_p + m_\phi\phi(\sigma_p) + m_\psi\psi(\sigma_p) = \sum_{\chi \in \mathrm{Irr}(CS)} m_\chi \chi(\sigma_p) = \chi_{CX}(\sigma_p) = 0.$$

Now we have

$$\begin{aligned}m_\psi &= \frac{\phi(\sigma_p)(n_S - 1) + n_p}{\phi(\sigma_p) - \psi(\sigma_p)}\\ &= \frac{\frac{1}{2}(a_{ppp} - a_{ppq} + \sqrt{(a_{ppp} - a_{ppq})^2 + 4(n_p - a_{ppq})})(n_S - 1) + n_p}{\sqrt{(a_{ppp} - a_{ppq})^2 + 4(n_p - a_{ppq})}}\\ &= \frac{1}{2}(n_S - 1 + \frac{(a_{ppp} - a_{ppq})(n_S - 1) + 2n_p}{\sqrt{(a_{ppp} - a_{ppq})^2 + 4(n_p - a_{ppq})}}).\end{aligned}$$

Thus, the claim follows from the fact that m_ψ is an integer.

(ii) Assume, by way of contradiction, that $a_{ppp} = 1 = a_{ppq}$. Then, by (i), $\sqrt{n_p - 1}$ divides n_p. Therefore, $n_p = 2$.

On the other hand, as we are assuming p to be symmetric, $1 \le a_{ppq}$ yields $1 \le a_{qpp}$; cf. Lemma 1.1.3(ii). Moreover, by the first equation of Lemma 1.1.1(i), we also have that $1 \le a_{1pp}$. Thus, by the second equation of Lemma 1.1.3(iii), $3 \le n_{p^*} = n_p$, contradiction.

(iii) Let us assume that $a_{ppp} = 0$ and that $a_{ppq} = 1$.

From the first equation of Lemma 1.1.1(i) we know that $a_{1pp} = 1$. Thus, as $a_{ppp} = 0$, the second equation of Lemma 1.1.3(iii) yields $a_{qpp} = n_{p^*} - 1$. Since $p^* = p$, this means that $a_{qp^*p} = n_p - 1$.

Now, as $a_{ppq} = 1$, Lemma 1.1.3(ii) yields $n_q = (n_p - 1)n_p$. Thus, as $\{1, p, q\} = S$,

$$n_S = n_p^2 + 1.$$

On the other hand, we know from (i) that $\sqrt{4n_p - 3}$ divides $n_S - 1 - 2n_p$. Therefore, $\sqrt{4n_p - 3}$ divides $n_p(n_p - 2)$.

Assume that $n_p \neq 2$, and set

$$m := \sqrt{4n_p - 3}.$$

Then

$$n_p = \frac{1}{4}(m^2 + 3)$$

and m divides

$$\frac{1}{16}(m^2 + 3)(m^2 - 5).$$

In particular, m divides 15, whence $m \in \{1, 3, 5, 15\}$.

Assume that $1 = n_p$. Then, by Lemma 1.5.1, s is thin. However, as we are assuming that $a_{ppq} = 1$, we have $q \in pp$, contradiction. Thus, $m \in \{3, 5, 15\}$.

Now the claim follows from $4n_p = m^2 + 3$ and $n_S = n_p^2 + 1$.

It seems to be a still open problem whether $n_S = 3250$ can actually occur in Theorem 9.5.3(iii).

9.6 Constrained Sets of Involutions

Let L be a set of involutions, and let us write ℓ instead of ℓ_L.

Recall that, for each element q in $\langle L \rangle$, $S_1(q, L)$ is our notation for the set of all elements p in $\langle L \rangle$ such that pq contains an element r with $\ell(r) = \ell(p) + \ell(q)$.

In accordance with Section 3.4 we shall write, for each element s in $\langle L \rangle$, $S_1(s)$ instead of $S_1(s, L)$.

Recall that L is called constrained if, for any two elements q in $\langle L\rangle$ and p in $S_1(q)$, $1 = |pq|$.

In this short section, the letter L stands for a constrained set of involutions.

Lemma 9.6.1 *Assume that $\langle L\rangle = S$ and that S has finite valency. Let p, q, and r be elements in S such that $r \in pq$ and $\ell_L(r) = \ell_L(p) + \ell_L(q)$. Then, we have the following.*

(i) *For each element s in S, $a_{pqs} = \delta_{sr}$.*

(ii) *We have $\sigma_p\sigma_q = \sigma_r$.*

PROOF. (i) From $r \in pq$ and $\ell(r) = \ell(p) + \ell(q)$ we obtain $a_{pqr} = 1$; cf. Lemma 3.5.2. On the other hand, as L is assumed to be constrained, $r \in pq$ yields $\{r\} = pq$. Thus, for each element s in $S \setminus \{r\}$, $a_{pqs} = 0$.

(ii) From (i) and Lemma 9.1.1(ii) we obtain

$$\sigma_p\sigma_q = \sum_{s\in S} a_{pqs}\sigma_s = \sum_{s\in S} \delta_{sr}\sigma_s = \sigma_r.$$

This proves (ii).

From Theorem 9.4.1 we know that, for each subset R of S with $C[R] = CS$, $\langle R\rangle = S$. The following corollary says that the converse of this theorem holds if R is constrained.

Corollary 9.6.2 *The set $O_\vartheta(L)$ is empty if and only if $\{1\} = O_\vartheta(\langle L\rangle)$.*

PROOF. Let us assume, by way of contradiction, that $\{1\} \neq O_\vartheta(\langle L\rangle)$, let us write ℓ instead of ℓ_L, and let us fix an element s in $O_\vartheta(\langle L\rangle) \setminus \{1\}$ such that $\ell(s)$ is as small as possible.

Since $1 \neq s$, there exist elements r in $\langle L\rangle$ and l in L such that $s \in rl$ and $\ell(s) = \ell(r) + 1$; cf. Lemma 3.1.2. From $\ell(s) = \ell(r) + 1$ and the minimal choice of s we obtain $1 = r$ or $r \notin O_\vartheta(\langle L\rangle)$.

If $1 = r$, $s \in L$, contradiction. Thus, $r \notin O_\vartheta(\langle L\rangle)$.

From $s \in rl$ and $\ell(s) = \ell(r) + 1$ we obtain $a_{rlq} = \delta_{qs}$ for each element q in $\langle L\rangle$; cf. Lemma 9.6.1(i). Thus, by Lemma 1.1.3(iv),

$$n_s = \sum_{q\in S} a_{rlq}n_q = n_r n_l.$$

contrary to $s \in O_\vartheta(\langle L\rangle)$ and $r \notin O_\vartheta(\langle L\rangle)$.

Corollary 9.6.3 *Assume that $\langle L\rangle = S$ and that S has finite valency. Then, for each field C we have $C[L] = CS$.*

PROOF. This is an immediate consequence of Lemma 9.6.1(ii).

10

Dihedral Closed Subsets

A closed subset of S is called *dihedral* if it is generated by a set of two involutions of S.

In this chapter, we investigate dihedral closed subsets of S. The letter L stands for a set of two (different) involutions of S. We shall look at $\langle L \rangle$. As earlier, we shall write ℓ instead of ℓ_L.

Recall that, for each element q in $\langle L \rangle$, $S_{-1}(q, L)$ is our notation for the set of all elements r in $\langle L \rangle$ such that there exists an element p in $\langle L \rangle$ with $r \in pq$ and $\ell(r) = \ell(p) + \ell(q)$. Keeping the notation introduced in Section 3.4 and used in Section 3.5 and Section 3.6 we shall write, for each element s in $\langle L \rangle$, $S_{-1}(s)$ instead of $S_{-1}(s, L)$.

In accordance with Section 3.5 we shall denote by $S_{-1}(L)$ the intersection of the two sets $S_{-1}(l)$ with $l \in L$.

It may happen that $S_{-1}(L)$ is empty. In this case, the structure of $\langle L \rangle$ is easy to describe. More challenging is it to look at the case where $S_{-1}(L)$ is not empty, and it is this case on which we shall focus in the present chapter. In fact, we are mainly interested in the case where $S_{-1}(L)$ contains exactly one element.

Later in this chapter, in Section 10.6, we shall assume additionally that $\langle L \rangle$ has finite valency. It is one of the main goals of this chapter to show that (in this case) L is a Coxeter set (defined in Section 3.6) or a Moore set. (The definition of a Moore set will be given in Section 10.6.) This goal will be achieved in Theorem 10.6.6.

A general theory of Coxeter sets will be developed in the last two chapters of this monograph. A more detailed investigation of the first case of Theorem 10.6.6 will be part of this more general approach.

If $\langle L \rangle = S$ and S has finite valency, the condition that $S_{-1}(L)$ has exactly one element is related to a natural condition about the scheme ring of S over an algebraically closed field of characteristic 0. In fact, we shall see that $S_{-1}(L)$

possesses exactly one element if there exists an algebraically closed field C of characteristic 0 such that $C[L] = CS$.

The results in this section show that the notion of a Coxeter set emerges naturally from the theory of dihedral closed subsets. On this occasion, it might be worthwhile to recall that the theory of Coxeter sets is equivalent to the theory of buildings in the sense of Jacques Tits.

In this chapter, we deal only with closed subsets generated by a set of two involutions. However, our approach indicates that the concepts and techniques developed in this chapter generalize meaningfully to closed subsets generated by an arbitrary finite set of involutions.

10.1 General Remarks

Let l be an element in L.

We set $R_0(l) := \{1\}$.

Let j be a positive integer, let $l_1, \ldots, l_j$ be elements in L, and let us assume that, for each integer i with $1 \leq i \leq j$, $l_i = l$ if and only if i is odd. In this case, we set

$$R_j(l) := l_1 \cdots l_j.$$

The following lemma is crucial for the remainder of this section.

Lemma 10.1.1 *Let i and j be positive integers such that $i \leq j$, let h and k be elements in L such that $h \neq k$, and let s be an element in $R_j(h)^*R_i(h)$. Then the following hold.*

(i) *If j is odd, there exists an integer n such that $j - i \leq n \leq j + i - 1$ and $s \in R_n(h)$.*

(ii) *If j is even, there exists an integer n such that $j - i \leq n \leq j + i - 1$ and $s \in R_n(k)$.*

Proof. Let us first look at the case where $i = 1$. In this case, our assumption says that $s \in R_j(h)^*h$. Thus, if j is odd,

$$s \in R_j(h)h \subseteq R_{j-1}(h) \cup R_j(h).$$

Similarly, if j is even,

$$s \in R_j(k)h \subseteq R_{j-1}(k) \cup R_j(k),$$

so that we are done in the case where $i = 1$.

Let us now assume that $2 \leq i$. Then

$$R_{j-1}(k)^*hhR_{i-1}(k) \subseteq R_{j-1}(k)^*R_{i-1}(k) \cup R_{j-1}(k)^*hR_{i-1}(k).$$

Thus, as $s \in R_j(h)^*R_i(h)$ and $R_j(h)^*R_i(h) = R_{j-1}(k)^*hhR_{i-1}(k)$,

$$s \in R_{j-1}(k)^*R_{i-1}(k) \cup R_{j-1}(k)^*hR_{i-1}(k).$$

If j is odd,

$$R_{j-1}(k)^*hR_{i-1}(k) = R_{j+i-1}(h).$$

If j is even,

$$R_{j-1}(k)^*hR_{i-1}(k) = R_{j+i-1}(k).$$

Thus, we may assume that $s \in R_{j-1}(k)^*R_{i-1}(k)$.

If j is odd, $j-1$ is even. Thus, by induction, there exists an integer n such that $j-i \leq n \leq j+i-3$ and $s \in R_n(h)$. If j is even, $j-1$ is odd. Thus, by induction, there exists an integer n such that $j-i \leq n \leq j+i-3$ and $s \in R_n(k)$.

Corollary 10.1.2 *Let i and j be positive integers, let h and k be elements in L such that $h \neq k$. Then we have the following.*

(i) *If $R_i(h) \cap R_j(k)$ is not empty, $1 \in R_{i+j}(h) \cup R_{i+j}(k)$.*

(ii) *Assume that $R_i(h) \cap R_j(h)$ is not empty and that $i \leq j$. Then there exists an integer n with $j-i \leq n \leq j+i-1$ and $1 \in R_n(h) \cup R_n(k)$.*

PROOF. (i) Let us assume that $R_i(h) \cap R_j(k)$ is not empty, and let s be an element in $R_i(h) \cap R_j(k)$. Referring to Lemma 1.3.2(iii) we then obtain

$$1 \in s^*s \subseteq R_i(h)^*R_j(k) \subseteq R_{i+j}(h) \cup R_{i+j}(k).$$

(ii) We are assuming that $R_i(h) \cap R_j(h)$ is not empty. Let s be an element in $R_i(h) \cap R_j(h)$. Then

$$1 \in s^*s \subseteq R_j(h)^*R_i(h),$$

so that the claim follows from Lemma 10.1.1.

Lemma 10.1.3 *Let n be a non-negative integer, and let s be an element in L^n. Then there exist elements i in $\{0, \ldots, n\}$ and l in L such that $s \in R_i(l)$.*

PROOF. The claim holds obviously if $n = 0$. Let us, therefore, assume that $1 \leq n$.

Since $s \in L^n$, there exists an element r in L^{n-1} such that $s \in rL$. From $r \in L^{n-1}$ we obtain, by induction, an integer i with $1 \leq i \leq n$ and an element l in L such that $r \in R_{i-1}(l)$.

If $1 = i$, we obtain from $r \in R_{i-1}(l)$ that $r \in R_0(l)$. It follows that $1 = r$. Thus, as $s \in rL$, $s \in L$, and we are done.

If $2 \leq i$, we obtain from $s \in rL$ and $r \in R_{i-1}(l)$ that

$$s \in R_{i-1}(l)L \subseteq R_{i-2}(l) \cup R_{i-1}(l) \cup R_i(l),$$

and that finishes our proof.

Corollary 10.1.4 *Let s be an element in $\langle L\rangle$, and let l be an element in L. Then $s \in S_{-1}(l)$ if and only if $s^* \in R_{\ell(s)}(l) \setminus \{1\}$.*

PROOF. We set $j := \ell(s)$, and we assume first that $s \in S_{-1}(l)$. Then $1 \neq s$ and, by definition, there exists an element r in $\langle L\rangle$ such that $s \in rl$ and $j = \ell(r) + 1$.

From $j = \ell(r) + 1$ we obtain $\ell(r) = j - 1$. Thus, $r \in L^{j-1}$. It follows that $r^* \in L^{j-1}$. Thus, there exist elements i in $\{1, \dots, j-1\}$ and k in L such that $r^* \in R_i(k)$; cf. Lemma 10.1.3.

From $r^* \in R_i(k)$ we obtain $r \in R_i(k)^*$. Thus, as $s \in rl$,

$$s \in R_i(k)^* l.$$

If $i = 0$, we obtain from $s \in R_i(k)^* l$ that $s = l$, and we are done. Thus, we may assume that $1 \leq i$.

If $k = l$, we obtain from $s \in R_i(k)^* l$ that $s \in R_i(k)^* k$. Thus, there exists a non-negative integer n such that $n \leq i$ and $s \in L^n$; cf. Lemma 10.1.1. It follows that $\ell(s) \leq n \leq i \leq j - 1$, contradiction.

If $k \neq l$, we obtain from $s \in R_i(k)^* l$ that $s \in R_{i+1}(l)^*$. Thus, $\ell(s) \leq i+1 \leq j$, so that $\ell(s) = i+1$. It follows that $s \in R_{\ell(s)}(l)^*$, and this implies $s^* \in R_{\ell(s)}(l)$.

Let us now, conversely, assume that $s^* \in R_{\ell(s)}(l) \setminus \{1\}$. Then, as $j = \ell(s)$, $s^* \in R_j(l) \setminus \{1\}$. Thus, there exists an element r in L^{j-1} such that $s \in rl$.

From r in L^{j-1} we obtain $\ell(r) \leq j - 1$. Thus, $\ell(s) = \ell(r) + 1$. Thus, as $s \in rl$, $s \in S_{-1}(l)$.

Lemma 10.1.5 *Assume that there exists a positive integer j and an element l in L such that $1 \in R_j(l)$. Let j be the smallest positive integer for which there exists an element l in L with $1 \in R_j(l)$.*
Then there exists an element s in $S_{-1}(L)$ such that $j = 2\ell(s)$.

PROOF. From $1 \in R_j(l)$ we obtain $4 \leq j$ and $1 \in R_2(l)R_{j-2}(l)$. By Lemma 1.3.2(ii), the latter condition implies $1 \in R_{j-2}(l)R_2(l)$.

Let us assume, by way of contradiction, that j is odd. Then, $j - 2$ is odd, so that $R_{j-2}(l)^* = R_{j-2}(l)$. Thus, as $1 \in R_{j-2}(l)R_2(l)$ and $2 \leq j - 2$, there exists an integer n with $j - 4 \leq n \leq j - 1$ and $1 \in R_n(l)$; cf. Lemma 10.1.1(i). This contradicts our (minimal) choice of j.

Thus, j is even, and that means that there exists an integer d such that $j = 2d$. It follows that $1 \in R_{2d}(h)$. Thus, there exists an element s in $R_d(h)$ such that $1 \in sR_d(k)^*$. It follows that $s \in R_d(h) \cap R_d(k)$.

From $s \in R_d(h)$ we obtain $\ell(s) \leq d$.

By Lemma 10.1.3, there exist elements i in $\{1, \ldots, \ell(s)\}$ and l in L such that $s \in R_i(l)$. Thus, as $s \in R_d(l)$,

$$1 \in s^* s \subseteq R_d(l)^* R_i(l).$$

Thus, as $\ell(s) \leq d$, there exists an integer n such that $d - i \leq n \leq d + i - 1$ and $s \in R_n(h) \cup R_n(k)$; cf. Lemma 10.1.1. From this we conclude that $i = d$. However, we have $i \leq \ell(s) \leq d$, so that $\ell(s) = d$. It follows that $s \in R_{\ell(s)}(h) \cap R_{\ell(s)}(k)$. Thus, by Corollary 10.1.4, $s^* \in S_{-1}(L)$.

Recall that L is called spherical if $S_{-1}(L)$ is not empty.

Lemma 10.1.6 *Assume L to be spherical, and let s be an element in $S_{-1}(L)$. Then, for each element l in L, $1 \in R_{2\ell(s)}(l)$.*

PROOF. We have fixed an element s in $S_{-1}(L)$. Thus, for each element l in L, s in $S_{-1}(l)$. Thus, for each element l in L, $s^* \in R_{\ell(s)}(l)$; cf. Corollary 10.1.4. Thus, by Corollary 10.1.2(i), $1 \in R_{2\ell(s)}(l)$.

From Lemma 10.1.5 together with Lemma 10.1.6 we obtain, in particular, that L is spherical if and only if there exists a positive integer j and an element l in L such that $1 \in R_j(l)$.

Lemma 10.1.7 *If $\langle L \rangle$ is finite, L is spherical.*

PROOF. Let l be an element in L, and let us assume $\langle L \rangle$ to be finite. Then there exist positive integers i and j such that $i+1 \leq j$ and $R_i(l) \cap R_j(l)$ is not empty. Thus, the claim follows from Corollary 10.1.2(ii) together with Lemma 10.1.5.

10.2 The Spherical Case

Throughout this section, the set L is assumed to be spherical.

We define d_L to be the smallest element in $\ell(S_{-1}(L))$. However, having fixed L for the remainder of this section, we shall write d instead of d_L.

Note that $2 \leq d$.

Let h and k be elements in L such that $h \neq k$. Since $S_{-1}(L)$ contains an element of length d, we obtain from Corollary 10.1.4 that $R_d(h) \cap R_d(k)$ is not empty. Of course, we also have that $R_0(h) \cap R_0(k)$ is not empty. The first part of the following lemma is a partial converse of this.

Lemma 10.2.1 *For any two integers i and j satisfying $0 \leq i \leq d$ and $0 \leq j \leq d$, we have the following.*

(i) *Let h and k be elements in L such that $h \neq k$ and $\emptyset \neq R_i(h) \cap R_j(k)$. Then $i = j \in \{0, d\}$.*

(ii) *Let l be an element in L with $\emptyset \neq R_i(l) \cap R_j(l)$. Then $i = j$.*

PROOF. (i) We are assuming that $R_i(h) \cap R_j(k)$ is not empty. Thus, by Corollary 10.1.2(i),

$$1 \in R_{i+j}(h) \cup R_{i+j}(k).$$

Thus, we must have $i + j = 0$ or $2d \leq i + j$; cf. Lemma 10.1.5.

In the first case, we are done. In the second case, we obtain from $0 \leq i \leq d$ and $0 \leq j \leq d$ that $i = j = d$.

(ii) The claim is obviously true if $i = 0$. Thus, we may assume that $1 \leq i \leq j$. In this case, we obtain from Corollary 10.1.2(ii) an integer n with $j - i \leq n \leq j + i - 1$ and an element k in L with $1 \in R_n(k)$.

From $i \leq d$, $j \leq d$, and $n \leq i + j - 1$ we obtain $n \leq 2d - 1$. Thus, by Lemma 10.1.5, $n = 0$. Thus, as $1 \leq i \leq j$ and $j - i \leq n$, $i = j$.

Lemma 10.2.2 *Let j be an integer such that $1 \leq j \leq d$, and let l be an element in L such that $|R_j(l)| = 1$. Then $|R_{j-1}(l)| = 1$.*

PROOF. We are assuming that $|R_j(l)| = 1$. Thus, there exists an element s in $R_j(l)$ such that $\{s\} = R_j(l)$. Let us fix elements p and q in $R_{j-1}(l)$. We shall see that $p = q$.

Let us first assume j to be odd. In this case, $j - 1$ is even, so that

$$pl \subseteq R_{j-1}(l)l = R_j(l) = \{s\}.$$

Similarly, $ql \subseteq \{s\}$, so that $pl = \{s\} = ql$. It follows that

$$q \in qll = pll \subseteq \{p\} \cup pl = \{p, s\}.$$

If $q = s$, $R_{j-1}(l) \cap R_j(l)$ is not empty, contrary to Lemma 10.2.1(ii). Therefore, $p = q$.

Similarly, the assumption that j is even leads to $p = q$.

Lemma 10.2.3 *For any two elements s in $\langle L \rangle$ and l in L, there exists a non-negative integer j such that $s \in R_j(l)$.*

PROOF. Let s be an element in $\langle L \rangle$. Then, by Lemma 3.1.1(i), there exists a non-negative integer n such that $s \in L^n$. Thus, by Lemma 10.1.3, there exists a non-negative integer i and an element l in L such that $s \in R_i(l)$.

Let j be a multiple of $2d$ such that $i \leq j$. From Lemma 10.1.6 we know that $1 \in R_{2d}(l)$. Thus, $1 \in R_j(l)$. It follows that

$$s \in 1s \subseteq R_j(l)^* R_i(l).$$

Now the claim follows from Lemma 10.1.1(ii).

Let s be an element in $\langle L \rangle$, and let l be an element in L. According to Lemma 10.2.3, there exists a non-negative integer j such that $s \in R_j(l)$. In the following, we shall denote by $\ell_l(s)$ the smallest non-negative integer j satisfying $s \in R_j(l)$.

Note that, for each element l in L, $\ell_l(1) = 0$ and $\ell_l(l) = 1$.

It follows immediately from the definition of ℓ that, for any two elements s in $\langle L \rangle$ and l in L, $\ell(s) \leq \ell_l(s)$. From Lemma 10.1.3 (together with Lemma 10.2.3) we obtain a little more, namely that

$$\ell(s) = \min\{\ell_l(s) \mid l \in L\}$$

for each element s in $\langle L \rangle$. We shall use this little observation frequently without further reference.

Lemma 10.2.4 *For each element s in $\langle L \rangle$, we have the following.*

(i) *Let l be an element in L such that $\ell_l(s)$ is odd. Then $\ell_l(s^*) \leq \ell_l(s)$.*

(ii) *Let h and k be elements in L such that $h \neq k$ and $\ell_h(s)$ is even. Then $\ell_k(s^*) \leq \ell_h(s)$.*

PROOF. (i) Set $j := \ell_l(s)$. Then $s \in R_j(l)$. It follows that $s^* \in R_j(l)^*$. On the other hand, we are assuming j to be odd, so that $R_j(l)^* = R_j(l)$. Thus, $s^* \in R_j(l)$, and it follows that $\ell_l(s^*) \leq j$.

(ii) Set $j := \ell_h(s)$. Then $s \in R_j(h)$. It follows that $s^* \in R_j(h)^*$. On the other hand, we are assuming j to be even, so that $R_j(h)^* = R_j(k)$. Thus, $s^* \in R_j(k)$, and it follows that $\ell_k(s^*) \leq j$.

Lemma 10.2.5 *Let p and q be elements in $\langle L \rangle$, and let l be an element in L. Then, if $q \in pL$ and $1 \neq p$, $\ell_l(q) \leq \ell_l(p) + 1$.*

PROOF. Set $j := \ell_l(p)$. Then, by definition, $p \in R_j(l)$. Moreover, as $1 \neq p$, $0 \neq j$. From $q \in pL$ and $p \in R_j(l)$ we obtain

$$q \in R_j(l)L \subseteq R_{j-1}(l) \cup R_j(l) \cup R_{j+1}(l).$$

It follows that $\ell_l(q) \leq j + 1$.

Lemma 10.2.6 *For each element s in $\langle L \rangle$, we have the following.*

(i) *Let h and k be elements in L such that $h \neq k$. Then, if $1 \neq s$, $2d \leq \ell_h(s) + \ell_k(s)$.*

(ii) *Let j be an integer with $0 \leq j \leq d$, and let l be an element in L such that $s \in R_j(l)$. Then $\ell(s) = j$.*

PROOF. (i) Set $i := \ell_h(s)$ and $j := \ell_k(s)$. Then $s \in R_i(h) \cap R_j(k)$. Thus, by Corollary 10.1.2(i), $1 \in R_{i+j}(h) \cup R_{i+j}(k)$.

On the other hand, we are assuming that $1 \neq s$, so that $0 \neq i$ and $0 \neq j$. Thus, by Lemma 10.1.5, $2d \leq i + j$.

(ii) We are assuming that $s \in R_j(l)$. Thus, $\ell_l(s) \leq j$. Set $i := \ell_l(s)$. Then $i \leq j$ and $s \in R_i(l)$. It follows that $s \in R_i(l) \cap R_j(l)$. Thus, as $0 \leq i \leq d$ and $0 \leq j \leq d$, we obtain from Lemma 10.2.1(ii) that $i = j$. Thus, $\ell_l(s) = j$.

The statement is obviously true if $1 = s$. Thus, we may assume that $1 \neq s$. Thus, as $\ell_l(s) = j$ and $j \leq d$, we obtain from (i) that $\ell(s) = \ell_l(s)$.

Lemma 10.2.7 *Let j be an integer with $1 \leq j \leq d$, and let l be an element in L. Then we have the following.*

(i) *If j is odd, $R_j(l) \subseteq S_{-1}(l)$.*

(ii) *Let h and k be elements in L such that $h \neq k$. Then, if j is even, $R_j(h) \subseteq S_{-1}(k)$.*

PROOF. (i) Assume j to be odd, and let s be an element in $R_j(l)$. Then there exists an element r in $R_{j-1}(l)$ such that $s \in rl$.

From $s \in R_j(l)$ and $j \leq d$ we obtain $\ell(s) = j$; cf. Lemma 10.2.6(ii). From $r \in R_{j-1}(l)$ we obtain $\ell(r) \leq j - 1$. Thus, as $s \in rl$, $s \in S_{-1}(l)$.

(ii) Assume j to be even, and let s be an element in $R_j(h)$. Then there exists an element r in $R_{j-1}(h)$ such that $s \in rk$.

From $s \in R_j(h)$ and $j \leq d$ we obtain $\ell(s) = j$; cf. Lemma 10.2.6(ii). From $r \in R_{j-1}(h)$ we obtain $\ell(r) \leq j - 1$. Thus, as $s \in rk$, $s \in S_{-1}(k)$.

Let us briefly look at the consequences of our results for $S_{-1}(L)$ and L.

Lemma 10.2.8 *We have $S_{-1}(L)^* = S_{-1}(L)$.*

PROOF. Let s be an element in $S_{-1}(L)$. Then, for each element l in L, $s^* \in R_{\ell(s)}(l)$; cf. Corollary 10.1.4. Thus, for each element l in L, $\ell_l(s^*) \leq \ell(s) = \ell(s^*)$, and that means that $\ell(s^*) = \ell_l(s^*)$. Thus, by Lemma 10.2.4, for each element l in L,

$$\ell_l(s) \leq \ell(s) = \ell(s^*).$$

Thus, by Corollary 10.1.4, $s^* \in S_{-1}(L)$.

Lemma 10.2.9 *We have $2d \leq |\langle L \rangle|$.*

PROOF. This follows from Lemma 10.2.1.

10.3 Arithmetic of the Length Function

Throughout this section, the set L is assumed to be spherical.

Let h and k be elements in L such that $h \neq k$. For each element s in $\langle L \rangle$, we define

$$\nu_h(s) := \ell_h(s) - \ell_k(s).$$

Note that, for any two elements s in $\langle L \rangle$ and l in L, $\ell(s) = \ell_l(s)$ if and only if $\nu_l(s) \leq 0$.

Lemma 10.3.1 *For each element l in L, we have*

$$S_{-1}(l) = \{s \in \langle L \rangle \mid \nu_l(s^*) \leq 0\} \setminus \{1\}.$$

PROOF. Let l be an element in L, and let s be an element in $\langle L \rangle$. Then we have $s^* \in R_{\ell(s)}(l)$ if and only if $\ell(s^*) = \ell_l(s^*)$, and that is equivalent to $\nu_l(s^*) \leq 0$. On the other hand, we know from Corollary 10.1.4 that $s \in S_{-1}(l)$ is equivalent to $s^* \in R_{\ell(s)}(l) \setminus \{1\}$.

As a consequence of Lemma 10.3.1 and Lemma 10.2.8 we obtain

$$S_{-1}(L) = \{s \in \langle L \rangle \mid \nu_l(s) = 0\} \setminus \{1\}$$

for each element l in L.

Lemma 10.3.2 *For any two elements s in $\langle L \rangle$ and l in L, we have the following.*

(i) *If $\ell(s)$ is odd, $0 \leq \nu_l(s^*)\nu_l(s)$.*

(ii) *If $\ell(s)$ is even, $\nu_l(s^*)\nu_l(s) \leq 0$.*

PROOF. (i) Assume first that $\nu_l(s) \leq 0$. Then $\ell(s) = \ell_l(s)$. Thus, assuming $\ell(s)$ to be odd we obtain from Lemma 10.2.4(i) that $\ell_l(s^*) \leq \ell(s^*)$. It follows that $\nu_l(s^*) \leq 0$.

Similarly, $0 \leq \nu_l(s)$ yields $0 \leq \nu_l(s^*)$.

(ii) Assume first that $\nu_l(s) \leq 0$. Then $\ell(s) = \ell_l(s)$. Thus, by Lemma 10.2.4(ii), $0 \leq \nu_l(s^*)$.

Similarly, $0 \leq \nu_l(s)$ yields $\nu_l(s^*) \leq 0$.

Lemma 10.3.3 *Let p and q be elements in $\langle L \rangle$ such that $q \in pL$. Then, for each element l in L, $0 \leq \nu_l(p)\nu_l(q)$.*

PROOF. Assume, by way of contradiction, that $\nu_l(p)\nu_l(q) \leq -1$. Then there exist elements h and k in L such that $\ell_h(p) \leq \ell_k(p) - 1$ and $\ell_k(q) \leq \ell_h(q) - 1$.

From $\nu_l(p)\nu_l(q) \leq -1$ we obtain $1 \neq p$ and $1 \neq q$. Thus, by Lemma 10.2.5,

$$\ell_h(p) \leq \ell_k(p) - 1 \leq \ell_k(q) \leq \ell_h(q) - 1 \leq \ell_h(p).$$

It follows that $\ell_h(p) = \ell_k(q)$. We set $j := \ell_h(p)$.

Since the claim is true for $p = 1$, we may assume that $1 \leq j$.

Since, by hypothesis, $q \in pL$, we have $q \in ph$ or $q \in pk$.

Let us assume first that $q \in ph$. If j is odd, $q \in ph \subseteq R_j(h)^*h$. Thus, by Lemma 10.1.1(i), $q \in R_{j-1}(h) \cup R_j(h)$, contrary to $j+1 \leq \ell_h(q)$. If j is even, $p \in qh \subseteq R_j(h)^*h$. Thus, by Lemma 10.1.1(ii), $p \in R_{j-1}(k) \cup R_j(k)$, contrary to $j+1 \leq \ell_k(p)$.

Assume now that $q \in pk$. If j is odd, $p \in qk \subseteq R_j(k)^*k$. Thus, by Lemma 10.1.1(i), $p \in R_{j-1}(k) \cup R_j(k)$, contrary to $j+1 \leq \ell_k(p)$. If j is even, $q \in pk \subseteq R_j(k)^*k$. Thus, by Lemma 10.1.1(ii), $q \in R_{j-1}(h) \cup R_j(h)$, contrary to $j+1 \leq \ell_h(q)$.

Lemma 10.3.4 *Let p, q, r, and s be elements in $\langle L\rangle \setminus S_{-1}(L)$ such that $p \in Lq \cap rL$ and $s \in Lr \cap qL$. Then $\ell(p) = \ell(r)$ if and only if $\ell(q) = \ell(s)$.*

Proof. Since p, q, r, $s \notin S_{-1}(L)$, we obtain from Lemma 10.3.3 that

$$1 \leq \nu_l(p^*)\nu_l(q^*),\ 1 \leq \nu_l(r^*)\nu_l(s^*),\ 1 \leq \nu_l(p)\nu_l(r),\ 1 \leq \nu_l(q)\nu_l(s)$$

for each element l in L.

Let l be an element in L, and let us assume that $\ell(p) = \ell(r)$. Then, by Lemma 10.3.2,

$$1 \leq \nu_l(p^*)\nu_l(p)\nu_l(r^*)\nu_l(r).$$

Thus, by the above four observations,

$$1 \leq \nu_l(q^*)\nu_l(q)\nu_l(s^*)\nu_l(s).$$

Thus, by Lemma 10.3.2, $\ell(q) - \ell(s)$ is even.

Now recall that we are assuming that $s \in qL$. Thus, $\ell(q) - \ell(s) \in \{-1, 0, 1\}$. It follows that $\ell(q) = \ell(s)$.

Lemma 10.3.5 *Let p, q, r, and s be elements in $\langle L\rangle \setminus S_{-1}(L)$, and let l be an element in L such that $p \in Lq \cap rl$ and $s \in Lr \cap ql$. Then $\ell(p) + 1 = \ell(r)$ if and only if $\ell(q) + 1 = \ell(s)$.*

Proof. Let us assume that $\ell(p) + 1 = \ell(r)$. Then, by Lemma 10.3.4, $\ell(q) \neq \ell(s)$. Thus, as we are assuming that $s \in qL$, we conclude that

$$\ell(s) \in \{\ell(q) - 1, \ell(q) + 1\}.$$

Let us assume, by way of contradiction, that $\ell(s) = \ell(q) - 1$.

From $s \in ql$ we obtain $q \in sl$; cf. Lemma 1.3.3(i). Thus, as we are assuming that $\ell(q) = \ell(s) + 1$, $q \in S_{-1}(l)$. Thus, by Lemma 10.3.1, $\nu_l(q^*) \leq -1$.

Similarly, we obtain from $p \in rl$ and $\ell(r) = \ell(p) + 1$ that $\nu_l(r^*) \leq -1$.

Now recall that, by Lemma 10.3.3, $1 \leq \nu_l(p^*)\nu_l(q^*)$. Thus, as $\nu_l(q^*) \leq -1$ and $\nu_l(r^*) \leq -1$,

$$1 \leq \nu_l(p^*)\nu_l(r^*).$$

From Lemma 10.3.3 we also obtain $1 \leq \nu_l(p)\nu_l(r)$. Thus,

$$1 \leq \nu_l(p^*)\nu_l(p)\nu_l(r^*)\nu_l(r).$$

But, as $\ell(p) + 1 = \ell(r)$, this contradicts Lemma 10.3.2.

Proposition 10.3.6 *Let p, q, and r be elements in $\langle L \rangle$ such that $p \in S_1(r)$ and $r \in S_1(q)$. Assume that, for each element s in $\langle L \rangle$ with*

$$\ell(r) + 1 \leq \ell(s) \leq \ell(p) + \ell(r) + \ell(q),$$

$s \notin S_{-1}(L)$. Then there exists an element s in prq such that $\ell(s) = \ell(p) + \ell(r) + \ell(q)$.

Proof. Let y be an element in X, and let z be an element in yr.

We are assuming that $r \in S_1(q)$. Thus, there exists an element u in rq such that $\ell(u) = \ell(r) + \ell(q)$.

From $u \in rq$ we obtain $r \in uq^*$; cf. Lemma 1.3.3(i). Thus, as $z \in yr$, $z \in yuq^*$. Thus, there exists an element w in yu such that $z \in wq^*$.

Set $n := \ell(q)$. Then $q \in L^n$. Thus, as $w \in zq$, there exist elements $z_0, \ldots, z_n$ in X such that $z_0 = z$, $z_n = w$, and, for each element j in $\{1, \ldots, n\}$, $z_j \in z_{j-1}L$; cf. Lemma 1.3.9.

We are assuming that $p \in S_1(r)$. Thus, there exists an element t in pr such that $\ell(t) = \ell(p) + \ell(r)$.

From $t \in pr$ we obtain $r \in p^*t$; cf. Lemma 1.3.3(ii). Thus, as $z \in yr$, $z \in yp^*t$. Thus, there exists an element v in yp^* such that $z \in vt$.

Set $m := \ell(p)$. Then, as before, we find elements $y_0, \ldots, y_m$ in X such that $y_0 = y$, $y_m = v$, and, for each element i in $\{1, \ldots, m\}$, $y_i \in Ly_{i-1}$.

For any two elements i in $\{0, \ldots, m\}$ and j in $\{0, \ldots, n\}$, we define s_{ij} to be the element in $\langle L \rangle$ which satisfies

$$z_j \in y_i s_{ij}.$$

Then $s_{00} = r$, $s_{m0} = t$, and $s_{0n} = u$. Moreover, for any two elements i in $\{1, \ldots, m\}$ and j in $\{1, \ldots, n\}$, we have

$$s_{ij} \in Ls_{i-1,j} \cap s_{i,j-1}L.$$

From $s_{00} = r$ and $s_{0n} = u$ we obtain $\ell(s_{00}) + n = \ell(s_{0n})$. Thus, we obtain from Lemma 3.4.1 that

$$\ell(s_{0j}) = j$$

for each element j in $\{1, \ldots, n\}$. Similarly, we have

$$\ell(s_{i0}) = i$$

for each element i in $\{1, \ldots, m\}$.

Our claim is that, for any two elements i in $\{1, \ldots, m\}$ and j in $\{1, \ldots, n\}$, $\ell(s_{ij}) = i + \ell(r) + j$.

By induction, we may assume that $\ell(s_{i-1,j-1}) = i+\ell(r)+j-2$, that $\ell(s_{i,j-1}) = i + \ell(r) + j - 1$, and that $\ell(s_{i-1,j}) = i + \ell(r) + j - 1$. Thus, by Lemma 10.3.5, $\ell(s_{ij}) = i + \ell(r) + j$.

Since we have that, for any two elements i in $\{1, \ldots, m\}$ and j in $\{1, \ldots, n\}$, $\ell(s_{ij}) = i + \ell(r) + j$,

$$\ell(s_{mn}) = m + \ell(r) + n = \ell(p) + \ell(r) + \ell(q),$$

and that finishes the proof.

10.4 Two Characteristic Subsets

Throughout this section, the set L is assumed to be spherical.

For each element l in L, we define

$$S_0(l) := \{s \in \langle L \rangle \mid \min \ell(sl) = \ell(s)\}.$$

Note that, for each element l in L, $1 \notin S_0(l)$.

Lemma 10.4.1 *Let l be an element in L, let q be an element in $S_0(l)$, and let p be an element in $\langle L \rangle$ such that $q \in S_{-1}(p)$. Then the following hold.*

(i) *We have $p \notin S_{-1}(l)$.*

(ii) *For each element r in $pl \setminus S_{-1}(l)$, $\ell(r) = \ell(p)$.*

PROOF. (i) Assume, by way of contradiction, that $p \in S_{-1}(l)$. Then, by Lemma 3.4.4(iii), $q \in S_{-1}(l)$. (We apply this lemma to p in place of q and l in place of p.) This contradicts our hypothesis that $q \in S_0(l)$.

(ii) Let r be an element in $pl \setminus S_{-1}(l)$, and let us assume, by way of contradiction, that $\ell(r) \neq \ell(p)$. Then, as $r \in pl$, $\ell(r) = \ell(p) - 1$ or $\ell(r) = \ell(p) + 1$.

In the first case, we obtain $p \in S_{-1}(l)$, contrary to (i). In the second case, we obtain $r \in S_{-1}(l)$, contrary to our hypothesis.

Let l be an element in L. Similarly to $S_{-1}(l)$, the set $S_0(l)$ can be described in terms of ν_l.

Lemma 10.4.2 *For each element l in L, we have*

$$S_0(l) = \{s \in \langle L \rangle \mid \nu_l(s^*) = 1\}.$$

PROOF. Let us first prove that, for each element s in $S_0(l)$, $\nu_l(s^*) = 1$. In order to do so we pick an element s in $S_0(l)$. From $s \in S_0(l)$ we obtain $s \notin S_{-1}(l)$. Thus, by Lemma 10.3.1, $1 \leq \nu_l(s^*)$.

Set $j := \ell(s)$. Then, as $s \in S_0(l)$ by definition, there exists an element r in sl such that $\ell(r) = j$. From $r \in sl$ we obtain $s^* \in lr^*$; cf. Lemma 1.3.3(iii). From $\ell(r) = j$ we obtain $\ell(r^*) = j$.

Let k be the element in $L \setminus \{l\}$.

Suppose that $\ell_l(r^*) = j$. Then $r^* \in R_j(l)$, whence $s^* \in lR_j(l)$. Thus, as $1 \leq j$, $s^* \in R_{j-1}(k) \cup R_j(l)$. If $s^* \in R_{j-1}(k)$, $\ell(s^*) \leq j-1$, contrary to $\ell(s) = j$. If $s^* \in R_j(l)$, $\ell_l(s^*) \leq j$, so that $\nu_l(s^*) \leq 0$, contrary to $1 \leq \nu_l(s^*)$.

Thus, we must have $\ell_k(r^*) = j$. It follows that $r^* \in R_j(k)$, whence $s^* \in lR_j(k) = R_{j+1}(l)$. Thus, $\ell_l(s^*) \leq j+1$. It follows that $\nu_l(s^*) \leq 1$.

Let us now prove that only elements s in $S_0(l)$ may satisfy $\nu_l(s^*) = 1$.

Let s be an element in $\langle L \rangle$ such that $\nu_l(s^*) = 1$. Set $j := \ell(s^*)$. Then, $\ell_l(s^*) = j+1$. In particular, $s^* \in R_{j+1}(l)$. Thus, there exists an element r in L^j such that $s^* \in lr$.

From $s^* \in lr$ we obtain $s \in r^*l$; cf. Lemma 1.3.2(iii). From $r \in L^j$ we obtain $r^* \in L^j$, so that $\ell(r^*) \leq j = \ell(s)$. Thus, $s \in S_{-1}(l)$ or $s \in S_0(l)$.

Since $\nu_l(s^*) = 1$, Lemma 10.3.1 tells us also that $s \notin S_{-1}(l)$. Therefore, we must have $s \in S_0(l)$.

From Lemma 3.4.5(ii) (together with Lemma 10.2.8) we know that, for each element s in $S_{-1}(L)$, $\{s\} = S_{-1}(s)$. The following lemma shows that, for any two elements l in L and s in $S_0(l)$, there exists an element q in $S_{-1}(L)$ such that s is 'almost' in $S_{-1}(q)$.

Proposition 10.4.3 *Let l be an element in L, and let s be an element in $S_0(l)$. Then there exist elements p in $\langle L \rangle$ and q in $S_{-1}(L)$ such that $s \in pq$ and $\ell(s) + 1 = \ell(p) + \ell(q)$.*

PROOF. We set $j := \ell_l(s^*)$. Then $s^* \in R_j(l)$. Note also that $2 \leq j$. Thus, there exist elements $l_1, \ldots, l_j$ in L such that

$$s^* \in l_1 \cdots l_j$$

and, for each element i in $\{1, \ldots, j\}$, $l_i = l$ if and only if i is odd.

Since $s^* \in l_1 \cdots l_j$, Lemma 1.3.5 yields elements $s_0, \ldots, s_j$ in $\langle L \rangle$ such that $s_0 = 1$, $s_j = s$, and, for each element i in $\{1, \ldots, j\}$,

$$s_i \in l_i s_{i-1}.$$

Since $s_1 = l$, $\ell_l(s_1^*) = 1$. Therefore, by induction, Lemma 10.2.5 yields that, for each element i in $\{1, \ldots, j\}$, $\ell_l(s_i^*) \leq i$. Similarly, as $s_j = s$, $\ell_l(s_j^*) = j$. Thus, by induction, Lemma 10.2.5 yields that, for each element i in $\{1, \ldots, j\}$, $i \leq \ell_l(s_i^*)$. It follows that, for each element i in $\{1, \ldots, j\}$,

$$\ell_l(s_i^*) = i.$$

From $s_1 = l$ we obtain $\nu_l(s_1^*) \leq -1$. On the other hand, as $s_j = s$, $s_j \in S_0(l)$, so that, according to Lemma 10.4.2, $\nu_l(s_j^*) = 1$.

We set

$$I := \{i \in \{1, \ldots, j\} \mid 0 \leq \nu_l(s_i^*)\}.$$

Since $\nu_l(s_j^*) = 1$, $j \in I$. Thus, I is not empty. We define $i := \min I$. Then $0 \leq \nu_l(s_i^*)$.

Since $\nu_l(s_1^*) \leq -1$, $1 \notin I$. Thus, $2 \leq i$ and $\nu_l(s_{i-1}^*) \leq -1$.

However, we have that $s_i^* \in s_{i-1}^* L$. Thus, by Lemma 10.3.3, $0 \leq \nu_l(s_i^*)\nu_l(s_{i-1}^*)$. Thus, as $0 \leq \nu_l(s_i^*)$ and $\nu_l(s_{i-1}^*) \leq -1$, $\nu_l(s_i^*) = 0$. This means that $s_i^* \in S_{-1}(L)$.

We set $q := s_i^*$. Then $\ell_l(q) = i$. Thus, as $q \in S_{-1}(L)$,

$$\ell(q) = i.$$

Since $\nu_l(s_i^*) = 0$ and $\nu_l(s_j^*) = 1$, $i \leq j - 1$. Since $\nu_l(s^*) = 1$ and $\ell_l(s^*) = j$, $\ell(s^*) = j - 1$. Thus,

$$\ell(s) = j - 1.$$

Note, finally, that $s \in l_j \cdots l_{i+1} q$. Thus, there exists an element p in $l_j \cdots l_{i+1}$ such that $s \in pq$.

From $p \in l_j \cdots l_{i+1}$ we obtain $\ell(p) \leq j - i$. From $s \in pq$ and $\ell(s) = j - 1$ we obtain $j - i - 1 \leq \ell(p)$. Thus,

$$j - i - 1 \leq \ell(p) \leq j - i.$$

Assume that $\ell(p) = j - i - 1$. Then $s \in S_{-1}(q)$. Thus, as $q \in S_{-1}(L)$, $q = s$; cf. Lemma 3.4.5(ii). This contradicts $q \in S_{-1}(L)$ and $s \in S_0(l)$.

Thus, we have $\ell(p) = j - i$. Now we obtain from $\ell(q) = i$ and $\ell(s) = j - 1$ that $\ell(s) + 1 = \ell(p) + \ell(q)$.

Recall that d_L is our notation for the smallest element in $\ell(S_{-1}(L))$. For the remainder of this section, we shall write d instead of d_L.

Corollary 10.4.4 *Let s be an element in $\langle L\rangle$ such that $\ell(s) \le d-1$. Then the following hold.*

(i) *For each element l in L, $s \notin S_0(l)$.*

(ii) *There exists an element l in L such that $s \notin S_{-1}(l)$.*

PROOF. (i) This is an immediate consequence of Proposition 10.4.3.

(ii) Since $\ell(s) \le d-1$, $s \notin S_{-1}(L)$. Thus, there exists an element l in L with $s \notin S_{-1}(l)$.

Let l be an element in L. We define

$$S_\infty(l) := \{s \in \langle L\rangle \mid \ell(sl) = \{\ell(s)\}\}.$$

Note that $\langle L\rangle$ is the disjoint union of the sets $S_{-1}(l)$, $S_\infty(l)$, and $S_1(l)$. Note also that

$$S_\infty(l) \subseteq S_0(l).$$

We shall use this latter observation frequently without further mentioning.

Lemma 10.4.5 *For any two elements l in L and s in $S_\infty(l)$, $s\langle l\rangle \subseteq S_\infty(l)$.*

PROOF. Let r be an element in $s\langle l\rangle$. Then, by Lemma 2.1.4, $r\langle l\rangle = s\langle l\rangle$. Thus, $|\ell(r\langle l\rangle)| = |\ell(s\langle l\rangle)| = 1$. It follows that $r \in S_\infty(l)$.

Lemma 10.4.6 *Let s be an element in $\langle L\rangle$ such that, for each element r in $\langle L\rangle$, $\ell(r) \le \ell(s)$. Then $s \in S_{-1}(L)$ or there exists an element l in L such that $s \in S_\infty(l)$.*

PROOF. Let l be an element in L. Then our hypothesis on $\ell(s)$ implies that, for each element r in $s\langle l\rangle$, $\ell(r) \in \{\ell(s), \ell(s)-1\}$. It follows that $s \in L_{-1}(l)$ or that $s \in S_\infty(l)$.

The following lemma is similar to Lemma 10.4.1(ii).

Lemma 10.4.7 *Let l be an element in L, and let q be an element in $S_\infty(l)$. Let p be an element in $\langle L\rangle$ such that $q \in Lp$ and $\ell(q) = 1 + \ell(p)$. Then, for each element r in pl, we have $r \in S_{-1}(L)$ or $\ell(r) = \ell(p)$.*

PROOF. Let r be an element in pl. Then, by Lemma 1.3.3(i), $p \in rl$. Similarly, as $q \in Lp$, $p \in Lq$; cf. Lemma 1.3.3(ii). Thus, $p \in Lq \cap rl$. Thus, by Lemma 1.3.4, $Lr \cap ql$ is not empty. Let s be an element in $Lr \cap ql$.

From $q \in Lp$ and $\ell(q) = 1 + \ell(p)$ we obtain $q \in S_{-1}(p)$. Thus, by Lemma 10.4.1(i), $p \notin S_{-1}(l)$. In particular, $p \notin S_{-1}(L)$.

Since $q \in S_\infty(l)$, $q \notin S_{-1}(l)$. In particular, $q \notin S_{-1}(L)$.

Since $q \in S_\infty(l)$ and $s \in ql$, $s \in S_\infty(l)$; cf. Lemma 10.4.5, Thus, $s \notin S_{-1}(l)$. In particular, $s \notin S_{-1}(L)$.

Since $q \in S_\infty(l)$ and $s \in ql$, $\ell(q) = \ell(s)$. Thus, by Lemma 10.3.4, $r \in S_{-1}(L)$ or $\ell(p) = \ell(r)$.

For each element l in L, we define

$$S^d_\infty(l) := \{s \in S_\infty(l) \mid \ell(s) = d\}.$$

It may happen that there exists an element l in L such that $S^d_\infty(l)$ is empty. Indeed, it will be an immediate consequence of Proposition 10.4.9(i) that, if both elements in L have finite valency, one of the two sets $S^d_\infty(l)$ with $l \in L$ must be empty.

Later, in Section 10.6, we shall see that, if $\langle L\rangle$ has finite valency and $S_{-1}(L)$ exactly one element, then the sets $S^d_\infty(l)$ (with $l \in L$) decide on the structure of $\langle L\rangle$.

Lemma 10.4.8 *Let l be an element in L, and let us assume that $S^d_\infty(l)$ is not empty. Then there exists an element s in $S_{-1}(L)$ such that $\ell(s) = d$ and, for each element r in $s\langle l\rangle \setminus S_{-1}(L)$, $\ell(r) = d - 1$.*

Proof. We are assuming that $S^d_\infty(l)$ is not empty. Let q be an element in $S^d_\infty(l)$. Then $\ell(q) = d$. Thus, as $2 \leq d$, there exists an element p in $\langle L\rangle$ such that $q \in Lp$ and $\ell(q) = 1 + \ell(p)$; cf. Lemma 3.1.2. Thus, by Lemma 10.4.7, $pl \cap S_{-1}(L)$ is not empty or $p \in S_\infty(l)$.

Since $\ell(q) = d$ and $\ell(q) = 1 + \ell(p)$, $\ell(p) = d - 1$. Thus, by Corollary 10.4.4(i), $p \notin S_0(l)$. In particular, $p \notin S_\infty(l)$. Thus, by the above dichotomy, the set $pl \cap S_{-1}(L)$ is not empty.

Let s be an element in $pl \cap S_{-1}(L)$.

From $s \in pl$ we obtain $\ell(s) \leq \ell(p) + 1$, from $s \in S_{-1}(L)$ we obtain $d \leq \ell(s)$. Thus, as $\ell(p) = d - 1$, $\ell(s) = d$.

From $p \in sl$ we obtain $s\langle l\rangle = p\langle l\rangle$; cf. Lemma 2.1.4.

Since $q \in Lp$, $\ell(q) = 1 + \ell(p)$, and $q \in S_\infty(l)$, we may apply Lemma 10.4.7. Thus, for each element r in $pl \setminus S_{-1}(L)$, $\ell(r) = \ell(p) = d - 1$.

Proposition 10.4.9 *Let h and k be elements in L such that $h \neq k$. Assume that n_h and n_k are finite and that $S^d_\infty(h)$ is not empty. Then the following hold.*

(i) *We have $n_h + 1 \leq n_k$.*

(ii) *The integer d is odd.*

PROOF. Let s be an element in $S^d_\infty(h)$, let y be an element in X, and let z be an element in ys. Let l be an element in L such that $\ell(s) = \ell_l(s)$. Then $\nu_l(s) \leq 0$.

For each element x in $z\langle h\rangle$, we set

$$W_x := yl \cap xL^{d-1}.$$

We shall first prove that, for each element x in $z\langle h\rangle$, W_x is not empty.

Let x be an element in $z\langle h\rangle$, and let us denote by r the element in $\langle L\rangle$ which satisfies $x \in yr$.

From $x \in z\langle h\rangle$ and $z \in ys$ we obtain $x \in ys\langle h\rangle$. Thus, as $x \in yr$, $r \in s\langle h\rangle$. Thus, as $\nu_l(s) \leq 0$, $\nu_l(r) \leq 0$; cf. Lemma 10.3.3. Thus, $\ell(r) = \ell_l(r)$.

On the other hand, as $s \in S^d_\infty(h)$ and $r \in s\langle h\rangle$, $\ell(r) = d$. Thus, $r \in lL^{d-1}$. Thus, as $x \in yr$, $x \in ylL^{d-1}$, and that implies that W_x is not empty.

We now shall prove that, for any two elements x and x' in $z\langle h\rangle$, $\emptyset \neq W_x \cap W_{x'}$ implies $x = x'$.

Let x and x' be elements in $z\langle h\rangle$ such that $W_x \cap W_{x'}$ is not empty, and let w be an element in $W_x \cap W_{x'}$.

Since $w \in xL^{d-1}$, $x \in wL^{d-1}$. Thus, there exists an element q in L^{d-1} such that $x \in wq$. Thus, as $w \in yl$, $x \in ylq$. Thus, there exists an element r in lq such that $x \in yr$. Then $r \in S_{-1}(q)$.

From $z \in x\langle h\rangle$ and $x \in yr$ we obtain $z \in yr\langle h\rangle$. Thus, as $z \in ys$, $s \in r\langle h\rangle$. Thus, as $s \in S_\infty(h)$, $r \in S_\infty(h)$. Thus, as $r \in S_{-1}(q)$, $q \notin S_{-1}(h)$; cf. Lemma 10.4.1(i).

Since $\ell(q) \leq d-1$, $q \notin S_0(h)$; cf. Corollary 10.4.4(i).

Let q' be the element in L^{d-1} such that $x' \in wq'$. Then, similarly $q' \notin S_{-1}(h)$ and $q' \notin S_0(h)$. However, as $x' \in x\langle z\rangle$, $q' \in q\langle h\rangle$. Thus, $x' = x$.

Now we conclude from

$$\bigcup_{x \in z\langle h\rangle} W_x \subseteq yl$$

that

$$n_h + 1 = |z\langle h\rangle| \leq |yl| = n_l.$$

It follows that $h \neq l$. (Recall that we are assuming that h and k have finite valency.) This means that $l = k$, whence $\ell(s) = \ell_k(s)$. This means that $0 \leq \nu_h(s)$.

On the other hand, we obtain from $s \in S_\infty(h)$ that $s \in S_0(h)$, and this implies $\nu_h(s^*) = 1$; cf. Lemma 10.4.2.

Thus, we have $0 \leq \nu_h(s^*)\nu_h(s)$. Now we obtain from Lemma 10.3.2(ii) that d is odd. (Recall that $s \in S_\infty(h)$, so that $s \notin S_{-1}(L)$.)

Assume that each of the two elements in L has finite valency. Then we obtain from Proposition 10.4.9(i) that one of the two sets $S^d_\infty(l)$ with l in L must be empty.

10.5 The Constrained Spherical Case

Throughout this section, the set L is assumed to be spherical. We investigate the case where L is constrained. Instead of d_L we shall write d.

Recall that L is called constrained if, for any two elements q in $\langle L\rangle$ and p in $S_1(q)$, $1 = |pq|$.

Lemma 10.5.1 *Assume L to be constrained, let j be an integer with $0 \leq j \leq d$, and let l be an element in L. Then $|R_j(l)| = 1$.*

PROOF. By definition, $R_0(l) = \{1\}$. Let us assume that $|R_{j-1}(l)| = 1$. We shall see that $|R_j(l)| = 1$.

Assuming that $|R_{j-1}(l)| = 1$ we obtain an element p in R_{j-1} such that $\{p\} = R_{j-1}(l)$. From $p \in R_{j-1}(l)$ and $j-1 \leq d$ we obtain $\ell(p) = j-1$; cf. Lemma 10.2.6(ii).

Let k be an element in L such that $R_j(l) = R_{j-1}(l)k$. Then, as $\{p\} = R_{j-1}(l)$, $R_j(l) = pk$.

Let q be an element in pk. Then $q \in R_j(l)$. Thus, as we are assuming that $j \leq d$, $\ell(q) = j$; cf. Lemma 10.2.6(ii). Thus, as $\ell(p) = j-1$ and $q \in pk$, $p \in S_1(k)$. Thus, as L is assumed to be constrained, $|pk| = 1$. Thus, as $R_j(l) = pk$, $|R_j(l)| = 1$.

Corollary 10.5.2 *Assume L to be constrained. Then the following hold.*

(i) *For each element l in L, $R_d(l) = S_{-1}(L)$.*

(ii) *We have $|S_{-1}(L)| = 1$.*

PROOF. (i) From Corollary 10.1.4 we know that, for each element l in L, $S_{-1}(L) \subseteq R_d(l)$. However, Lemma 10.5.1 tells us that, for each element l in L, $|R_j(l)| = 1$. Thus, the claim follows from the hypothesis that $S_{-1}(L)$ is not empty.

(ii) Considering Lemma 10.5.1 this follows from (i).

Lemma 10.5.3 *Assume L to be constrained. Then, for each element s in $\langle L\rangle$, $\ell(s) \leq d$.*

PROOF. It is enough to show that there is no element in $\langle L\rangle$ of length $d+1$.

By way of contradiction, we assume that $\langle L\rangle$ contains an element s such that $\ell(s) = d+1$. Then, by Lemma 3.1.2, there exist elements r in $\langle L\rangle$ and l in

L such that $s \in rl$ and $\ell(s) = \ell(r) + 1$. From $s \in rl$ and $\ell(s) = \ell(r) + 1$ we obtain $r \in S_1(l)$.

From $d + 1 = \ell(s)$ and $\ell(s) = \ell(r) + 1$ we obtain $\ell(r) = d$. Thus, there exists an element l in L such that $s \in R_d(l)$; cf. Lemma 10.1.3. Thus, by Corollary 10.5.2(i), $r \in S_{-1}(L)$, contradiction.

Let j be an element in $\{0, \ldots, d\}$, and let l be an element in L. From Lemma 10.5.1 we know that $|R_j(l)| = 1$. In the following, we shall denote by $r_j(l)$ the uniquely determined element in $R_j(l)$.

Theorem 10.5.4 *The following statements are equivalent.*

(a) *We have* $2d = |\langle L \rangle|$.

(b) *The set* L *is constrained.*

(c) *We have* $\{r_j(l) \mid l \in L,\ j \in \{0, \ldots, d\}\} = \langle L \rangle$.

PROOF. (a) $\Rightarrow$ (b) Let q be an element in $\langle L \rangle$, and let p be an element in $S_1(q)$. We have to show that $1 = |pq|$.

Since $p \in S_1(q)$, there exists an element r in pq such that $\ell(r) = \ell(p) + \ell(q)$. Since we are assuming that $2d = |\langle L \rangle|$, $\ell(r) \leq d$; cf. Lemma 10.2.1. Thus, $1 = |pq|$.

(b) $\Rightarrow$ (c) Assume L to be constrained. Then, for each element s in $\langle L \rangle$, $\ell(s) \leq d$; cf. Lemma 10.5.3. Thus, the desired equation follows from Lemma 10.1.3.

(c) $\Rightarrow$ (a) By definition, we have that, for any two elements h and k in L, $r_0(h) = r_0(k)$. From Corollary 10.5.2 we also know that, for any two elements h and k in L, $r_d(h) = r_d(k)$. Thus, the equation in (c) yields $|\langle L \rangle| \leq 2d$, so that we obtain the desired equation in (a) from Lemma 10.2.1.

Theorem 10.5.5 *If* L *is constrained,* L *satisfies the exchange condition.*

PROOF. Let us assume L to be constrained, let us fix elements h, k in L and s in $S_1(k)$, and let us assume that $h \in S_1(s)$. We have to show that $hs = sk$ or that $hs \subseteq S_1(k)$.

From $h \in S_1(s)$ we obtain an element t in hs such that $\ell(t) = 1 + \ell(s)$. Similarly, as $s \in S_1(k)$, there exists an element u in sk such that $\ell(u) = \ell(s) + 1$.

From $t \in hs$, $h \in S_1(s)$, and the hypothesis that L is constrained we obtain $\{t\} = hs$. Similarly, we obtain $\{u\} = sk$.

Assume first that $\ell(s) = d - 1$. Then $\ell(t) = d = \ell(u)$. Thus, by Lemma 10.5.1, $t = u$. Thus, as $\{t\} = hs$ and $\{u\} = sk$, $hs = sk$.

Assume now that $\ell(s) \leq d - 2$, and let us fix an element r in tk. Then $r \in hsk = hu$. Thus, as $s \in ht \cap uk$ and $\ell(s) + 1 = \ell(u)$, $\ell(t) + 1 = \ell(r)$; cf. Lemma 10.3.5. It follows that $hs = \{t\} \subseteq S_1(k)$.

10.6 Dihedral Closed Subsets of Finite Valency

Throughout this section, $\langle L \rangle$ is assumed to have finite valency.

Assuming $\langle L \rangle$ to have finite valency we obtain that $\langle L \rangle$ is finite, so that, according to Lemma 10.1.7, L is spherical.

Instead of d_L we shall write d.

Lemma 10.6.1 *Let h and k be elements in L such that $h \neq k$, and assume $\langle L \rangle$ to have finite valency. Let R be the set of all elements s of S such that $\ell(s) \leq d$. Then the following hold.*

(i) *Assume d to be odd, and let c be the integer satisfying $d = 2c + 1$. Then*

$$(n_h + 1)((n_k + 1)\sum_{j=0}^{c-1}(n_h n_k)^j + (n_h n_k)^c) \leq n_R$$

with equality if and only if L is constrained.

(ii) *Assume d to be even, and let c be the integer satisfying $d = 2c$. Then*

$$(n_h + 1)(n_k + 1)\sum_{j=0}^{c-1}(n_h n_k)^j \leq n_R$$

with equality if and only if L is constrained.

Proof. For any two elements j in $\{0, \ldots, d\}$ and l in L, we have $R_j(l) \subseteq R$. Moreover, the sets

$$R_1(h), \ \ldots, \ R_d(h), \ R_0(k), \ \ldots, \ R_{d-1}(k)$$

are pairwise disjoint; cf. Lemma 10.2.1. Thus,

$$\sum_{j=1}^{d} n_{R_j(h)} + \sum_{j=0}^{d-1} n_{R_j(k)} \leq n_R.$$

Let us now fix an integer j with $0 \leq j \leq d - 1$ and compute $n_{R_j(h)}$.

If j is odd and i the integer satisfying $j = 2i + 1$, we have

$$n_{R_j(h)} = (n_h n_k)^i n_h.$$

If j is even and i the integer satisfying $j = 2i$, we have

$$n_{R_j(h)} = (n_h n_k)^i.$$

The values $n_{R_j(k)}$ are computed similarly. This proves the lemma.

We are assuming $\langle L\rangle$ to have finite valency. As a consequence, $\langle L\rangle$ is a finite set. In particular, $\ell(\langle L\rangle)$ has a maximal element. In the following, we shall denote this element by $\max \ell(\langle L\rangle)$.

For the remainder of this section, we shall focus on the case where

$$|S_{-1}(L)| = 1.$$

In this case, we shall obtain precise information about the structure of $\langle L\rangle$.

Let l be an element in L. Recall that $S_\infty(l)$ is our notation for the set of all elements s in $\langle L\rangle$ such that $\ell(sl) = \{\ell(s)\}$. Recall also that $S^d_\infty(l)$ denotes the set of all elements s in $S_\infty(h)$ with $\ell(s) = d$.

Proposition 10.6.2 *If* $|S_{-1}(L)| = 1$, $d = \max \ell(\langle L\rangle)$.

PROOF. Let us assume that $|S_{-1}(L)| = 1$, and let us denote by m the element in $S_{-1}(L)$.

Since $\{m\} = S_{-1}(L)$, $\ell(m) = d$.

We set $e := \max \ell(\langle L\rangle)$. Then $d \leq e$, and our claim is that $d = e$. In order to prove $d = e$ we assume, by way of contradiction, that $d + 1 \leq e$.

Let s be an element in $\langle L\rangle$ such that $\ell(s) = e$. Then, as $\ell(m) = d \leq e - 1$, $m \neq s$. Thus, as $\{m\} = S_{-1}(L)$, $s \notin S_{-1}(L)$. Thus, by Lemma 10.4.6, there exists an element h in L such that

$$s \in S_\infty(h).$$

From $s \in S_\infty(h)$ we obtain $s \in S_0(h)$. Thus, by Proposition 10.4.3, there exists an element p in $\langle L\rangle$ such that

$$s \in pm$$

and $e + 1 = \ell(p) + d$.

Since $\ell(p) = e - d + 1$, $\ell(p^*) = e - d + 1$. Thus, there exist elements $l_1, \ldots,$ l_{e-d+1} in L such that

$$p^* \in l_1 \cdots l_{e-d+1}$$

and, for each element i in $\{2, \ldots, e - d + 1\}$, $l_{i-1} \neq l_i$.

Since $p^* \in l_1 \cdots l_{e-d+1}$ and $y \in xp^*$, there exist elements

$$y_0, \ \ldots, \ y_{e-d+1}$$

in X such that $y_{e-d+1} \in y_0 p^*$ and, for each element i in $\{1, \ldots, e - d + 1\}$,

$$y_i \in y_{i-1}l_i;$$

cf. Lemma 1.3.9.

From $\{m\} = S_{-1}(L)$ we conclude that $m^* = m$; cf. Lemma 10.2.8. Thus, as $s \in pm$, $p \in sm$; cf. Lemma 1.3.3(i). Thus, as $y_0 \in y_{e-d+1}p$, $y_0 \in y_{e-d+1}sm$. Thus, there exists an element z_0 in $y_{e-d+1}s$ such that $y_0 \in z_0 m$.

For each element i in $\{0, \ldots, e-d+1\}$, we denote by s_{i0} the element in $\langle L \rangle$ which satisfies

$$z_0 \in y_i s_{i0}.$$

(This implies $s_{00} = m$ and $s_{e-d+1,0} = s$.)

Let i be an element in $\{1, \ldots, e-d+1\}$. Then, as $z_0 \in y_i s_{i0}$ and $y_i \in y_{i-1}l_i$, $z_0 \in y_{i-1}l_i s_{i0}$. Thus, as $z_0 \in y_{i-1}s_{i-1,0}$,

$$s_{i-1,0} \in l_i s_{i0}.$$

For each element i in $\{1, \ldots, e-d+1\}$, we set

$$s_{0i} := s_{i0}^*.$$

Our first claim is now that there exist elements

$$z_1, \ \ldots, \ z_{e-d+1}$$

in X such that, for each element i in $\{1, \ldots, e-d+1\}$,

$$z_i \in y_0 s_{0i} \cap z_{i-1}l_i.$$

Recall that $m^* = m$. Thus, as $m = s_{00}$, $s_{00}^* = s_{00}$. Thus, as $s_{00} \in l_1 s_{10}$, $s_{00} \in s_{01}l_1$; cf. Lemma 1.3.2(iii). Thus, as $z_0 \in y_0 s_{00}$, $z_0 \in y_0 s_{01} l_1$. Thus, there exists an element z_1 in $y_0 s_{01} \cap z_0 l_1$. This proves our claim for $i = 1$.

Let us now fix an element i in $\{2, \ldots, e-d+1\}$, and let us assume, by induction hypothesis, that there exists an element z_{i-1} in X with

$$z_{i-1} \in y_0 s_{0,i-1} \cap z_{i-2}l_{i-1}.$$

Since $s_{i-1,0} \in l_i s_{i0}$, $s_{0,i-1} \in s_{0i}l_i$. Thus, as $z_{i-1} \in y_0 s_{0,i-1}$, $z_{i-1} \in y_0 s_{0i} l_i$. Thus, there exists an element z_i in $y_0 s_{0i} \cap z_{i-1}l_i$, so that we have shown our first claim.

From $s_{00} \in l_1 s_{10}$ and $m = s_{00}$ we obtain $m \in l_1 s_{10}$. Thus, as $m^* = m$, $s_{01} \in ml_1$; cf. Lemma 1.3.3(iii). Thus, as $m \in S_{-1}(L)$, $\ell(s_{01}) \leq d$. Thus, for each element i in $\{1, \ldots, e-d+1\}$,

$$\ell(s_{0i}) = d - 1 + i;$$

cf. Lemma 3.4.1. It follows that, for each element $i \in \{1, \ldots, e-d+1\}$,

$$\ell(s_{i0}) = d - 1 + i.$$

(Recall that, for each element i in $\{1, \ldots, e-d+1\}$, $s_{i0} := s_{0i}^*$.)
Recall that $s_{e-d+1,0} = s \in S_\infty(h)$. Thus, as $s_{e-d+1,0} \in S_{-1}(s_{10})$, $s_{10} \in S_\infty^d(h)$; cf. Lemma 10.4.1(ii). In particular, $S_\infty^d(h)$ is not empty, so that, according to Proposition 10.4.9(ii), d is odd. Moreover, Lemma 10.4.8 yields

$$\ell(m\langle h\rangle \setminus \{m\}) = \{d-1\}.$$

Since $s_{10} \in S_\infty^d(h)$, $s_{10} \neq m$. Thus, as $s_{01} \in s_{00}l_1 \subseteq m\langle l_1\rangle$ and $\ell(s_{01}) = d$, the last equation implies

$$h \neq l_1.$$

For any two elements i and j in $\{1, \ldots, e-d+1\}$, we define s_{ij} to be the element in $\langle L\rangle$ which satisfies

$$z_j \in y_i s_{ij}.$$

Suppose that $s_{11} = m$. Then, as $h \neq l_1$, $s_{12} \in m\langle h\rangle$. Thus, by the above equation, $s_{12} = m$ or $\ell(s_{12}) = d - 1$. If $s_{12} = m$, $s_{21} \in S_{-1}(L)$, $s_{20} \in s_{21}L$, $\ell(s_{21}) = d$, and $\ell(s_{20}) = d + 1$, contradiction. If $\ell(s_{12}) = d - 1$, $\ell(s_{02}) \leq d$, contradiction. Therefore, we must have

$$s_{11} \neq m.$$

Assume that $\ell(s_{11}) = d-1$. Since d is odd, $d-1$ is even. Thus, $s_{11} \in R_{d-1}(l_1)$ or $s_{11}^* \in R_{d-1}(l_1)$. In the first case,

$$s_{01} \in l_1 s_{11} \subseteq l_1 R_{d-1}(l_1) \subseteq R_{d-2}(h) \cup R_{d-1}(l_1),$$

which yields $\ell(s_{01}) \leq d - 1$, contradiction. In the second case, we obtain, similarly, the contradiction $\ell(s_{10}) \leq d-1$. Thus, we have $\ell(s_{11}) \neq d-1$. From $\ell(s_{10}) = d$ and $s_{11} \in s_{10}L$ we now conclude that

$$\ell(s_{11}) \in \{d, d+1\}.$$

Assume that $\ell(s_{11}) = d+1$. Then $\{s_{11}^*, s_{11}\} \subseteq S_{-1}(l_1)$. Therefore, by Lemma 10.3.1, $\nu_{l_1}(s_{11}^*)$, $\nu_{l_1}(s_{11}) \leq 0$. It follows that $0 \leq \nu_{l_1}(s_{11}^*)\nu_{l_1}(s_{11})$. On the other hand, as $d + 1$ is even, Lemma 10.3.2(ii) yields $\nu_{l_1}(s_{11}^*)\nu_{l_1}(s_{11}) \leq 0$. Thus, $\nu_{l_1}(s_{11}^*)\nu_{l_1}(s_{11}) = 0$, which means that $s_{11} \in S_{-1}(L)$. It follows that $s_{11} = m$. This contradiction yields

$$\ell(s_{11}) = d.$$

We claim that, more generally, for each element i in $\{1, \ldots, e-d+1\}$,

$$\ell(s_{i1}) = d - 1 + i.$$

The case $i = 1$ has just been proven.

By induction, we may assume that $\ell(s_{i-1,1}) = d - 1 + i - 1$. We have that $\ell(s_{i-1,0}) = d - 1 + i - 1$ and that $\ell(s_{i0}) = d - 1 + i$. Thus, as

$$s_{i-1,0} \in Ls_{i0} \cap s_{i-1,1}L$$

and

$$s_{i1} \in Ls_{i-1,1} \cap s_{i0}L,$$

$\ell(s_{i1}) = d - 1 + i$; cf. Lemma 10.3.4.

It follows that $\ell(s_{e-d+1,1}) = e$.

Let p' be the element in S satisfying $y_1 \in y_{e-d+1}p'$. Then $\ell(p') = e - d$ and $p' \in S_1(s_{11})$. Similarly, let q' be the element in S satisfying $z_{e-d+1} \in y_1 q'$. Then $\ell(q') = e - d$ and $s_{11} \in S_1(q')$.

From $p' \in S_1(s_{11})$ and $s_{11} \in S_1(q')$ we obtain an element t in $p's_{11}q'$ such that

$$\ell(t) = \ell(p') + \ell(s_{11}) + \ell(q');$$

cf. Proposition 10.3.6. Thus, as $\ell(p') = e - d$, $\ell(s_{11}) = d$, and $\ell(q') = e - d$, we obtain $\ell(t) = 2e - d$. It follows that

$$\ell(s) + 1 = e + 1 \leq 2e - d = \ell(t),$$

contrary to the (maximal) choice of s.

Corollary 10.6.3 *Assume that $|S_{-1}(L)| = 1$. Then, for each element h in L, there exists an element k in L such that $R_d(h) \subseteq S_{-1}(L) \cup S^d_\infty(k)$.*

PROOF. Let h be an element in L, and let s be an element in $R_d(h)$. Then, by Lemma 10.2.6(ii), $\ell(s) = d$. Thus, by Proposition 10.6.2, $\ell(s) = \max \ell(\langle L \rangle)$, so that the claim follows from Lemma 10.4.6.

Proposition 10.6.4 *Assume that $|S_{-1}(L)| = 1$ and that L is not constrained. Then there exist elements h and k in L such that $h \neq k$, $R_d(h) = S_{-1}(L)$, $R_d(k) = S_{-1}(L) \cup S^d_\infty(h)$, and $S^d_\infty(h) \subseteq S_{-1}(L)k$.*

PROOF. Let us assume that, for each element l in L, $S^d_\infty(l)$ is empty. Then, for each element l in L, $S_{-1}(L) = R_d(l)$; cf. Corollary 10.6.3. Thus, we obtain from Lemma 10.2.2 and Lemma 10.2.1 that $2d = |\langle L \rangle|$. From $2d = |\langle L \rangle|$ we obtain that L is constrained; cf. Theorem 10.5.4.

Thus, we may assume that there exists an element h in L such that $S^d_\infty(h)$ is not empty. Then, by Proposition 10.4.9(ii), d is odd. Thus, for each element l in L, $R_d(l) \subseteq S_{-1}(l)$; cf. Lemma 10.2.7(i).

Let us denote by k the element in $L \setminus \{h\}$.

From $R_d(h) \subseteq S_{-1}(h)$ we obtain that $R_d(h) \cap S_\infty(h)$ is empty, from $R_d(k) \subseteq S_{-1}(k)$ that $R_d(k) \cap S_\infty(k)$ is empty. Thus, by Corollary 10.6.3, $R_d(h) \subseteq S_{-1}(L) \cup S_\infty^d(k)$ and

$$R_d(k) \subseteq S_{-1}(L) \cup S_\infty^d(h).$$

From Proposition 10.4.9(i) we obtain that $S_\infty^d(k)$ is empty. Thus, as $R_d(h) \subseteq S_{-1}(L) \cup S_\infty^d(k)$ and $S_{-1}(L) \subseteq R_d(h)$,

$$R_d(h) = S_{-1}(L).$$

In order to prove that $S_\infty^d(h) \subseteq S_{-1}(L)k$ we fix an element in $S_\infty^d(h)$ and call it s.

From $s \in S_\infty^d(h)$ we obtain $s^* \in R_d(h) \cup R_d(k)$. Thus, as $R_d(k) \subseteq S_{-1}(L) \cup S_\infty^d(h)$ and $R_d(h) \subseteq S_{-1}(L)$, $s^* \in S_\infty^d(h)$. From $s^* \in S_\infty^d(h)$ we obtain $s^* \in S_0(h)$. Thus, by Proposition 10.4.3, there exists an element l in L such that $s^* \in lS_{-1}(L)$. Thus, by Lemma 1.3.2(iii), $s \in S_{-1}(L)l$.

From $s \in S_\infty^d(h)$ we obtain $s \notin S_{-1}(L)h$. Thus, $s \in S_{-1}(L)k$.

Proposition 10.6.5 *Assume that $|S_{-1}(L)| = 1$. Then we have the following.*

(i) *For each element s in $\langle L \rangle$, there exist elements j in $\{0, \ldots, d\}$ and l in L such that $s \in R_j(l)$.*

(ii) *Let j be an element in $\{0, \ldots, d-1\}$, and let l be an element in L. Then $|R_j(l)| = 1$.*

Proof. (i) Let s be an element in $\langle L \rangle$. Set $j := \ell(s)$, and let l be an element in L such that $\ell(s) = \ell_l(s)$. Then $s \in R_j(l)$ and, by Proposition 10.6.2, $j \leq d$.

(ii) According to Lemma 10.5.1, we may assume that L is not constrained. By Lemma 10.2.2, it suffices to show that $|R_{d-1}(l)| = 1$.

Let r be an element in $R_{d-1}(l)$. Then, by Lemma 10.2.6(ii), $\ell(r) = d - 1$. Thus, by Corollary 10.4.4, there exists an element s in rL such that $\ell(s) = d$.

We are assuming that $|S_{-1}(L)| = 1$. Thus, there exists an element m in $S_{-1}(L)$ such that $\{m\} = S_{-1}(L)$. We wish to prove that $r \in mL$.

If $s = m$, this follows from $s \in rL$. Therefore, we assume that $s \neq m$.

If $s \neq m$, there exist elements h and k in L such that $h \neq k$, $s \in S_\infty(h)$, and $s \in mk$; cf. Proposition 10.6.4. (Recall that $\ell(s) = d$ and that L is not constrained.) From $s \in S_\infty(h)$, $r \in sL$, and $\ell(r) + 1 = \ell(s)$ we conclude that $r \in sk$. Thus, as $s \in mk$, $r \in mk$; cf. Lemma 2.1.4. In particular, $r \in mL$.

Since r has been chosen arbitrarily in $R_{d-1}(l)$, we have shown that

$$R_{d-1}(l) \subseteq mL.$$

Now Corollary 10.4.4(i) yields $|R_{d-1}(l)| = 1$.

Recall that, according to Theorem 10.5.4, L is constrained if and only if $2d = |\langle L\rangle|$. In this case, L satisfies the exchange condition; cf. Theorem 10.5.5.

Assume that $|S_{-1}(L)| = 1$. We call L a *Moore set* if there exist elements h and k in L such that $h \neq k$, $R_d(h) = S_{-1}(L)$, $R_d(k) = S_{-1}(L) \cup S^d_\infty(h)$, and $S^d_\infty(h) \subseteq S_{-1}(L)k$.

Assume that L is a Coxeter set. Then, by Corollary 10.5.2(ii), $|S_{-1}(L)| = 1$. By definition, this equation also holds if L is a Moore set.

The following theorem is the main result of this chapter. (Recall that $\langle L\rangle$ is assumed to have finite valency.)

Theorem 10.6.6 *Assume that $|S_{-1}(L)| = 1$. Then L is a Coxeter set or a Moore set.*

PROOF. If L is constrained, L is a Coxeter set; cf. Theorem 10.5.5. If L is not constrained, we obtain from Proposition 10.6.4 elements h and k in L such that $h \neq k$, $R_d(h) = S_{-1}(L)$, $R_d(k) = S_{-1}(L) \cup S^d_\infty(h)$, and $S^d_\infty(h) \subseteq S_{-1}(L)k$.

Much more can be said in either of the two cases of Theorem 10.6.6.

The first case of Theorem 10.6.6 will be investigated in more detail in Section 12.4. In Theorem 12.4.6, for instance, we shall see that, if L is a Coxeter set, $\langle L\rangle$ is thin or

$$d \in \{2, 3, 4, 6, 8, 12\}.$$

More details about the valencies of the elements in L will be given in each of the six cases in the subsequent six theorems of Section 12.4.

Similar investigations have been made about the second case of Theorem 10.6.6. In fact, in this case, one can show that

$$d \in \{3, 5\}.$$

However, the proof of this latter fact is beyond the scope of this monograph. It is based on modular representation theory of finite schemes and follows from [13; Theorem 3], [6; Theorem 2], and [14; Theorem 1].

In the introduction to this chapter, we mentioned a representation theoretic condition which implies $|S_{-1}(L)| = 1$. Let us now, at the end of this section, look at this condition.

Theorem 10.6.7 *Assume that there exists an algebraically closed field C of characteristic 0 such that $C[L] = C\langle L\rangle$. Then $|S_{-1}(L)| = 1$.*

PROOF. We may assume that $\langle L\rangle = S$, so that $C[L] = CS$. Let h and k be elements in L such that $h \neq k$. We first shall prove that

$$|S| \leq |S/\langle h\rangle| + |S/\langle k\rangle|$$

and

$$||S/\langle h\rangle| - |S/\langle k\rangle|| \leq 1.$$

For each element i in $\{1, 2\}$, we define M_i to be the sum of the i-dimensional submodules of the CS-module CS. Then, by Lemma 9.4.5,

$$CS = M_1 \oplus M_2.$$

Let l be an element in L. Set $\sigma_l := \sigma_l^C$, and define

$$C_{\langle l\rangle} := \{\sigma \in CS \mid \sigma\sigma_l = n_l\sigma\}.$$

Let σ be an element in $C_{\langle l\rangle}$. Then there exist elements σ_1 in M_1 and σ_2 in M_2 such that $\sigma = \sigma_1 + \sigma_2$. Since $\sigma \in C_{\langle l\rangle}$,

$$\sigma\sigma_l = n_l\sigma = n_l(\sigma_1 + \sigma_2) = n_l\sigma_1 + n_l\sigma_2.$$

On the other hand,

$$\sigma\sigma_l = (\sigma_1 + \sigma_2)\sigma_l = \sigma_1\sigma_l + \sigma_2\sigma_l.$$

Thus, as $CS = M_1 \oplus M_2$, we obtain $\sigma_1\sigma_l = n_l\sigma_1$ and $\sigma_2\sigma_l = n_l\sigma_2$. Since $\sigma \in CS$ has been chosen arbitrarily, we have shown that

$$C_{\langle l\rangle} = (M_1 \cap C_{\langle l\rangle}) \oplus (M_2 \cap C_{\langle l\rangle}).$$

Let i be an element in $\{1, 2\}$. We set

$$b_i := \dim_C(M_i),$$

and, for each element l in L, we define

$$a_i^l := \dim_C(M_i \cap C_{\langle l\rangle}).$$

Since $CS = M_1 \oplus M_2$, this notation yields

$$|S| = b_1 + b_2.$$

Moreover, for each element l in L,

$$|S/\langle l\rangle| = a_1^l + a_2^l.$$

(Recall that $C_{\langle l\rangle} = (M_1 \cap C_{\langle l\rangle}) \oplus (M_2 \cap C_{\langle l\rangle})$, and use Theorem 9.4.7.) From Corollary 9.4.4(ii) we obtain

$$\frac{1}{2}b_2 = a_2^l$$

for each element l in L. From Corollary 9.4.4(ii) we also deduce that

$$b_1 \leq \sum_{l \in L} a_1^l.$$

(Recall that CS possesses a 1-dimensional right ideal which affords the character 1_{CS}.) Thus, we have

$$|S| \leq \sum_{l \in L} (a_1^l + a_2^l) = \sum_{l \in L} |S/\langle l \rangle|.$$

From Corollary 9.4.4(ii) we finally deduce that, for each element l in L, $a_1^l \in \{1, 2\}$. Thus,

$$|\,|S/\langle h \rangle| - |S/\langle k \rangle|\,| = |a_1^h - a_1^k| \leq 1.$$

Let us now define

$$C := \bigcup_{l \in L} S/\langle l \rangle,$$

$$r := \{(s\langle h \rangle, s\langle k \rangle) \mid s \in S\},$$

and $\Gamma := (C, r)$. Then, by Theorem 9.4.1, the graph Γ is connected.

Note that Γ is not a tree. Thus,

$$|C| \leq |r|.$$

Moreover, the definition of r yields

$$|r| \leq |S|.$$

Finally, as $|S| \leq |S/\langle h \rangle| + |S/\langle k \rangle|$, we have

$$|S| \leq |C|.$$

It follows that

$$|C| = |r| = |S|.$$

In particular, for each element s in S, $s\langle h \rangle \cap s\langle k \rangle = \{s\}$. Therefore, we conclude that $|S_{-1}(L)| = 1$.

Theorem 10.6.6 (together with Corollary 10.5.2(ii)) says that Coxeter sets as well as Moore sets are characterized by the equation $|S_{-1}(L)| = 1$.

The following corollary is a representation theoretic characterization of Coxeter sets and Moore sets. It was proved first in [42; Theorem B]. Here, we obtain it as a corollary of Theorem 10.6.6 and Theorem 10.6.7.

Corollary 10.6.8 *Assume that there exists an algebraically closed field C of characteristic 0 with $C[L] = C\langle L \rangle$. Then L is a Coxeter set or a Moore set.*

11

Coxeter Sets

In this chapter, we start to look more thoroughly at Coxeter sets. In order to do so we first recall the definition of a Coxeter set.

We fix a set of involutions of S and call it L. Instead of ℓ_L we shall write ℓ.

Recall that, for each element q in $\langle L \rangle$, $S_1(q, L)$ is our notation for the set of all elements p in $\langle L \rangle$ such that pq contains an element r satisfying $\ell(r) = \ell(p) + \ell(q)$. In accordance with Section 3.4 we shall write, for each element s in $\langle L \rangle$, $S_1(s)$ instead of $S_1(s, L)$.

Recall that L is called constrained if, for any two elements q in $\langle L \rangle$ and p in $S_1(q)$, $1 = |pq|$. Recall also that L is said to satisfy the exchange condition if, for any three elements h, k in L and s in $S_1(k)$, $h \in S_1(s)$ implies $hs \subseteq sk \cup S_1(k)$. Recall, finally, that L is called a Coxeter set if L is constrained and satisfies the exchange condition.

For the remainder of this chapter, we assume L to be a Coxeter set.

It is the goal of this chapter to provide global information about the structure of $\langle L \rangle$.

In the first section, we compile a few general facts about closed subsets of $\langle L \rangle$ generated by subsets of L. Such closed subsets are called *parabolic*. The section can be viewed as a continuation of Section 3.6.

In the second section, we show that, if L is a finite set and does not contain thin elements, the closed subset generated by L is a direct product of simple closed subsets each of which is generated by the elements of L which it contains.

In the third section of this chapter, we investigate faithful maps defined on subsets of $x\langle L \rangle$ with $x \in X$. Recall that a map χ from a subset Y of X to X is called faithful if, for any three elements v, w in Y and s in S, $w \in vs$ implies $w\chi \in v\chi s$.

The goal of the last of the four sections of this chapter is the proof of a specific extension theorem (Theorem 11.4.6) for Coxeter sets.

By an extension theorem we mean a theorem which guarantees that, given subsets Y and Z of X, a faithful map χ from Y of X extends faithfully to a map from Z of X.

In Theorem 11.4.6, we deal with a case where the set Z contains only one element which is not in Y. This theorem will be crucial in the proof of Theorem 12.3.4.

11.1 Parabolic Subsets

Throughout this section, the letter K stands for a subset of L.

Lemma 11.1.1 *Let p be an element in $S_1(K)$, and let q be an element in $p\langle K\rangle$. Then $q^* \in S_{-1}(p^*)$.*

PROOF. We are assuming that $q \in p\langle K\rangle$. Thus, there exists an element s in $\langle K\rangle$ such that $q \in ps$.

We are assuming that $p \in S_1(K)$. Thus, by Theorem 3.6.4, $p \in S_1(\langle K\rangle)$. Thus, as $s \in \langle K\rangle$, $p \in S_1(s)$. Thus, as $q \in ps$ and L is assumed to be constrained, $\ell(q) = \ell(p) + \ell(s)$.

From $q \in ps$, we obtain $q^* \in s^*p^*$. From $\ell(q) = \ell(p) + \ell(s)$ we obtain $\ell(q^*) = \ell(s^*) + \ell(p^*)$. Thus, $q^* \in S_{-1}(p^*)$.

Lemma 11.1.2 *We have $\langle L \setminus K\rangle \subseteq S_1(\langle K\rangle)$.*

PROOF. Let l be an element in $L \setminus K$. Then, as L is assumed to satisfy the exchange condition, $l \in S_1(K)$. Thus, by Theorem 3.6.4, $l \in S_1(\langle K\rangle)$. Thus, by Lemma 3.4.4(i), $\langle K\rangle \subseteq S_1(l)$.

Since l has been chosen arbitrarily in $L \setminus K$, we have shown that $\langle K\rangle \subseteq S_1(L \setminus K)$. Thus, by Theorem 3.6.4, $\langle K\rangle \subseteq S_1(\langle L \setminus K\rangle)$. Thus, by Lemma 3.4.4(i), $\langle L \setminus K\rangle \subseteq S_1(\langle K\rangle)$.

The following lemma generalizes Lemma 3.4.8 for Coxeter sets.

Lemma 11.1.3 *For each subset H of L, $(\langle H\rangle \cap S_1(K))\langle H \cap K\rangle = \langle H\rangle$.*

PROOF. From Lemma 11.1.2 we know that $\langle H\rangle \subseteq S_1(K \setminus H)$. Thus,

$$\langle H\rangle \cap S_1(H \cap K) = \langle H\rangle \cap S_1(K).$$

On the other hand, we obtain from Lemma 3.4.8 that

$$\langle H\rangle \cap S_1(H \cap K)\langle H \cap K\rangle = \langle H\rangle,$$

and according to Lemma 2.2.1(ii), the left hand side of this equation is equal to $(\langle H\rangle \cap S_1(H \cap K))\langle H \cap K\rangle$.

Lemma 11.1.4 *Let p and q be elements in $\langle L\rangle$, and let l be an element in $L \cap S_1(p) \cap S_1(q)$. Then, if $lp \subseteq lq\langle K\rangle$, $p \in q\langle K\rangle$.*

Proof. Let us assume, by way of contradiction, that $lp \subseteq lq\langle K\rangle$ and that $p \notin q\langle K\rangle$. Assuming $lp \subseteq lq\langle K\rangle$ we must have $p \in q\langle K\rangle$ or $p \in lq\langle K\rangle$. Thus, as $p \notin q\langle K\rangle$, $p \in lq\langle K\rangle$.

From Lemma 3.4.8 we know that $S_1(K)\langle K\rangle = \langle L\rangle$. Thus, as $q \in \langle L\rangle$, there exists an element u in $S_1(K)$ such that $q \in u\langle K\rangle$. From $u \in S_1(K)$ and $q \in u\langle K\rangle$ we obtain $q^* \in S_{-1}(u^*)$; cf. Lemma 11.1.1. On the other hand, we are assuming that $l \in S_1(q)$, and that means that $q^* \in S_1(l)$; cf. Lemma 3.4.4(i). Thus, by Lemma 3.4.4(ii), $u^* \in S_1(l)$, and this implies $l \in S_1(u)$; cf. Lemma 3.4.4(i).

Now recall that L is assumed to satisfy the exchange condition. Thus, we obtain from $l \in S_1(u)$ and $u \in S_1(K)$ that $lu \subseteq uK$ or $lu \subseteq S_1(K)$. However, as $p \in lq\langle K\rangle$, $q\langle K\rangle = u\langle K\rangle$, and $p \notin q\langle K\rangle$, we cannot have $lu \subseteq uK$. Thus, $lu \subseteq S_1(K)$.

Similarly, we obtain an element t in $S_1(K)$ such that $p \in t\langle K\rangle$, $l \in S_1(t)$, and $lt \subseteq S_1(K)$. Thus, referring to Theorem 3.6.4 we now obtain $lt = lu$. (Note that $lt\langle K\rangle = lu\langle K\rangle$.) Thus, by Lemma 3.5.1, $t = u$. Thus, as $p \in t\langle K\rangle$ and $q \in u\langle K\rangle$, $p \in q\langle K\rangle$, contradiction.

Lemma 11.1.5 *Let l be an element in L, and let s be an element in $S_1(K)$. Assume there exists an element r in $s\langle K\rangle$ such that $l \in S_1(r) \setminus \langle K\rangle^{r^*}$. Then $ls \subseteq S_1(K)$.*

Proof. From $r \in s\langle K\rangle$ we obtain $r\langle K\rangle = s\langle K\rangle$. According to Lemma 1.3.2(iii), this implies $\langle K\rangle r^* = \langle K\rangle s^*$. Thus, by Lemma 1.3.7, $\langle K\rangle^{r^*} = \langle K\rangle^{s^*}$. Thus, as we are assuming that $l \notin \langle K\rangle^{r^*}$, $l \notin \langle K\rangle^{s^*}$. Thus, by definition, $s^*l \not\subseteq \langle K\rangle s^*$. Thus, by Lemma 1.3.2(iii), $ls \not\subseteq s\langle K\rangle$.

From $r \in s\langle K\rangle$ and $s \in S_1(K)$ we obtain $r^* \in S_{-1}(s^*)$; cf. Lemma 11.1.1. Thus, by Lemma 3.4.4(iv), $S_1(r) \subseteq S_1(s)$. Thus, as we are assuming that $l \in S_1(r)$, $l \in S_1(s)$. Thus, as L is assumed to satisfy the exchange condition, we have $ls \subseteq S_1(K)$.

Lemma 11.1.6 *For any two elements y and z in X, $|yS_1(K) \cap z\langle K\rangle| = 1$.*

Proof. Let v and w be elements in $yS_1(K) \cap z\langle K\rangle$. Since v, $w \in z\langle K\rangle$, $w \in v\langle K\rangle$. Thus, there exists an element s in $\langle K\rangle$ such that $w \in vs$.

Since $v \in yS_1(K)$, there exists an element p in $S_1(K)$ such that $v \in yp$.

From $w \in vs$ and $v \in yp$ we obtain $w \in yps$. Thus, there exists an element q in ps such that $w \in yq$. Since $p \in S_1(K)$, $s \in \langle K\rangle$, and $q \in ps$, $\ell(q) = \ell(p) + \ell(s)$; cf. Theorem 3.6.4.

From $w \in yS_1(K)$ we, similarly, obtain $\ell(p) = \ell(q) + \ell(s^*)$. Thus, $\ell(s) = 0$. It follows that $1 = s$. Thus, as $w \in vs$, $v = w$.

Lemma 11.1.7 *Let x be an element in X, let l be an element in L, and let y and z be elements in $x\langle l\rangle$ such that $y \neq z$. Then, for each subset H of L, $y\langle H\rangle \cap z\langle K\rangle \subseteq x(\langle H\rangle \cup \langle K\rangle)$.*

PROOF. Let w be an element in $y\langle H\rangle \cap z\langle K\rangle$. Since $w \in z\langle K\rangle$, $z \in w\langle K\rangle$. Thus, as $w \in y\langle H\rangle$, $z \in y\langle H\rangle\langle K\rangle$. Thus, as $z \in yl$, $l \in \langle H\rangle\langle K\rangle \subseteq \langle H \cup K\rangle$. Thus, as $l \in L$, $l \in H \cup K$; cf. Lemma 3.6.2(ii).

Since $w \in y\langle H\rangle$ and $y \in x\langle l\rangle$, $w \in x\langle l\rangle\langle H\rangle$. Similarly, $w \in x\langle l\rangle\langle K\rangle$. Thus, as $l \in H \cup K$, we must have $w \in x\langle H\rangle$ or $w \in x\langle K\rangle$. Thus, $w \in x(\langle H\rangle \cup \langle K\rangle)$.

11.2 Direct Products

In this section, we shall see that $\langle L\rangle$ is the direct product of simple closed subsets each of which is generated by the elements of L which it contains. As a consequence, we shall also look at the case where $\langle L\rangle$ is simple.

Theorem 11.2.1 *Let H and K be subsets of L. Then the following hold.*

(i) *We have $\langle H \cap K\rangle = \langle H\rangle \cap \langle K\rangle$.*

(ii) *If $\langle H\rangle = \langle K\rangle$, then $H = K$.*

PROOF. (i) It is clear that $\langle H \cap K\rangle \subseteq \langle H\rangle \cap \langle K\rangle$. Thus, we just have to show that $\langle H\rangle \cap \langle K\rangle \subseteq \langle H \cap K\rangle$.

Let us assume, by way of contradiction, that $\langle H\rangle \cap \langle K\rangle \not\subseteq \langle H \cap K\rangle$. Among the elements in $\langle H\rangle \cap \langle K\rangle$ which are not in $\langle H \cap K\rangle$ we fix an element s such that $\ell(s)$ is as small as possible.

From $s \notin \langle H \cap K\rangle$ we obtain $1 \neq s$. From $1 \neq s$ we obtain elements r in $\langle L\rangle$ and l in L such that $s \in rl$ and $\ell(s) = \ell(r) + 1$; cf. Lemma 3.1.2. From $s \in rl$ and $\ell(s) = \ell(r) + 1$ we obtain $s \in S_{-1}(l)$.

Assume that $l \notin H$. Then, as $s \in \langle H\rangle$, $s \in S_1(l)$; cf. Lemma 11.1.2. This contradiction yields $l \in H$.

Similarly, we obtain from $s \in \langle K\rangle$ that $l \in K$. It follows that $l \in H \cap K$.

From $s \in rl$ we obtain $r \in sl$; cf. Lemma 1.3.3(i). Thus, as $s \in \langle H\rangle \cap \langle K\rangle$ and $l \in H \cap K$, $r \in \langle H\rangle \cap \langle K\rangle$. Thus, the (minimal) choice of s yields $r \in \langle H \cap K\rangle$. Thus, as $s \in rl$ and $l \in H \cap K$, $s \in \langle H \cap K\rangle$, contradiction.

(ii) This follows from Lemma 3.6.2(ii).

Theorem 11.2.2 *Assume that L does not contain thin elements. Then $K \mapsto \langle K\rangle$ is a bijective map from the power set of L to the set of all closed subsets of $\langle L\rangle$.*

PROOF. From Corollary 3.5.4(ii) we know that, for each closed subset T of $\langle L\rangle$, there exists a subset K of L such that $\langle K\rangle = T$. This shows that the map in question is surjective.

Injectivity follows from Theorem 11.2.1(ii).

Recall that, by Lemma 3.5.5(iii), L does not contain thin elements if and only if $\{1\} = O_\vartheta(\langle L\rangle)$.

Let P and Q be subsets of S, and let us assume Q to be not empty. Recall that $C_P(Q)$ is our notation for the set of all elements p in P such that, for each element q in Q, $qp = pq$.

Theorem 11.2.3 *For each subset K of L, the following conditions are equivalent.*

(a) *For any two elements k in K and l in $L \setminus K$, $kl = lk$.*

(b) *We have $\langle L \setminus K\rangle \subseteq C_{\langle L\rangle}(\langle K\rangle)$.*

(c) *The closed subset $\langle K\rangle$ is normal in $\langle L\rangle$.*

PROOF. (a) $\Rightarrow$ (b) Let p be an element in $\langle K\rangle$, and let q be an element in $\langle L \setminus K\rangle$. We have to show that $pq = qp$.

If $1 = p$ or $1 = q$, the claim is obvious. Thus, we may assume that $1 \neq p$ and that $1 \neq q$. From $1 \neq p$ we obtain elements t in $\langle K\rangle$ and h in K such that $p \in th$ and $\ell(p) = \ell(t) + 1$; cf. Lemma 3.1.2. Similarly, we obtain elements k in $L \setminus K$ and u in $\langle L \setminus K\rangle$ such that $q \in ku$ and $\ell(q) = 1 + \ell(u)$.

From $p \in th$ and $\ell(p) = \ell(t) + 1$ we obtain $\{p\} = th$. From $q \in ku$ and $\ell(q) = 1 + \ell(u)$ we obtain $\{q\} = ku$. We are assuming that $hk = kh$, and, by induction, we may assume that $tk = kt$, that $hu = uh$, and that $tu = ut$. Thus,

$$pq = thku = tkhu = ktuh = kuth = qp.$$

(b) $\Rightarrow$ (c) Let s be an element in $\langle L\rangle$. We have to show that $\langle K\rangle s \subseteq s\langle K\rangle$.

There is nothing to show if $1 = s$. Therefore, we assume that $1 \neq s$. From $1 \neq s$ we obtain elements r in $\langle L\rangle$ and l in L such that $s \in rl$ and $\ell(s) = \ell(r) + 1$; cf. Lemma 3.1.2. From $s \in rl$ and $\ell(s) = \ell(r) + 1$ we obtain $r \in S_1(l)$. Thus, as L is assumed to be constrained, we conclude that $\{s\} = rl$.

By induction, we have $\langle K\rangle r \subseteq r\langle K\rangle$, and from (b) we obtain $\langle K\rangle l \subseteq l\langle K\rangle$. Thus, as $\{s\} = rl$,

$$\langle K\rangle s = \langle K\rangle rl \subseteq r\langle K\rangle l \subseteq rl\langle K\rangle = s\langle K\rangle.$$

(c) $\Rightarrow$ (a) Let k be an element in K and let l be an element in $L \setminus K$. We have to show that $kl = lk$.

Since $k \in K$ and $l \in L \setminus K$, $k \in S_1(l)$; cf. Lemma 11.1.2. Thus, there exists an element s in kl such that $\ell(s) = \ell(k) + \ell(l) = 2$. Since L is assumed to be constrained, we obtain from $k \in S_1(l)$ and $s \in kl$ that $\{s\} = kl$.

Since $\langle K \rangle$ is assumed to be normal in $\langle L \rangle$, we have $kl \subseteq l\langle K \rangle$. Thus, $s \in l\langle K \rangle$. Thus, there exists an element r in $\langle K \rangle$ such that $s \in lr$.

From $l \in L \setminus K$ and $r \in \langle K \rangle$ we obtain $l \in S_1(r)$; cf. Lemma 11.1.2. From $l \in S_1(r)$ and $s \in lr$ we obtain $\ell(s) = 1 + \ell(r)$ and $\{s\} = lr$.

From $\ell(s) = 2$ and $\ell(s) = 1 + \ell(r)$ we obtain $\ell(r) = 1$. Thus, $r \in L$.

From $s \in lr$ we obtain $r \in ls$; cf. Lemma 1.3.3(ii). Thus, as $s \in kl$, $r \in lkl$. In particular, $r \in \langle k, l \rangle$. Thus, as $r \in L$, $r \in \{k, l\}$; cf. Lemma 3.6.2(ii).

From $kl = \{s\} = lr$ we obtain $r \neq l$. Thus, $r = k$. Thus, as $kl = \{s\} = lr$, $kl = lk$.

Recall that a closed subset T of S different from $\{1\}$ is called simple if $\{1\}$ and T are the only normal closed subsets of T.

Theorem 11.2.4 *Assume that L is finite and does not contain thin elements. Then L contains subsets $L_1, \ldots, L_n$ such that $\{L_1, \ldots, L_n\}$ is a partition of L,*

$$\langle L_1 \rangle \times \ldots \times \langle L_n \rangle = \langle L_1 \rangle,$$

and, for each element i in $\{1, \ldots, n\}$, $\langle L_i \rangle$ is simple.

Proof. Assume that $\langle L \rangle$ is not simple. Then, by definition, $\langle L \rangle$ possesses a normal closed subset T with $\{1\} \neq T \neq \langle L \rangle$.

Since we are assuming that L has no thin element, we obtain from Theorem 11.2.2 a subset K of L such that $\langle K \rangle = T$. Since $\{1\} \neq T$, K is not empty, and that means that $L \setminus K \neq L$. Since $T \neq \langle L \rangle$, $K \neq L$, cf. Theorem 11.2.1(ii).

Since T is normal in $\langle L \rangle$ and $\langle K \rangle = T$, $\langle K \rangle$ is normal in $\langle L \rangle$. Thus, by Theorem 11.2.3, $\langle L \setminus K \rangle$ is normal in $\langle L \rangle$.

From Theorem 11.2.1(i) we obtain $\{1\} = \langle K \rangle \cap \langle L \setminus K \rangle$. Thus, by definition, $\langle K \rangle \times \langle L \setminus K \rangle = \langle L \rangle$.

Now the claim follows by induction.

Theorem 11.2.4 tells us that, in order to investigate Coxeter sets without thin elements, it is enough to look at Coxeter sets which generate a simple closed subset.

The following lemma is a result about simple closed subsets generated by a Coxeter set. It will be needed in the proof of Lemma 12.3.1.

Lemma 11.2.5 *Assume that $3 \leq |L|$ and that $\langle L \rangle$ is simple. Let l be an element in L. Then there exists an element s in $S_1(l)$ such that, for each subset K of L with $|K| = 2$, $\langle K \rangle s \subseteq S_1(l)$.*

PROOF. Since $\{l\} \neq L$ and $\langle L\rangle$ is assumed to be simple, we find an element k in $L \setminus C_S(l)$; cf. Theorem 11.2.3.

Since L is assumed to have at least three elements and $\langle L\rangle$ is assumed to be simple, we find an element h in $L\setminus\{k,l\}$ such that $h \notin C_S(\langle k,l\rangle)$; cf. Theorem 11.2.3.

Let s be the element in hk, let K be a subset of L with $|K| = 2$, and let q be an element in $\langle L\rangle$ such that $\langle K\rangle \subseteq S_1(q)$ and $\langle K\rangle s = \langle K\rangle q$; cf. Lemma 3.4.8.

From $s \in \langle K\rangle q$ and $\langle K\rangle \subseteq S_1(q)$ we obtain $s \in S_{-1}(q)$. From $s \in hk$ and $l \notin \{h,k\}$ we obtain $s \in S_1(l)$. Thus, $S_{-1}(q) \cap S_1(l)$ is not empty. Thus, by Lemma 3.4.4(ii), $q \in S_1(l)$.

Note also that $hkl \not\subseteq \langle K\rangle s$. Thus, by Lemma 11.1.5, $\langle K\rangle \in S_1(ql)$. Thus, by Lemma 3.4.6(ii), $\langle K\rangle q \subseteq S_1(l)$. Thus, as $\langle K\rangle q = \langle K\rangle s$, $\langle K\rangle s \subseteq S_1(l)$.

11.3 Faithful Maps

Throughout this section, the letter K stands for a subset of L.

Lemma 11.3.1 *Let y be an element in X, let z be an element in $yS_1(K)$, and let χ be a map from $\{y\} \cup z\langle K\rangle$ to X. Then, if $\chi|_{\{y,z\}}$ and $\chi|_{z\langle K\rangle}$ are faithful, so is χ.*

PROOF. Let x be an element in $z\langle K\rangle$. We shall be done if we succeed in showing that $\chi|_{\{y,x\}}$ is faithful.

Since $x \in z\langle K\rangle$, there exists an element q in $\langle K\rangle$ such that $x \in zq$. Since we are assuming that $z \in yS_1(K)$, there exists an element p in $S_1(K)$ such that $z \in yp$. Since $p \in S_1(K)$, $p \in S_1(\langle K\rangle)$; cf. Theorem 3.6.4. Thus, as $q \in \langle K\rangle$, $p \in S_1(q)$. Thus, as $\chi|_{\{y,z\}}$ and $\chi|_{\{z,x\}}$ are faithful, $\chi|_{\{y,x\}}$ is faithful; cf. Lemma 6.6.1.

Lemma 11.3.2 *Let x be an element in X, let H be a subset of L, and let χ be a map from $x(\langle H\rangle \cup \langle K\rangle)$ to X. Then, if $\chi|_{x\langle H\rangle}$ and $\chi|_{x\langle K\rangle}$ are faithful, so is χ.*

PROOF. Let y be an element in $x\langle H\rangle$. Then, $x \in y\langle H\rangle$. Thus, by Lemma 11.1.3, $x \in y(\langle H\rangle \cap S_1(K))\langle K\rangle$. Thus, there exists an element z in $y(\langle H\rangle \cap S_1(K))$ such that $x \in z\langle K\rangle$.

Since $z \in y\langle H\rangle$, $\chi|_{\{y,z\}}$ is faithful. Since $x \in z\langle K\rangle$ and $\chi|_{x\langle K\rangle}$ is faithful, $\chi|_{z\langle K\rangle}$ is faithful. Thus, as $z \in yS_1(K)$, $\chi|_{\{y\}\cup x\langle K\rangle}$ must be faithful; cf. Lemma 11.3.1.

Lemma 11.3.3 *Let s be an element in $\langle L\rangle$, and let l be an element in L with $l \in S_1(s) \setminus \langle K\rangle^{s^*}$. Let x be an element in X, y an element in xl, and z an element in ys. Finally, let χ be a map from $\{x,y\} \cup z\langle K\rangle$ to X. Then, if $\chi|_{\{x,y\}}$ and $\chi|_{\{y\}\cup z\langle K\rangle}$ are faithful, so is χ.*

PROOF. By Lemma 3.4.8, there exists an element r in $S_1(K)$ such that $s \in r\langle K\rangle$. Thus, as we are assuming that $l \in S_1(s)\backslash\langle K\rangle^{s^*}$, $lr \subseteq S_1(K)$; cf. Lemma 11.1.5.

Since $z \in ys$ and $s \in r\langle K\rangle$, $z \in yr\langle K\rangle$. Thus, there exists an element w in yr such that $z \in w\langle K\rangle$. Since $w \in yr$ and $y \in xl$, $w \in xlr$. Thus, as $lr \subseteq S_1(K)$, $w \in xS_1(K)$.

Since r in $S_1(K)$ and $s \in r\langle K\rangle$, $s^* \in S_{-1}(r^*)$. Thus, by Lemma 3.4.4(iv), $S_1(s) \subseteq S_1(r)$. Thus, as as $l \in S_1(s)$, $l \in S_1(r)$.

By hypothesis, $\chi|_{\{x,y\}}$ and $\chi|_{\{y,w\}}$ are faithful. (Note that $w \in z\langle K\rangle$.) Thus, as $y \in xl$, $w \in yr$, and $l \in S_1(r)$, we obtain that $\chi|_{\{x,w\}}$ is faithful; cf. Lemma 6.6.1. On the other hand, as $\chi|_{z\langle K\rangle}$ is assumed to be faithful, we obtain from $z\langle K\rangle = w\langle K\rangle$ that $\chi|_{w\langle K\rangle}$ is faithful. Thus, as $w \in xS_1(K)$, $\chi|_{\{x\}\cup w\langle K\rangle}$ is faithful; cf. Lemma 11.3.1. Thus, as $w\langle K\rangle = z\langle K\rangle$, $\chi|_{\{x\}\cup z\langle K\rangle}$ is faithful.

Lemma 11.3.4 *Let H be a subset of L, and let s be an element in $S_1(H)$ such that $\langle H\rangle \subseteq \langle K\rangle^s$. Let y be an element in X, let z be an element in $ys\langle H\rangle$, and let z' be an element in $z\langle H\rangle$.*
Then, for each faithful map χ from $y\langle K\rangle \cup \{z\}$ to X, there exists at most one faithful map χ' from $y\langle K\rangle \cup \{z'\}$ to X such that $\chi'|_{y\langle K\rangle} = \chi|_{y\langle K\rangle}$ and $z'\chi' \in z\chi\langle H\rangle$.

PROOF. Since we are assuming that $z' \in z\langle H\rangle$ and $z \in ys\langle H\rangle$, $z' \in ys\langle H\rangle$. On the other hand, we are assuming that $\langle H\rangle \subseteq \langle K\rangle^s$, and that means that $s\langle H\rangle \subseteq \langle K\rangle s$. Thus, as $z' \in ys\langle H\rangle$, $z' \in y\langle K\rangle s$. Thus, there exists an element y' in $y\langle K\rangle$ with $z' \in y's$.

Let us fix a faithful map χ' from $y\langle K\rangle \cup \{z'\}$ to X which satisfies $\chi'|_{y\langle K\rangle} = \chi|_{y\langle K\rangle}$ and $z'\chi' \in z\chi\langle H\rangle$.

Since $y' \in y\langle K\rangle$, χ' is defined on y'. Moreover, we have $z' \in y's$. Thus, as χ' is assumed to be faithful, we have $z'\chi' \in y'\chi's$. On the other hand, as we are assuming that $\chi'|_{y\langle K\rangle} = \chi|_{y\langle K\rangle}$, we have $y'\chi' = y'\chi$. Thus, $z'\chi' \in y'\chi s$. Thus, as $s \in S_1(H)$, $z'\chi' \in y'\chi S_1(H)$.

Thus, as $z'\chi' \in z\chi\langle H\rangle$, the claim follows from Lemma 11.1.6.

11.4 The Extension Theorem

The goal of this section is the proof of Theorem 11.4.6. In this theorem, we focus on faithful maps from certain subsets Y of X to X which extend faithfully to a subset of X containing Y and one additional element. The main idea of the proof of this theorem is the use of Corollary 11.4.3 in the proof of Proposition 11.4.5.

Theorem 11.4.6 plays a crucial role in the proof of Lemma 12.3.1.

In this section, we fix a nonempty set of subsets of L and call it $\mathcal{K}$.

Lemma 11.4.1 *We have* $\langle\bigcap_{K\in\mathcal{K}} K\rangle = \bigcap_{K\in\mathcal{K}}\langle K\rangle$.

PROOF. Let us denote by H the intersection of the elements in $\mathcal{K}$ and by T the intersection of the sets $\langle K\rangle$ where $K \in \mathcal{K}$. We have to show that $\langle H\rangle = T$.

By way of contradiction, we assume that $\langle H\rangle \neq T$. Then, as $\langle H\rangle \subseteq T$, $T \not\subseteq \langle H\rangle$. We pick an element t in $T \setminus \langle H\rangle$, and we do this in such a way that $\ell(t)$ is as small as possible.

Since $1 \in \langle H\rangle$ and $t \notin \langle H\rangle$, $1 \neq t$. Thus, by Lemma 3.1.2, there exist elements s in $\langle L\rangle$ and l in L such that $t \in sl$ and $\ell(t) = \ell(s) + 1$. It follows that $t \in S_{-1}(l)$. Thus, as $t \in T$, $l \in H$; cf. Lemma 11.1.2. Thus, as $t \in sl$ and $t \in T$, $s \in T$. Thus, as $\ell(t) = \ell(s) + 1$, the minimal choice of t forces $s \in \langle H\rangle$. Thus, as $t \in sl$ and $l \in H$, $t \in \langle H\rangle$, contradiction.

Here is a generalization of Lemma 11.4.1.

Lemma 11.4.2 *For each element* s *in* $\langle L\rangle$, $s\langle\bigcap_{K\in\mathcal{K}} K\rangle = \bigcap_{K\in\mathcal{K}} s\langle K\rangle$.

PROOF. Let us denote by R the set of all elements in $\langle L\rangle$ which do not satisfy the equation in question. By way of contradiction, we assume that R is not empty. We pick an element r in R, and we do this in such a way that $\ell(r)$ is as small as possible.

By Lemma 11.4.1, $1 \notin R$. Thus, as $r \in R$, $1 \neq r$. Thus, by Lemma 3.1.2, there exist elements l in L and q in $\langle L\rangle$ such that $r \in lq$ and $\ell(r) = 1 + \ell(q)$.

Let us denote by H the intersection of the elements in $\mathcal{K}$. By Q we shall denote the intersection of the sets $r\langle K\rangle$ with $K \in \mathcal{K}$. Then, as $r \in R$, $r\langle H\rangle \neq Q$. Thus, as $r\langle H\rangle \subseteq Q$, $Q \not\subseteq r\langle H\rangle$. Thus, we find an element s in Q such that $s \notin r\langle H\rangle$.

Let us first assume that $s^* \in S_{-1}(l)$. Then, there exists an element p in $\langle L\rangle$ such that $s \in lp$ and $\ell(s) = 1 + \ell(p)$. Thus, $l \in S_1(p)$. Moreover, since $r \in lq$ and $\ell(r) = 1+\ell(q)$, $l \in S_1(q)$. On the other hand, for each element K in $\mathcal{K}$, we have $s \in r\langle K\rangle$. Thus, for each element K in $\mathcal{K}$, $p \in q\langle K\rangle$; cf. Lemma 11.1.4. Thus, as $\ell(r) = 1 + \ell(q)$, the minimal choice of r forces $p \in q\langle H\rangle$. It follows that $s \in lp \subseteq lq\langle H\rangle = r\langle H\rangle$, contrary to the choice of s.

Let us now assume that $s^* \notin S_{-1}(l)$. Then, by Lemma 3.4.7, $s^* \in S_1(l)$. Thus, by Lemma 3.4.4(i), $l \in S_1(s)$. Thus, there exists an element t in $\langle L\rangle$ such that $\{t\} = ls$. Since $s \in Q$, we have that, for each element K in $\mathcal{K}$, $s \in r\langle K\rangle$. Thus, for each element K in $\mathcal{K}$,

$$q \in lr \subseteq ls\langle K\rangle = t\langle K\rangle,$$

and this is equivalent to $t \in q\langle K\rangle$. Thus, as $\ell(r) = 1+\ell(q)$, the minimal choice of r forces $t \in q\langle H\rangle$. Thus, $s \in lt \subseteq lq\langle H\rangle = r\langle H\rangle$, and this contradicts our choice of s.

Corollary 11.4.3 *For each element s in $\langle L\rangle$, $\langle\bigcap_{K\in\mathcal{K}} K\rangle^s = \bigcap_{K\in\mathcal{K}}\langle K\rangle^s$.*

PROOF. Let us denote by H the intersection of the elements in $\mathcal{K}$. Then, for each element K in $\mathcal{K}$, $\langle H\rangle^s \subseteq \langle K\rangle^s$.

Conversely, let r be an element in $\langle L\rangle$ such that, for each element K in $\mathcal{K}$, $r \in \langle K\rangle^s$. Then, for each element K in $\mathcal{K}$, $sr \subseteq \langle K\rangle s$. Thus, by Lemma 11.4.2, $sr \subseteq \langle H\rangle s$, and this means that $r \in \langle H\rangle^s$.

Let y and z be elements in X, and let n be the smallest non-negative integer n with $z \in yL^n$. Recall from Section 6.6 that $D(y,z)$ is our notation for the union of the sets $yL^i \cap zL^j$ which satisfy $i+j=n$.

For the remainder of this section, we denote by V the union of the sets $\langle K\rangle$ with $K \in \mathcal{K}$. We also fix two elements y in X and z in $y\langle L\rangle$.

Proposition 11.4.4 *Let x be an element in $D(y,z)$. Then, each faithful map χ from $\{y\}\cup zV$ to X extends faithfully to a bijective map from $\{x,y\}\cup zV$ to $\{x\chi, y\chi\}\cup z\chi V$.*

PROOF. Let p (respectively q) denote the uniquely determined element in $\langle L\rangle$ which satisfies $x \in yp$ (respectively $z \in xq$). Then $z \in ypq$. Thus, there exists an element s in pq such that $z \in ys$.

Let χ be a faithful map from $\{y\}\cup zV$ to X. Then, as $z \in ys$, $z\chi \in y\chi s$. Thus, as $s \in pq$, $z\chi \in y\chi pq$. Therefore, there exists an element v in $y\chi p$ such that $z\chi \in vq$. We set $x\chi := v$. Then, $\chi|_{\{y,x,z\}}$ is faithful.

Now we pick an element K in $\mathcal{K}$. Then, by Lemma 3.4.8, $z \in xS_1(K)\langle K\rangle$. Thus, there exists an element w in $xS_1(K)$ such that $z \in w\langle K\rangle$. Thus, by Theorem 3.6.4, $w \in D(x,z)$. On the other hand, as $w \in z\langle K\rangle \subseteq zV$, $\chi|_{\{y,w,z\}}$ is faithful. Thus, as $x \in D(y,z)$ and $\chi|_{\{y,x,z\}}$ is faithful, $\chi|_{\{x,w\}}$ is faithful; cf. Lemma 6.6.2. Thus, as $w \in xS_1(K)$ and $z \in w\langle K\rangle$, $\chi|_{\{x\}\cup z\langle K\rangle}$ is faithful; cf. Lemma 11.3.1.

Now the claim follows from the fact that K has been chosen arbitrarily in $\mathcal{K}$.

Proposition 11.4.5 *Let x be an element in X such that $y \in D(x,z)$. Then, if $y \in xL$, each faithful map from $\{y\}\cup zV$ to X extends faithfully to a bijective map from $\{x,y\}\cup zV$ to $\{x\chi, y\chi\}\cup z\chi V$.*

PROOF. Let us denote by s the uniquely determined element in $\langle L\rangle$ which satisfies $z \in ys$ and by l the uniquely determined element in L which satisfies $y \in xl$. Then, as $y \in D(x,z)$, $l \in S_1(s)$.

Let χ be a faithful map from $\{y\}\cup zV$ to X, and let us denote by $\mathcal{H}$ the set of all elements K of $\mathcal{K}$ which satisfy $l \in \langle K\rangle^{s^*}$.

Assume first that $\mathcal{H}$ is empty. In this case, we pick an arbitrary element in $y\chi l$, and we denote this element by $x\chi$. By Lemma 11.3.3, the extension of χ defined in this way, must be faithful.

Let us now assume that $\mathcal{H}$ is not empty, and let us denote by J the intersection of the elements in $\mathcal{H}$. Then, by Corollary 11.4.3, $l \in \langle J\rangle^{s^*}$, and that means that $ls \subseteq s\langle J\rangle$.

On the other hand, we have $z \in ys$ and $y \in xl$, whence $z \in xls$. Thus, as $ls \subseteq s\langle J\rangle$, $z \in xs\langle J\rangle$. Thus, there exists an element w in xs such that $z \in w\langle J\rangle$. From $x \in yl$, $w \in xs$, and $l \in S_1(s)$ we obtain $x \in D(y, w)$.

Let us denote by V' the union of the sets $\langle H\rangle$ with $H \in \mathcal{H}$. Then, as $w \in z\langle J\rangle$, $wV' = zV'$. Thus, as $x \in D(y, w)$, $\chi|_{\{y\}\cup zV'}$ extends faithfully to a bijective map from $\{x, y\} \cup zV'$ to $\{x\chi, y\chi\} \cup z\chi V'$; cf. Proposition 11.4.4. Thus, by Lemma 11.3.3, χ is faithful also on $\{x, y\} \cup zV$.

Here is the *Extension Theorem.*

Theorem 11.4.6 *Let x be an element in X. Then, each faithful map χ from $yV \cup \{z\}$ to X extends faithfully to a bijective map from $yV \cup \{x, z\}$ to $y\chi V \cup \{x\chi, z\chi\}$.*

PROOF. Let us denote by s the uniquely determined element in $\langle L\rangle$ such that $x \in zs$. Clearly, we may assume that $1 \neq s$. Thus, by Lemma 3.1.2, there exist elements r in $\langle L\rangle$ and l in L such that $s \in rl$ and $\ell(s) = \ell(r) + 1$. From $x \in zs$ and $s \in rl$ we obtain $x \in zrl$. Thus, there exists an element w in zr such that $x \in wl$.

Let χ be a faithful map from $yV \cup \{z\}$ to X. Then, by induction, χ extends faithfully to $yV \cup \{w, z\}$.

From Proposition 11.4.4 and Proposition 11.4.5 we obtain that $\chi|_{yV\cup\{w\}}$ extends faithfully to $yV \cup \{w, x\}$. However, as $\ell(s) = \ell(r) + 1$ and $x \in zs$, $x\chi \in z\chi s$.

For the remaining three results of this section, we shall now fix the letter s to denote the uniquely determined element in $\langle L\rangle$ which satisfies $z \in ys$.

For each element K in $\mathcal{K}$, we fix a subset K' of L such that, if r denotes the uniquely determined element in $s\langle K'\rangle \cap S_1(K')$, $\langle K'\rangle \subseteq \langle K\rangle^r$.

We denote by V' the union of the sets $\langle K'\rangle$ with $K \in \mathcal{K}$.

Corollary 11.4.7 *Each faithful map χ from $yV \cup \{z\}$ to X extends faithfully to a bijective map from $yV \cup zV'$ to $y\chi V \cup z\chi V'$.*

PROOF. Let χ be a faithful map from $yV \cup \{z\}$ to X. Then, by Theorem 11.4.6, χ extends to $yV \cup zV'$ in such a way that, for each element x in zV', $\chi|_{yV\cup\{x,z\}}$ is faithful.

Let us now fix an element K in $\mathcal{K}$, and let us pick two elements v and w in $z\langle K'\rangle$. By Theorem 11.4.6, $\chi|_{y\langle K\rangle\cup\{v\}}$ extends faithfully to a bijective map from $y\langle K\rangle \cup \{v, w\}$ to $y\chi\langle K\rangle \cup \{v\chi, w\chi\}$. By Lemma 11.3.4, this extension coincides with $\chi|_{y\langle K\rangle\cup\{v,w\}}$. Thus, $\chi|_{\{v,w\}}$ is faithful.

Since v and w have been chosen arbitrarily in $z\langle K'\rangle$, we have shown that $\chi|_{z\langle K'\rangle}$ is faithful. Thus, the claim follows from Lemma 11.3.2.

Corollary 11.4.8 *Each faithful map from yV to X extends faithfully to a bijective map from $yV \cup zV'$ to $y\chi V \cup z\chi V'$.*

PROOF. Let χ be a faithful map from yV to X. Then, by Theorem 11.4.6, χ extends faithfully to $yV \cup \{z\}$. Thus, the claim follows from Corollary 11.4.7.

Corollary 11.4.9 *Let z' be an element in $ys \cap zV'$. Then, for each faithful map χ from $yV \cup zV'$ to X, there exists a faithful map χ' from $yV \cup z'V'$ to X which coincides with χ on yV and on $zV' \cap z'V'$.*

PROOF. We are assuming that $z' \in ys$. Thus, there exists a faithful map χ' from $yV \cup z'V'$ to X which coincides with χ on yV and on z'; cf. Corollary 11.4.7. We shall show that χ and χ' coincide on $zV' \cap z'V'$. In order to do so, we pick an element x in $zV' \cap z'V'$. Since $x \in z'V'$, there exists an element K in $\mathcal{K}$ such that $x \in z'\langle K'\rangle$.

Since $\langle K\rangle \subseteq V$, the (faithful) maps $\chi|_{y\langle K\rangle\cup\{x\}}$ and $\chi'|_{y\langle K\rangle\cup\{x\}}$ coincide on $y\langle K\rangle$. Moreover, as $x \in z'\langle K'\rangle$, $x\chi' \in z'\chi'\langle K'\rangle = z'\chi\langle K'\rangle$. Finally, as $x \in z'\langle K'\rangle$, $x\chi \in z'\chi\langle K'\rangle$. Thus, by Lemma 11.3.4, $x\chi = x\chi'$.

12

Spherical Coxeter Sets

This final chapter is the second part of our investigation of Coxeter sets. It deals with spherical Coxeter sets.

Let L be a set of involutions of S.

Recall that L is called a Coxeter set if L is constrained and satisfies the exchange condition. Recall also that L is called spherical if $S_{-1}(L)$ is not empty. Recall, finally, that a closed subset T of S is called faithfully embedded in S if, for any two elements y in X and z in yT, each faithful map χ from $\{y, z\}$ to X extends to a bijective map from yT to $y\chi T$.

The first goal of this chapter is to show that $\langle L \rangle$ is faithfully embedded in S if L is a spherical Coxeter set having at least three elements none of them thin. The corresponding Schur groups turn out to have a Tits system. The situation will be completely described in the corresponding recognition theorem (Theorem 12.3.4).

In the first section of this chapter, we focus on specific characteristics of spherical Coxeter sets such as maximal elements and conjugation. The second section is devoted to an extension theorem for spherical Coxeter sets. Our approach to this theorem (which follows the line of [46]) is partially inspired by a geometrical reasoning provided by Jacques Tits in [37].

In the third section of this chapter, we apply results from the two previous sections in order to prove the above-mentioned recognition theorem for spherical Coxeter sets of cardinality at least 3.

In Section 12.4, we shall look closer at the case where L is a spherical Coxeter set consisting of two elements.[1] Assuming $\langle L \rangle = S$ we shall see that

$$|S| \in \{4, 6, 8, 12, 16, 24\}$$

if S is not thin and has finite valency. This is the scheme theoretic version of a well-known theorem on generalized polygons due to Walter Feit and

[1] We promised towards the end of Section 10.6 to look at this case.

Graham Higman. Our proof is the scheme-theoretic version of the proof given by Robert Kilmoyer and Louis Solomon in [29].

In Section 12.5, we shall look closer at the individual cases which we obtained in Section 12.4.

We finish this chapter by looking at the case where L is a Coxeter set consisting of two elements and $\langle L\rangle$ has finite valency and is the kernel of a semidirect product. This is a case which, geometrically, has been investigated by Udo Ott and Stanley Payne.

12.1 Elements of Maximal Length

In this section, the letter L stands for a spherical Coxeter set. Instead of ℓ_L we shall write ℓ.

Assuming L to be a spherical Coxeter set we obtain from Lemma 3.6.8 that $S_{-1}(L)$ contains exactly one element. In the following, we shall denote this element by m_L.

Recall from Section 10.2 that we denote by d_L the smallest element in $\ell(S_{-1}(L))$. Thus, as $\{m_L\} = S_{-1}(L)$,

$$\ell(m_L) = d_L.$$

From this equation we obtain $\ell(s) \leq d_L - 1$ for each element s in $\langle L\rangle \setminus \{m_L\}$; cf. Lemma 3.6.7. In particular, as $(m_L)^* \in \langle L\rangle$ and $\ell((m_L)^*) = \ell(m_L)$,

$$(m_L)^* = m_L.$$

Let s be an element in $\langle L\rangle$. Then, as $m_L \in S_{-1}(L)$, $m_L \in S_{-1}(s)$; cf. Theorem 3.6.6. Thus, by definition, $\langle L\rangle$ contains an element r with $m_L \in rs$ and $\ell(m_L) = \ell(r) + \ell(s)$. Moreover, Lemma 3.5.1 says that $\langle L\rangle$ contains at most one such element. Thus, there exists exactly one such element. For the remainder of this section, we shall denote this element by $s^{(L)}$.

Lemma 12.1.1 *For any two elements p and q in $\langle L\rangle$, we have the following.*

(i) *If $p \neq q$, $p^{(L)} \neq q^{(L)}$.*

(ii) *Let r be an element in pq, and assume that $\ell(r) = \ell(p) + \ell(q)$. Then*

$$q^{(L)} \in r^{(L)}p$$

and $\ell(q^{(L)}) = \ell(r^{(L)}) + \ell(p)$.

PROOF. (i) Let p and q be elements in $\langle L\rangle$ such that $p^{(L)} = q^{(L)}$. Then, by definition, $m_L \in p^{(L)}p$, $\ell(m_L) = \ell(p^{(L)}) + \ell(p)$, $m_L \in p^{(L)}q$, and $\ell(m_L) = \ell(p^{(L)}) + \ell(q)$. Thus, by Lemma 3.5.1, $p = q$.

(ii) By definition, we have $m_L \in r^{(L)}r$ and $\ell(m_L) = \ell(r^{(L)}) + \ell(r)$. By hypothesis, we have $r \in pq$ and $\ell(r) = \ell(p) + \ell(q)$. Thus, by Lemma 3.4.2, there exists an element s in $r^{(L)}p$ such that $m_L \in sq$, $\ell(s) = \ell(r^{(L)}) + \ell(p)$, and $\ell(m_L) = \ell(s) + \ell(q)$.

From $m_L \in sq$ and $\ell(m_L) = \ell(s) + \ell(q)$ we obtain $s = q^{(L)}$. Thus, the claim follows from $s \in r^{(L)}p$ and $\ell(s) = \ell(r^{(L)}) + \ell(p)$.

For each element s in $\langle L \rangle$, we shall write $s^{[L]}$ instead of $s^{(L)(L)}$.

Lemma 12.1.2 *For any two elements p and q in $\langle L \rangle$, we have the following.*

(i) *If $p \neq q$, $p^{[L]} \neq q^{[L]}$.*

(ii) *Let r be an element in pq, and assume that $\ell(r) = \ell(p) + \ell(q)$. Then*

$$r^{[L]} \in p^{[L]}q^{[L]}$$

and $\ell(r^{[L]}) = \ell(p^{[L]}) + \ell(q^{[L]})$.

PROOF. (i) This follows from Lemma 12.1.1(i). (Apply this lemma twice.)

(ii) This follows from Lemma 12.1.1(ii). (Apply this lemma three times.)

Lemma 12.1.3 *For each element s in $\langle L \rangle$, the following hold.*

(i) *We have $\ell(s^{[L]}) = \ell(s)$.*

(ii) *We have $s^{[L][L]} = s$.*

PROOF. (i) This follows immediately from the definition of $s^{[L]}$.

(ii) Let s be an element in $\langle L \rangle$. Then, by definition, $m_L \in s^{[L](L)}s^{[L]}$. Thus, as $(m_L)^* = m_L$, $m_L \in (s^{[L]})^*(s^{[L](L)})^*$; cf. Lemma 1.3.2(iii). Thus, as $m_L \in s^{(L)}s$, the set

$$(s^{(L)})^*(s^{[L]})^* \cap ss^{[L](L)}$$

is not empty; cf. Lemma 1.3.4. Thus, as $\{m_L\} = (s^{(L)})^*(s^{[L]})^*$, $m_L \in ss^{[L](L)}$. However, by definition, we have

$$m_L \in s^{[L][L]}s^{[L](L)}.$$

Moreover, by (i), we have $\ell(s^{[L][L]}) = \ell(s)$. Thus, the claim follows from Lemma 3.5.1.

Recall that, for each subset R of $\langle L \rangle$, $S_1(R)$ is our notation for the intersection of the sets $S_1(r)$ with $r \in \{1\} \cup R$.

Lemma 12.1.4 *For each subset K of L, we have $(m_K)^{(L)} \in S_1(K)$ and $(m_K)^{(L)} \in m_L\langle K \rangle$.*

PROOF. Let k be an element in K. Then $m_K \in S_{-1}(k)$. Thus, by Lemma 3.4.4(iv), $S_1(m_K) \subseteq S_1(k)$. Thus, as $(m_K)^{(L)} \in S_1(m_K)$, $(m_K)^{(L)} \in S_1(k)$. Since k has been chosen arbitrarily in K, we have shown that $(m_K)^{(L)} \in S_1(K)$.

By definition, $m_L \in (m_K)^{(L)} m_K$. Thus, as $m_K \in \langle K \rangle$, $m_L \in (m_K)^{(L)}\langle K \rangle$, so that the second claim follows from Lemma 2.1.4.

For each nonempty subset R of $\langle L \rangle$, we define $R^{[L]}$ to be the set of all elements $r^{[L]}$ with $r \in R$.

Lemma 12.1.5 *For each nonempty subset K of L, the following hold.*

(i) *We have $m_{K^{[L]}} = (m_K)^{[L]}$.*

(ii) *We have $\langle K \rangle \subseteq \langle K^{[L]} \rangle^{((m_K)^{(L)})}$.*

PROOF. (i) From Lemma 12.1.2 we obtain by induction that $s \mapsto s^{[L]}$ is a bijective map from $\langle K \rangle$ to $\langle K^{[L]} \rangle$. Thus, the claim follows from Lemma 12.1.3(i).

(ii) From Lemma 1.3.6(iii) together with Lemma 3.1.1(i) we know that it is enough to show that $K \subseteq \langle K^{[L]} \rangle^{((m_K)^{(L)})}$.

In order to show this we pick an element k in K. From the first statement of Lemma 12.1.4 we know that

$$(m_K)^{(L)} \in S_1(k).$$

Thus, there exists an element p in $(m_K)^{(L)} k$ such that $\ell(p) = \ell((m_K)^{(L)}) + 1$. Assume that $K^{[L]} \subseteq S_1(p)$. Then, by Theorem 3.6.4, $\langle K^{[L]} \rangle \subseteq S_1(p)$. Thus, as $m_{K^{[L]}} \in \langle K^{[L]} \rangle$, $m_{K^{[L]}} \in S_1(p)$. Thus, by (i),

$$(m_K)^{[L]} \in S_1(p).$$

Thus, there exists an element q in $(m_K)^{[L]} p$ such that $\ell(q) = \ell((m_K)^{[L]}) + \ell(p)$. Thus, as $\ell(p) = \ell((m_K)^{(L)}) + 1$,

$$\ell(q) = \ell((m_K)^{[L]}) + \ell((m_K)^{(L)}) + 1 = \ell(m_L) + 1,$$

contrary to $q \in \langle L \rangle$ and $\ell(m_L) = d_L$.

Thus, as L is assumed to satisfy the exchange condition, there exists an element h in $K^{[L]}$ such that

$$h(m_K)^{(L)} = (m_K)^{(L)} k.$$

Thus, as $h \in \langle K^{[L]} \rangle$, $(m_K)^{(L)} k \subseteq \langle K^{[L]} \rangle (m_K)^{(L)}$, and that means that $k \in \langle K^{[L]} \rangle^{((m_K)^{(L)})}$.

12.2 Faithful Maps

In this section, the letter L stands for a spherical Coxeter set. Instead of m_L we shall write m.

In the previous section, we did not refer to the set X. We shall do this now. In fact, all results of this section deal with faithful maps.

Lemma 12.2.1 *Let y be an element in X, let z be an element in ym, let l be an element in L, and let χ be a faithful map from $yl^{[L]} \cup \{y, z\}$ to X. Then there exists at most one faithful map from $y\langle l^{[L]}\rangle \cup z\langle l\rangle$ to X which coincides with χ on $yl^{[L]} \cup \{y, z\}$.*

PROOF. By the first statement of Lemma 12.1.4, $l^{(L)} \in S_1(l)$, and, by Lemma 12.1.5(ii), $\langle l\rangle \subseteq \langle l^{[L]}\rangle^{l^{(L)}}$. Thus, the claim follows from Lemma 11.3.4 (applied to $\{l\}$, $\{l^{[L]}\}$, and $l^{(L)}$ instead of H, K, and s).

Lemma 12.2.2 *Let y be an element in X, let z be an element in ym, and let K be a subset of L. Assume that K does not contain thin elements.*
Let χ be a faithful map from $yK^{[L]}\cup\{y, z\}$ to X. Then there exists at most one faithful map from $y\langle K^{[L]}\rangle\cup z\langle K\rangle$ to X which coincides with χ on $yK^{[L]}\cup\{y, z\}$.

PROOF. Let χ be a faithful map from $y\langle K^{[L]}\rangle \cup z\langle K\rangle$ to X. From Lemma 12.2.1 we know that the values of χ on zK are uniquely determined by the values on $y\langle K\rangle^{[L]} \cup \{y, z\}$.

Let us now prove that the values of χ on zK^2 are determined by the values of χ on $yK^{[L]} \cup \{y, z\}$. In order to show this we pick an element w in zK^2. Since $w \in zK^2$, there exists an element v in zK such that $w \in vK$. Since $v \in zK$, there exists an element h in K such that $v \in zh$. Since $w \in vK$, there exists an element l in K such that $w \in vl$.

We are assuming that K does not contain thin elements. Thus, $h^{[L]}$ is not thin. Thus, $zm\cap vm\cap y\langle h^{[L]}\rangle$ is not empty. Let x be an element in $zm\cap vm\cap y\langle h^{[L]}\rangle$.

By hypothesis, χ is uniquely defined on $yh^{[L]}$. In particular, χ is uniquely defined on x. Thus, as $x \in zm$, χ is uniquely defined on $x\langle l^{[L]}\rangle$; cf. Lemma 12.2.1. Thus, as $v \in xm$, χ is uniquely defined on $v\langle l\rangle$; cf. Lemma 12.2.1 once again.

Induction now finishes the proof.

For the remainder of this section, we assume that L does not contain thin elements.

We fix a positive integer n such that $n \leq |L|$, and we denote by $\mathcal{K}$ the set of all subsets K of L with $1 \leq |K| \leq n$. The union of the sets $\langle K\rangle$ with $K \in \mathcal{K}$ will be denoted by V.

From Lemma 12.1.3(i) we know that, for each element K of $\mathcal{K}$, $K^{[L]} \subseteq L$. Moreover, Lemma 12.1.2(i) tells us that, for each element K of $\mathcal{K}$, $|K^{[L]}| = |K|$. Thus, for each element K of $\mathcal{K}$, $K^{[L]} \in \mathcal{K}$. In particular, $V^{[L]} = V$.

The fact that $V^{[L]} = V$ together with Lemma 12.1.4 and Lemma 12.1.5(ii) allows us to apply Corollary 11.4.7, Corollary 11.4.8, and Corollary 11.4.9 to V instead of V'.

For any two elements y in X and z in ym, we define $\Xi_{\{y,z\}}$ to be the set of all faithful maps from $yV \cup zV$ to X.

As a consequence of Lemma 12.2.2 we obtain the following.

Lemma 12.2.3 *Let y be an element in X, and let z be an element in ym. Then each map in $\Xi_{\{y,z\}}$ is uniquely determined by its action on $yL \cup \{y, z\}$.*

Let x be an element in X, let y be an element in xm, let z be an element in $xm \cap yV$, let ϕ be an element in $\Xi_{\{x,y\}}$, and let ψ be an element in $\Xi_{\{x,z\}}$.

The maps ϕ and ψ will be called *close to each other* if they coincide on xV and on $yV \cap zV$.

Let K be an element in $\mathcal{K}$, let y' be an element in X, let y be an element in $y'm$, let z' be an element in $y'\langle K^{[L]}\rangle$, and let z be an element in $z'm \cap y\langle K\rangle$. Let ϕ be an element in $\Xi_{\{y',y\}}$, and let ψ be an element in $\Xi_{\{z',z\}}$.

The faithful maps ϕ and ψ will be called *K-compatible* if, for each element x in $ym \cap zm \cap y'\langle K^{[L]}\rangle$, there exist maps η in $\Xi_{\{x,y\}}$ close to ϕ and ζ in $\Xi_{\{x,z\}}$ close to η and ψ.

We define Ξ to be the union of the sets $\Xi_{\{y,z\}}$ with $y \in X$ and $z \in ym$.

For the remainder of this section, we assume that $2 \leq n$.

Lemma 12.2.4 *K-compatibility is an equivalence relation on Ξ.*

PROOF. By Corollary 11.4.9, K-compatibility is reflexive. That K-compatibility is symmetric follows immediately from the definition of K-compatibility. Let us prove transitivity.

Let (x', x), (y', y), (z', z) be elements in m such that y, $z \in x\langle K\rangle$ and y', $z' \in x'\langle K^{[L]}\rangle$.

Let χ_x be an element in $\Xi_{\{x',x\}}$, let χ_z be an element in $\Xi_{\{z',z\}}$, and let χ_y be an element in $\Xi_{\{y',y\}}$ which is K-compatible with both χ_x and χ_z. We have to show that χ_x and χ_z are K-compatible. By induction, we may assume that $z \in yK$. Thus, there exists an element k in K such that $z \in yk$.

In order to show that χ_x and χ_z are K-compatible, we pick an element u' in $xm \cap zm \cap x'\langle K^{[L]}\rangle$. Then, as we are assuming that L does not contain thin elements, there exist elements v' in $xm \cap ym \cap u'\langle k^{[L]}\rangle$ and w' in $ym \cap zm \cap u'\langle k^{[L]}\rangle$. (Recall that $z \in yk$.)

Since χ_x and χ_y are assumed to be compatible, there exist maps ρ in $\Xi_{\{v',x\}}$ close to χ_x and σ in $\Xi_{\{v',y\}}$ close to ρ and χ_y.

Similarly, as χ_y and χ_z are assumed to be compatible, there exist maps ζ in $\Xi_{\{w',y\}}$ close to χ_y and η in $\Xi_{\{w',z\}}$ close to ζ and χ_z.

By Corollary 11.4.9, there exists an element ϕ in $\Xi_{\{u',x\}}$ close to χ_x. Similarly, we find an element ψ in $\Xi_{\{u',z\}}$ close to ϕ. We shall be done if we succeed in showing that ψ is close to χ_z.

From Corollary 11.4.9 we obtain an element $\tilde{\eta}$ in $\Xi_{\{w',z\}}$ close to ψ.

We claim that $\tilde{\eta} = \eta$, and in order to see this we pick an element w in $w'L$. Then, as we are assuming that $2 \leq n$,

$$w\tilde{\eta} = w\psi = w\phi = w\rho = w\sigma = w\zeta = w\eta.$$

Since w has been chosen arbitrarily in $w'L$, we have that $\tilde{\eta}|_{w'L} = \eta|_{w'L}$. Thus, as

$$z\tilde{\eta} = z\psi = z\phi = z\rho = z\sigma = z\zeta = z\eta,$$

$\tilde{\eta} = \eta$; cf. Lemma 12.2.3.

The maps ψ and $\tilde{\eta}$ coincide on z' and on zV. The same is true for η and χ_z. Thus, ψ and χ_z coincide on z' and on zV. Thus, they coincide on $u'V \cap z'V$ and on zV. Thus, ψ is close to χ_z.

Let (y, z) be an element in m, and let K be an element in $\mathcal{K}$. From Corollary 11.4.9 we know that each K-equivalence class intersects $\Xi_{\{y,z\}}$ in at most one element. The following lemma shows that, if Ξ is not empty, each K-equivalence class intersects $\Xi_{\{y,z\}}$ in at least one element.

Lemma 12.2.5 *Let (y', y) and (z', z) be elements in m, let K be an element in $\mathcal{K}$ such that $z' \in y'\langle K^{[L]}\rangle$ and $z \in y\langle K\rangle$, and let χ be an element in $\Xi_{\{y',y\}}$. Then there exists an element in $\Xi_{\{z',z\}}$ which is K-compatible with χ.*

PROOF. By Lemma 12.2.4, we may assume that $z \in yK$. In this case, there exists an element k in K such that $z \in yk$.

Since $z \in yk$, we may assume that $z' \in y'k^{[L]}$. Since we are assuming that L does not contain thin elements, there exists an element x in $ym \cap zm \cap y'\langle k^{[L]}\rangle$.

By Corollary 11.4.9, there exists a map ϕ in $\Xi_{\{x,v\}}$ close to χ. Similarly, we find a map ψ in $\Xi_{\{x,z\}}$ close to ϕ. Finally, referring to Corollary 11.4.9 a third time we obtain a map χ' in $\Xi_{\{z',z\}}$ close to ψ.

The maps χ and ϕ coincide on $y'V \cap xV$, ϕ and ψ coincide on xV, and ψ and χ' coincide on $xV \cap z'V$. On the other hand, we know from Lemma 11.1.7 that $y'V \cap z'V \subseteq xV$. Thus, χ and χ' coincide on $y'V \cap z'V$.

The maps χ and ϕ coincide on yV, ϕ and ψ coincide on $yV \cap zV$, and ψ and χ' coincide on zV. Thus, χ and χ' coincide on $yV \cap zV$.

In order to show that χ and χ' are K-compatible, we pick an element t in $z'\langle K^{[L]}\rangle \cap ym \cap zm$. By Corollary 11.4.9, there exists a map ϕ in $\Xi_{\{t,y\}}$ close to χ. Similarly, there exists a map ψ in $\Xi_{\{t,z\}}$ close to ϕ.

We claim that ψ and χ' coincide on zL. In order to see this we pick an element x in zL. Then, as $z \in zK$, $x\psi = x\phi = x\chi = x\chi'$. Thus, as $z'\psi = z'\phi = z'\chi = z'\chi_w$, we conclude that ψ and χ' are close to each other.

We call $\mathcal{K}$-*compatibility* the smallest equivalence relation on Ξ which contains K-compatibility for each K in $\mathcal{K}$.

Let (y', y) and (z', z) be elements in m, let ϕ be an element in $\Xi_{\{y',y\}}$, and let ψ be an element in $\Xi_{\{z',z\}}$. Then, if ϕ and ψ are $\mathcal{K}$-compatible, $z' \in y'\langle L\rangle$ and $z \in y\langle L\rangle$.

In the following lemma, we shall write ℓ instead of ℓ_L.

Lemma 12.2.6 *Let (y, z) be an element in m. Then, for each element χ in Ξ there exists at most one element in $\Xi_{\{y,z\}}$ which is $\mathcal{K}$-compatible with χ.*

PROOF. Let us denote by R the set of all elements s in $\langle L\rangle$ such that zs contains an element contradicting our claim. By way of contradiction, we assume that R is not empty. We pick an element r in R such that $\ell(r)$ is as small as possible.

Since $r \in R$, $1 \neq r$. Thus, by Lemma 3.1.2, there exist elements q in $\langle L\rangle$ and l in L such that $r \in ql$ and $\ell(r) = \ell(q) + 1$.

Since $r \in R$, there exists an element x in zr such that $\Xi_{\{x\}}$ contains at least two (different) elements χ_x and χ'_x which are $\mathcal{K}$-compatible with χ.

From $x \in zr$ and $r \in ql$ we obtain $x \in zql$. Thus, there exists an element u in zq such that $x \in ul$.

By Lemma 12.2.5, there exists an element χ_u in $\Xi_{\{u\}}$ which is $\mathcal{K}$-compatible with χ_x. Similarly, we find an element χ'_u in $\Xi_{\{u\}}$ which is $\mathcal{K}$-compatible with χ'_x. Both, χ_u and χ'_u are $\mathcal{K}$-compatible with χ. Thus, the choice of r forces $\chi_u = \chi'_u$. Thus, by Lemma 12.2.4, $\chi_x = \chi'_x$.

Proposition 12.2.7 *Let y be an element in X, let z be an element in ym. Then each faithful map χ from $yV \cup \{z\}$ to X extends to a faithful map from $y\langle L\rangle$ to $y\chi\langle L\rangle$.*

PROOF. Let χ be a faithful map from $yV \cup \{z\}$ to X. Then, by Corollary 11.4.7, χ extends faithfully to a map χ_z from $yV \cup zV$ to X. Thus, by Lemma 12.2.6, there exists, for each element w in $x\langle L\rangle$, a uniquely determined element χ_w in $\Xi_{\{w\}}$ which is $\mathcal{K}$-compatible with χ_z.

For each element w in $x\langle L\rangle$, we set $w\chi := w\chi_w$. Then we obtain $wl\chi \subseteq w\chi l$ for any two elements w in $x\langle L\rangle$ and l in L. Thus, by Lemma 6.6.3, χ is a faithful map from $x\langle L\rangle$ to X.

Let v be an element in xV. Then, by definition, there exists an element K in $\mathcal{K}$ such that $v \in x\langle K\rangle$. Thus, by induction, $\chi_v|_{x\langle K\rangle} = \chi|_{x\langle K\rangle}$. This shows that χ extends χ_x.

12.3 The Main Theorem

In this section, we look at spherical Coxeter sets containing at least three elements none of them thin. We shall apply Proposition 12.2.7 in order to prove that closed subsets of S generated by such Coxeter sets are faithfully embedded in S. We also establish the corresponding recognition theorem.

Let y and z be elements in X, and let n be the smallest non-negative integer n with $z \in yL^n$. Recall from Section 6.6 that we $D(y, z)$ is our notation for the union of the sets $yL^i \cap zL^j$ which satisfy $i + j = n$.

Lemma 12.3.1 *Let L be a spherical Coxeter set. Assume that L has at least three elements none of them thin. Let l be an element in L, let s be an element in $S_1(l)$, let y be an element in X, and let z be an element in ys.*
Let us denote by G the Schur group of $\langle L\rangle$ with respect to y. Then G_{yz} acts transitively on zl.

PROOF. Let v and w be elements in zl. We have to find an element in G_{yz} with $vg = w$. Without loss of generality we may assume that $s = l^{(L)}$.

Let us denote by V the union of the sets $\langle K\rangle$ with $K \subseteq L$ and $|K| \leq 2$. Then, by Lemma 11.2.5, there exists an element x in $D(y, z)$ such that $|K| = 2$,

$$xV \cap D(y, z) = xV \cap D(y, v)$$

and

$$xV \cap D(y, z) = xV \cap D(y, v).$$

Let us denote by χ the map on $xV \cup \{v\}$ which is the identity on xV and maps v to w. Then χ is faithful. Thus, as L is assumed to have no thin element, we obtain from Proposition 12.2.7 an element g in G which is the identity on xV and maps v to w.

Since y, $z \in D(y, z)$, we have $g \in G_{yz}$ and $vg = w$.

Proposition 12.3.2 *Let L be a spherical Coxeter set. Assume that L has at least three elements none of them thin. Let x be an element in X, and let us denote by G the Schur group of $\langle L\rangle$ with respect to x. Then we have the following.*

(i) *Let x be an element in X, and let s be an element in $\langle L\rangle$. Then G_x acts transitively on xs.*

(ii) *The group G acts transitively on $x\langle L\rangle$.*

PROOF. (i) Let us denote by R the set of all elements r in $\langle L \rangle$ such that G_x does not act transitively on xr. By way of contradiction, we assume that R is not empty. We pick an element r in R such that $\ell(r)$ is as small as possible.

Since $r \in R$, $1 \neq r$. Thus, by Lemma 3.1.2, there exist elements q in $\langle L \rangle$ and l in L such that $r \in ql$ and $\ell(r) = \ell(q) + 1$.

Let y and z be elements in xr. From $y \in xr$ and $r \in ql$ we obtain $y \in xql$. Thus, there exists an element v in xq such that $y \in vl$. Similarly, as $z \in xr$ and $r \in ql$, there exists an element w in xq such that $z \in wl$.

Since $q \in \langle L \rangle$ and $\ell(r) = \ell(q) + 1$, the minimal choice of r yields $q \notin R$. Thus, as v, $w \in xq$, there exists an element e in G_x such that $ve = w$. Since e is faithful, we obtain from $y \in vl$ and $w = ve$ that

$$ye \in vle \subseteq vel = wl;$$

cf. Lemma 6.1.1(i). Thus, as $z \in wl$, there exists an element f in G_{xw} such that $yef = z$; cf. Lemma 12.3.1. Setting $g := ef$ we, therefore, obtain $g \in G_x$ and $yg = z$. This contradiction finishes the proof of (i).

(ii) Let y be an element in X, and let z be an element in yL. Then there exists an element l in L such that $z \in yl$. Since L is assumed to have no thin element, there exists an element x in X with $y \in xl$ and $z \in xl$. Thus, by (i), there exists an element g in G_x with $yg = z$. (We apply (i) to l in place of s.)

What we have seen so far is that, for any two elements y and z in X with $z \in yL$, there exists an element g in G with $yg = z$. Thus, the claim follows from Lemma 6.3.4.

Theorem 12.3.3 *Let L be a spherical Coxeter set. Assume that L has at least three elements none of them thin. Then $\langle L \rangle$ is schurian.*

PROOF. Referring to Theorem 6.3.1 this follows immediately from Proposition 12.3.2.

A scheme which is generated by a Coxeter set L is called a *Coxeter scheme with respect to L.*

If S is a Coxeter scheme with respect to L, $|L|$ is called the *rank* of S.

If $\{1\} = S$, S is a Coxeter scheme with respect to the empty set.

If S consists of two elements and s denotes the non-identity element of S, s is an involution and S is a Coxeter scheme (of rank 1) with respect to $\{s\}$.

We shall now see how to obtain Coxeter schemes from thin schemes.

Assume S to be thin, let T be a closed subset of S, and let J be a subset of S such that $\langle T \cup J \rangle = S$.

Assume that $T \cap \langle J \rangle$ is normal in $\langle J \rangle$. Assume that, for each element j in J, $j^2 \in T$ and

$$T \neq TjTjT.$$

Assume that, for any two elements $j \in J$ and s in $\langle J \rangle$,

$$TsTjT \subseteq TsjT \cup TsT.$$

Then (T, J) is called a *Tits system* for S.

Theorem 12.3.4 *Let L be a set of involutions of S such that $\langle L \rangle = S$. Assume that L is a spherical Coxeter set consisting of at least three elements none of them thin. Then there exists a finite thin scheme $\bar{S}$ with a Tits system $(\bar{T}, \bar{J})$ such that $S \cong \bar{S}//\bar{T}$.*

PROOF. This is a consequence of Theorem 12.3.3.

The following theorem is the converse of Theorem 6.5.4. It says that Tits systems give rise to Coxeter sets. It makes Theorem 6.5.4 to be one of our recognition theorems.

Theorem 12.3.5 *Assume that S is thin and possesses a Tits system (T, J). Then $S//T$ is a Coxeter scheme with respect to $J//T$.*

PROOF. We are assuming that, for each element j in J, $T \neq TjTjT$. Thus, for each element j in J, j^T is an involution in $S//T$.

From $\langle T \cup J \rangle = S$ we obtain that $S//T$ is generated by $J//T$; cf. Lemma 4.2.2(i).

The facts that $J//T$ is constrained and satisfies the exchange condition follow from the hypothesis that $TsTjT \subseteq TsjT \cup TsT$ for any two elements j in J and s in $\langle J \rangle$.

12.4 Coxeter Schemes of Finite Valency and Rank 2

Throughout this section, the letter L stands for a set of involutions. We assume that S is a Coxeter scheme with respect to L, that $|L| = 2$, and that S has finite valency.

Since S is assumed to have finite valency, S is finite. Thus, by Lemma 10.1.7, L is spherical, and that means that $S_{-1}(L)$ is not empty.

Keeping the notation introduced in Section 10.2 we write d_L to denote the smallest element in $\ell(S_{-1}(L))$. However, having fixed L for the remainder of this section, we shall write d instead of d_L.

Recall that, by Theorem 10.5.4, $2d = |S|$.

The main result of this section is Theorem 12.4.6, a result due to Walter Feit and Graham Higman. This theorem says that

$$d \in \{2, 3, 4, 6, 8, 12\}$$

if S is not thin. After that, we shall give a few arithmetical details in each of the individual cases.

The main idea of the proof of Theorem 12.4.6 is to apply Theorem 9.1.7(ii) to an algebraically closed field C of characteristic 0. According to Lemma 9.2.5, the left hand side of the equation in Theorem 9.1.7(ii) is algebraic over the smallest unitary subring Z of C and the right hand side of that equation is in the smallest subfield of C. Thus, according to Lemma 8.2.5, both sides must be in Z, and one obtains an integral divisibility condition.

Group theoretic proofs based on this type of reasoning have been given repeatedly by William Burnside. (An example is his proof of the solvability of finite groups of order divisible by at most two primes.) Assuming Burnside would have known and would have been interested in the notion of a finite scheme he might have given a proof of Theorem 12.4.6 long time before Walter Feit and Graham Higman gave their proof.

Lemma 12.4.1 *Let C be an algebraically closed field of characteristic 0. Then the following hold.*

(i) *Each irreducible character of CS is linear or has degree 2.*

(ii) *If d is odd, CS has two linear characters and $\frac{d-1}{2}$ irreducible characters of degree 2.*

(iii) *If d is even, CS has four linear characters and $\frac{d}{2} - 1$ irreducible characters of degree 2.*

Proof. (i) Since S is assumed to be a Coxeter scheme with respect to L, L is constrained and $\langle L \rangle$. Thus, by Corollary 9.6.3, $C[L] = CS$, so that the claim follows from Lemma 9.4.5.

(ii), (iii) Let us denote by λ_1 the number of linear characters of CS and by λ_2 the number of irreducible characters of degree 2. Since $2d = |S|$, we obtain from (i) together with Corollary 8.6.5 that $2d = \lambda_1 + 4\lambda_2$.

From Corollary 9.6.3 we know that $C[L] = CS$. Thus, each linear character of CS is determined by its values on $\{\sigma_l \mid l \in L\}$. Thus, by Corollary 9.4.4(ii), $\lambda_1 \leq 4$. Since 1_{CS} is linear, we also obtain $1 \leq \lambda_1$.

Our claims are obvious consequences of $2d = \lambda_1 + 4\lambda_2$ and $1 \leq \lambda_1 \leq 4$.

For the remainder of this section, we set

$$n := \prod_{l \in L} \sqrt{n_l}.$$

Let l be an element in L, and let j be a positive integer. Recall from Section 10.1 that $R_j(l)$ is our notation for the complex product $l_1 \cdots l_j$ of elements in

L which satisfy $l_1 = l$ and, for each integer i with $1 \le i \le j$, $l_i = l$ if and only if i is odd. We also write $R_0(l)$ instead of $\{1\}$.

Recall from Lemma 10.5.1 that, for any two elements l in L and j in $\{0, \ldots, d\}$, $|R_j(l)| = 1$. Recall also (from Section 10.5) that, for any two elements l in L and j in $\{0, \ldots, d\}$, $r_j(l)$ is our notation for the element in $R_j(l)$.

From Theorem 10.5.4 we know that

$$\{r_j(l) \mid l \in L,\ j \in \{0, \ldots, d\}\} = S.$$

From Corollary 10.5.2 we also know that, for any two elements h and k in L, $r_d(h) = r_d(k)$. Clearly, by definition, we also have that, for any two elements h and k in L, $r_0(h) = r_0(k)$.

Let us now denote by C_d the set of all elements c in $C \setminus \{-1, 1\}$ satisfying $c^d = 1$.

In the following proposition, we completely compute the values of all non-linear irreducible characters of CS, provided that C is an algebraically closed field of characteristic 0.

Proposition 12.4.2 *Let C be an algebraically closed field of characteristic 0, and let χ be an irreducible character of CS of degree 2. Then there exists an element c in C_d such that the following hold.*

(i) *Let h and k be elements in L with $h \neq k$. Then, for each non-negative integer i with $2i + 1 \le d$,*

$$\chi(\sigma_{r_{2i+1}(h)}) = \frac{n^i}{c - c^{-1}}((n_h - 1)(c^{i+1} - c^{-(i+1)}) + (n_k - 1)\frac{n_h}{n}(c^i - c^{-i})).$$

(ii) *Let l be an element in L. Then, for each non-negative integer i with $2i \le d$,*

$$\chi(\sigma_{r_{2i}(l)}) = n^i(c^i + c^{-i}).$$

PROOF. Let M be an irreducible CS-module such that $\chi_M = \chi$, let h and k be elements in L such that $h \neq k$, and let us denote by a and b the characteristic roots of $d_M(\sigma_h \sigma_k)$. (Recall from Section 8.1 that $d_M(\sigma_h \sigma_k)$ is our notation for the endomorphism of M induced by $\sigma_h \sigma_k$.)

First of all, we have $\det(d_M(\sigma_h \sigma_k)) = ab$. On the other hand, for each element l in L, -1 and n_l are the characteristic roots of $d_M(\sigma_l)$; cf. Corollary 9.4.4(ii). Thus, $\det(d_M(\sigma_h)) = -n_h$ and $\det(d_M(\sigma_k)) = -n_k$. It follows that $ab = n^2$.

Set

$$c := \frac{a}{n}.$$

Let us first prove that $c^2 \neq 1$. In order to do this, we assume, by way of contradiction, that $c^2 = 1$. Then $a = b$. Thus, $\sigma_h\sigma_k$ induces a scalar multiplication on M, contrary to the irreducibility of M. Thus, $c \notin \{-1, 1\}$.

Set $e := d$ if d is odd and $e := \frac{d}{2}$ if d is even. Define $\sigma := (\sigma_h\sigma_k)^e$.

From Corollary 10.5.2 we know that $r_d(h) = r_d(k)$. Thus, $\sigma\sigma_h = \sigma_h\sigma$ and $\sigma\sigma_k = \sigma_k\sigma$. On the other hand, we know from Corollary 9.6.3 that $C[L] = CS$. Thus, $d_M(\sigma) \in \mathrm{End}_{CS}(M)$. Therefore, by Lemma 8.4.3 and Lemma 8.2.6, σ induces a scalar multiplication on M. It follows that $a^e = b^e$. Therefore, as $c^2 = \frac{a}{b}$, we obtain $c^d = 1$. Thus, as $c \notin \{-1, 1\}$, $c \in C_d$.

From $ab = n^2$ and the definition of c we obtain $cb = n$. Thus,

$$a + b = n(c + c^{-1}).$$

It follows that

$$d_M(\sigma_h\sigma_k)^2 = (a + b)d_M(\sigma_h\sigma_k) - ab\sigma_1 = n(c + c^{-1})d_M(\sigma_h\sigma_k) - n^2\sigma_1.$$

Thus, for each element j in $\{4, \ldots, d\}$,

$$d_M(\sigma_{r_j(h)}) = n(c + c^{-1})d_M(\sigma_{r_{j-2}(h)}) - n^2 d_M(\sigma_{r_{j-4}(h)});$$

see Lemma 9.6.1(ii).

If $i \in \{0, 1\}$, the equations can be computed explicitly. Let us, therefore, pick an integer i such that $2 \leq i \leq \frac{d}{2}$. We assume that the equations in question hold for $i - 1$ and $i - 2$. Then the last equation yields

$$\begin{aligned}
\chi(\sigma_{r_{2i+1}(h)}) &= n(c + c^{-1})\frac{n^{i-1}}{c - c^{-1}}(n_h - 1)(c^i - c^{-i}) \\
&\quad + n(c + c^{-1})\frac{n^{i-1}}{c - c^{-1}}(n_k - 1)\frac{n_h}{n}(c^{i-1} - c^{-(i-1)}) \\
&\quad - n^2\frac{n^{i-2}}{c - c^{-1}}(n_h - 1)(c^{i-1} - c^{-(i-1)}) \\
&\quad - n^2\frac{n^{i-2}}{c - c^{-1}}(n_k - 1)\frac{n_h}{n}(c^{i-2} - c^{-(i-2)}) \\
&= \frac{n^i}{c - c^{-1}}((n_h - 1)(c^{i+1} - c^{-(i+1)}) + (n_k - 1)\frac{n_h}{n}(c^i - c^{-i}))
\end{aligned}$$

and

$$\chi(\sigma_{r_{2i}(l)}) = n^i(c + c^{-1})(c^{i-1} + c^{-(i-1)}) - n^i(c^{i-2} + c^{-(i-2)}) = n^i(c^i + c^{-i}).$$

This finishes the proof of the proposition.

Let C be an algebraically closed field of characteristic 0, and let χ be an irreducible character of CS of degree 2. We define C_χ to be the set of all elements in C_d satisfying conditions (i) and (ii) of Proposition 12.4.2.

According to Proposition 12.4.2, the set C_χ is not empty.

In Proposition 12.4.2, we have computed all values of all irreducible characters of CS of degree 2. This allows us to compute, for each of these characters, the left hand side of the equation given in Theorem 9.1.7(ii). We shall do this in the following proposition.

Proposition 12.4.3 *Let C be an algebraically closed field of characteristic 0, let χ be an irreducible character of CS of degree 2, and let c be an element in C_χ. Then we have the following.*

(i) *If d is odd,*

$$\sum_{s\in S}\frac{1}{n_{s^*}}\chi(\sigma_{s^*})\chi(\sigma_s) = 2d\Big(1-\frac{(n-1)^2}{n(c+c^{-1}-2)}\Big).$$

(ii) *Assume that d is even, and let h and k be elements in L such that $h \neq k$. Then*

$$\sum_{s\in S}\frac{1}{n_{s^*}}\chi(\sigma_{s^*})\chi(\sigma_s) =$$

$$2d\Big(1-\frac{1}{(c-c^{-1})^2}\Big(\frac{(n_h-1)^2}{n_h}+\frac{(n_k-1)^2}{n_k}\Big)-\frac{(n_h-1)(n_k-1)(c+c^{-1})}{n(c-c^{-1})^2}\Big).$$

Proof. (i) Let us assume that d is odd. Then there exists an integer e such that $2e+1=d$. Moreover, for each element l in L, $n_l = n$; cf. Lemma 10.6.1(i). It follows that, for each element l in L,

$$\sum_{s\in S}\frac{1}{n_{s^*}}\chi(\sigma_{s^*})\chi(\sigma_s) = \sum_{i=0}^{e}\Big(2\frac{\chi(\sigma_{r_{2i}(l)})^2}{n^{2i}}+2\frac{\chi(\sigma_{r_{2i+1}(l)})^2}{n^{2i+1}}\Big)-\frac{\chi(\sigma_{r_d(l)})^2}{n^d}-4.$$

Let us now compute the fractions on the right hand side of this equation. In order to do so we fix an element l in L.

From Proposition 12.4.2(ii) together with the first equation of Lemma 8.7.2 we obtain

$$\sum_{i=0}^{e}2\frac{\chi(\sigma_{r_{2i}(l)})^2}{n^{2i}} = \sum_{i=0}^{e}2(c^i+c^{-i})^2 = 2(d+2).$$

From Proposition 12.4.2(i) we obtain

$$\chi(\sigma_{r_{2i+1}(l)}) = \frac{n^i}{c-c^{-1}}(n-1)(c^{i+1}-c^{-(i+1)}+c^i-c^{-i})$$

for any two elements l in L and i in $\{0,\ldots,\frac{d-1}{2}\}$. From this equation together with the second equation of Lemma 8.7.2 we obtain

$$\sum_{i=0}^{e} 2\frac{\chi(\sigma_{r_{2i+1}(l)})^2}{n^{2i+1}} = \frac{2(n-1)^2}{n(c-c^{-1})^2}\sum_{i=0}^{e}(c^{i+1}-c^{-(i+1)}+c^{i}-c^{-i})^2$$
$$= -\frac{2(n-1)^2}{n(c-c^{-1})^2}d(c+c^{-1}+2).$$

Since $c^{e+1}=c^{-e}$ and $c^{-(e+1)}=c^{e}$, the above equation also yields

$$\frac{\chi(\sigma_{r_d(l)})^2}{n^d} = \frac{(n-1)^2}{n(c-c^{-1})^2}(c^{e+1}-c^{-(e+1)}+c^{e}-c^{-e})^2 = 0.$$

Thus,

$$\sum_{s\in S}\frac{1}{n_{s^*}}\chi(\sigma_{s^*})\chi(\sigma_s) = 2(d+2) - \frac{2(n-1)^2}{n(c-c^{-1})^2}d(c+c^{-1}+2) - 4$$
$$= 2d(1-\frac{(n-1)^2}{n(c+c^{-1}-2)}).$$

(ii) Assume that d is even, and let h and k be elements in L with $h \neq k$. Since d is assumed to be even, there exists an integer e such that $2e = d$. Thus,

$$\sum_{s\in S}\frac{1}{n_{s^*}}\chi(\sigma_{s^*})\chi(\sigma_s)$$
$$= \sum_{i=0}^{e-1}(2\frac{\chi(\sigma_{r_{2i}(h)})^2}{(n_h n_k)^i} + \frac{\chi(\sigma_{r_{2i+1}(h)})^2}{(n_h n_k)^i n_h} + \frac{\chi(\sigma_{r_{2i+1}(k)})^2}{(n_k n_h)^i n_k}) + \frac{\chi(\sigma_{r_d(h)})^2}{(n_h n_k)^e} - 4$$

Let us now compute the fractions on the right hand side of this equation. Referring to Proposition 12.4.2(ii) and Lemma 8.7.1 we obtain

$$\sum_{i=0}^{e-1} 2\frac{\chi(\sigma_{r_{2i}(h)})^2}{(n_h n_k)^i} = \sum_{i=0}^{e-1} 2(c^{i}+c^{-i})^2$$
$$= \sum_{i=0}^{e-1} 2(c^{2i}+2+c^{-2i})$$
$$= 4e + 2\sum_{i=0}^{e-1}(c^{2i}+c^{-2i}) = 2d.$$

From Proposition 12.4.2(ii) we also obtain

$$\frac{\chi(\sigma_{r_d(h)})^2}{(n_h n_k)^e} = (c^{e}+c^{-e})^2 = c^{2e}+2+c^{-2e} = c^{d}+2+c^{-d} = 4.$$

Finally, referring to Proposition 12.4.2(i) and Lemma 8.7.3 we obtain

$$\sum_{i=0}^{e-1}\left(\frac{\chi(\sigma_{r_{2i+1}(h)})^2}{(n_h n_k)^i n_h}+\frac{\chi(\sigma_{r_{2i+1}(k)})^2}{(n_k n_h)^i n_k}\right)$$
$$=\sum_{i=0}^{e-1}\frac{1}{n_h(c-c^{-1})^2}((n_h-1)(c^{i+1}-c^{-(i+1)})+(n_k-1)\frac{n_h}{n}(c^i-c^{-i}))^2$$
$$+\sum_{i=0}^{e-1}\frac{1}{n_k(c-c^{-1})^2}((n_k-1)(c^{i+1}-c^{-(i+1)})+(n_h-1)\frac{n_k}{n}(c^i-c^{-i}))^2$$
$$=\sum_{i=0}^{e-1}\frac{1}{(c-c^{-1})^2}\left(\frac{(n_h-1)^2}{n_h}+\frac{(n_k-1)^2}{n_k}\right)((c^{i+1}-c^{-(i+1)})^2+(c^i-c^{-i})^2)$$
$$+\sum_{i=0}^{e-1}\frac{4(n_h-1)(n_k-1)}{n(c-c^{-1})^2}(c^{i+1}-c^{-(i+1)})(c^i-c^{-i})$$
$$=-2d\left(\frac{1}{(c-c^{-1})^2}\left(\frac{(n_h-1)^2}{n_h}+\frac{(n_k-1)^2}{n_k}\right)+\frac{(n_h-1)(n_k-1)(c+c^{-1})}{n(c-c^{-1})^2}\right),$$

and that proves the desired equation.

Lemma 12.4.4 *Let C be an algebraically closed field of characteristic 0, and let c be an element in C_d. Then there exists an irreducible character χ of CS of degree 2 such that $c \in C_\chi$.*

PROOF. If d is odd, $|C_d| = d-1$, and if d is even, $|C_d| = d-2$. Thus, by Lemma 12.4.1(ii), (iii), $\frac{1}{2}|C_d|$ is the number of irreducible characters of CS of degree 2.

Let χ be an irreducible character of CS of degree 2, and let c be an element in C_χ. Then, looking at the equations of Proposition 12.4.2, we see that $c^{-1} \in C_\chi$. Thus, for each of the $\frac{1}{2}|C_d|$ irreducible characters χ of CS of degree 2, we have that

$$\{c^{-1}, c\} \subseteq C_\chi \subseteq C_d.$$

This proves the lemma.

Lemma 12.4.5 *Let C be an algebraically closed field of characteristic 0, let χ be an irreducible character of CS of degree 2, and let c be an element in C_χ. Let Q be the smallest subfield of C.*
Assume that S is not thin, that d is even, and that $4 \le d$. Then $(c-c^{-1})^2 \in Q$.

PROOF. By Lemma 12.4.4, CS possesses an irreducible character χ' of degree 2 with $-c \in C_{\chi'}$. Adding the values computed for χ and χ' in Proposition 12.4.3(ii) we obtain from Theorem 9.1.7(ii) that

$$2d\left(2-\frac{2}{(c-c^{-1})^2}\left(\frac{(n_h-1)^2}{n_h}+\frac{(n_k-1)^2}{n_k}\right)\right) \in Q.$$

Thus, the claim follows from the fact that S is assumed to be thin.

The following two theorems are due to Walter Feit and Graham Higman; cf. [10; Theorem 1].

Theorem 12.4.6 *If S is not thin, $d \in \{2, 3, 4, 6, 8, 12\}$.*

PROOF. Let C be an algebraically closed field of characteristic 0. Let us write Q to denote the smallest subfield of C, and let Z denote the smallest unitary subring of C.

Suppose first that d is odd. Then $3 \leq d$. Thus, by Lemma 12.4.1(ii), CS possesses at least one irreducible character χ of degree 2. Let c be an element in C_χ. Then we obtain from Proposition 12.4.3(i) and Theorem 9.1.7(ii) that

$$2d(1 - \frac{(n-1)^2}{n(c + c^{-1} - 2)}) \in Q.$$

Now recall that S is assumed not to be thin. Thus, by Lemma 3.1.5(i), $1 \neq n$, and this implies $c + c^{-1} \in Q$.

On the other hand, by Lemma 9.2.5 together with Theorem 8.2.4, $c + c^{-1} \in I_C(Z)$. Thus, by Lemma 8.2.5, $c + c^{-1} \in Z$. Now Lemma 8.7.4 yields $d = 3$.

Suppose now that d is even and that $4 \leq d$. Then, according to Lemma 12.4.1(iii), CS possesses at least one irreducible character χ of degree 2.

Let c be an element in C_χ. Then, by Lemma 12.4.5, $(c - c^{-1})^2 \in Q$. It follows that $c^2 + c^{-2} \in Q$.

On the other hand, by Lemma 9.2.5 together with Theorem 8.2.4, $c^2 + c^{-2} \in I_C(Z)$. Thus, by Lemma 8.2.5, $c^2 + c^{-2} \in Z$. Now Lemma 8.7.4 yields $d \in \{4, 6, 8, 12\}$.

Theorem 12.4.7 *If $\{1\} = O_\vartheta(S)$, we have the following.*

(i) *If $d = 6$, n is an integer.*

(ii) *If $d = 8$, $\sqrt{2}n$ is an integer.*

(iii) *We have $d \neq 12$.*

PROOF. Let C be an algebraically closed field of characteristic 0. Let us write Q to denote the smallest subfield of C, and let Z denote the smallest unitary subring of C.

Let χ be an irreducible character of CS of degree 2, and let c be an element in C_χ. Then, by Lemma 12.4.5, $(c - c^{-1})^2 \in Q$.

Now recall that $S \setminus \{1\}$ is assumed to have no thin element. Thus, $1 \neq n_h$ and $1 \neq n_k$. Thus, as $(c - c^{-1})^2 \in Q$, we obtain from Proposition 12.4.3(ii) and Theorem 9.1.7(ii) that

$$\frac{c + c^{-1}}{n} \in Q.$$

(i) Assume that $d = 6$. Then $(c + c^{-1})^2 = 1$. Thus, we must have that $n \in Q$. Therefore, by Lemma 8.2.5, $n \in Z$.

(ii) Assume that $d = 8$. By Lemma 12.4.4, we may assume that $(c+c^{-1})^2 = 2$. Thus, we must have that $\sqrt{2}n \in Q$. Therefore, by Lemma 8.2.5, $\sqrt{2}n \in Z$.

(iii) Assume finally that $d = 12$. In this case, we may assume that $(c+c^{-1})^2 = 3$ or that $(c + c^{-1})^2 = 1$.

If $(c + c^{-1})^2 = 3$, $\sqrt{3}n \in Q$. Then, by Lemma 8.2.5, $\sqrt{3}n \in Z$.

Suppose now that $(c+c^{-1})^2 = 1$. Thus, we must have that $n \in Q$. Therefore, by Lemma 8.2.5, $n \in Z$, contrary to $\sqrt{3}n \in Z$.

As a consequence of Theorem 12.4.7(ii) one obtains that the valencies of the two elements in L must be different if $d = 8$.

12.5 Valencies and Multiplicities

In this section, we continue our investigation on Coxeter schemes of finite valency and rank 2 which we started in the previous section. The letter L stands for a set of involutions. We assume that S is a Coxeter scheme with respect to L, that $|L| = 2$, and that S has finite valency.

Since S is assumed to have finite valency, S is finite. Thus, by Lemma 10.1.7, L is spherical, and that means that $S_{-1}(L)$ is not empty.

Keeping the notation of the previous section we write d_L to denote the smallest element in $\ell(S_{-1}(L))$. However, having fixed L for the remainder of this section, we shall write d instead of d_L.

In Theorem 12.4.6, we saw that $d \in \{2, 3, 4, 6, 8, 12\}$. If $d = 2$, S is the direct product of two closed subsets each of them generated by an involution. We shall now look separately at each of the individual cases $d = 3$, $d = 4$, $d = 6$, $d = 8$, and $d = 12$. In order to do this, we fix an algebraically closed field of characteristic 0 and call it C.

From Lemma 12.4.1(ii), (iii) and Corollary 9.4.4(ii) we know that there exists a linear character st such that, for each element l in L, $st(\sigma_l) = -1$. The linear character st is called the *Steinberg character* of CS.

Theorem 12.5.1 *If $d = 3$, the following hold.*

(i) *We have $n_S = (n + 1)(n^2 + n + 1)$.*

(ii) *We have $m_{st} = n^3$.*

(iii) *The multiplicity of the only non-linear irreducible character of CS is $n(n + 1)$.*

PROOF. (i) Applying Theorem 9.1.7(ii) to 1_{CS} in place of ϕ and ψ we obtain

$$\frac{n_S}{m_{1_{CS}}} = \sum_{s \in S} \frac{1}{n_{s^*}} 1_{CS}(\sigma_{s^*}) 1_{CS}(\sigma_s) = 1 + 2n + 2n^2 + n^3.$$

Thus, the claim follows from Lemma 9.1.8(ii).

(ii) Applying Theorem 9.1.7(ii) to st in place of ϕ and ψ we obtain from (i) that

$$\frac{n_S}{m_{st}} = \sum_{s \in S} \frac{1}{n_{s^*}} st(\sigma_{s^*}) st(\sigma_s) = 1 + \frac{2}{n} + \frac{2}{n^2} + \frac{1}{n^3} = \frac{n_S}{n^3}.$$

(iii) Let us denote by χ the only non-linear irreducible character of CS. Then, referring to Lemma 9.1.8(ii), Lemma 12.4.1(i), and (ii) we obtain

$$\chi_{CX}(\sigma_1) = m_{1_{CS}} 1_{CS}(\sigma_1) + m_{st} st(\sigma_1) + m_\chi \chi(\sigma_1) = 1 + n^3 + m_\chi \chi(\sigma_1).$$

On the other hand, by (i),

$$\chi_{RX}(\sigma_1) = n_S = 1 + 2n + 2n^2 + n^3.$$

Therefore, as $\chi(\sigma_1) = 2$, $m_\chi = n(n+1)$.

It is a well-known open question, whether or not $d = 3$ implies that n is a prime power.

For the remainder of this section, d will be even, and we assume that $4 \leq d$.

Since C is assumed to be algebraically closed, C_d contains an element c such that, for each integer i with $1 \leq i \leq d - 1$, $c^i \neq 1$. We fix one of these elements. For each integer i with $1 \leq i \leq \frac{d}{2} - 1$, we define χ_i to be the non-linear irreducible character of CS which satisfies $c^i \in C_{\chi_i}$; cf. Lemma 12.4.4.

Let us fix elements h and k in L such that $h \neq k$. Then, by Lemma 12.4.1(iii) and Corollary 9.4.4(ii), there exists a linear character λ_h such that

$$\lambda_h(\sigma_h) = n_h, \quad \lambda_h(\sigma_k) = -1$$

and there exists a linear character λ_k such that

$$\lambda_k(\sigma_h) = -1, \quad \lambda_k(\sigma_k) = n_k.$$

Without loss of generality, we assume that $n_h \leq n_k$.

Theorem 12.5.2 *Assume that $d = 4$. Then the following hold.*

(i) *We have $n_S = (n_h + 1)(n_k + 1)(n_h n_k + 1)$.*

(ii) *We have* $m_{st} = n_h^2 n_k^2$ *and*

$$m_{\lambda_h} = \frac{n_k^2(n_h n_k + 1)}{n_h + n_k},$$
$$m_{\lambda_k} = \frac{n_h^2(n_h n_k + 1)}{n_h + n_k},$$
$$m_{\chi_1} = \frac{n_h n_k(n_h + 1)(n_k + 1)}{n_h + n_k}.$$

PROOF. (i) Applying Theorem 9.1.7(ii) to 1_{CS} in place of ϕ and ψ we obtain

$$\frac{n_S}{m_{1_{CS}}} = \sum_{s\in S} \frac{1}{n_{s^*}} 1_{CS}(\sigma_{s^*})1_{CS}(\sigma_s)$$
$$= 1 + n_h + n_k + 2n_h n_k + n_h n_k^2 + n_h^2 n_k + n_h^2 n_k^2.$$

Thus, the claim follows from Lemma 9.1.8(ii).

(ii) Applying Theorem 9.1.7(ii) to st in place of ϕ and ψ we obtain from (i) that

$$\frac{n_S}{m_{st}} = \sum_{s\in S} \frac{1}{n_{s^*}} st(\sigma_{s^*})st(\sigma_s)$$
$$= \frac{1}{1} + \frac{1}{n_h} + \frac{1}{n_k} + 2\frac{1}{n_h n_k} + \frac{1}{n_h n_k^2} + \frac{1}{n_h^2 n_k} + \frac{1}{n_h^2 n_k^2} = \frac{n_S}{n_h^2 n_k^2}.$$

The values of m_{λ_h}, m_{λ_k}, and m_{χ_1} are obtained from (i) and Proposition 12.4.3(ii) by applying Theorem 9.1.7(ii) to m_{λ_h}, m_{λ_k}, and m_{χ_1} in place of ϕ and ψ.

From Theorem 12.5.2(ii) we obtain, in particular, that $n_h + n_k$ divides $n_h n_k(n_h + 1)(n_k + 1)$.

Assume that $d = 4$ and that $\{1\} = O_\vartheta(S)$. Then, if S is isomorphic to a known example, we have

$$(n_h, n_k) \in \{(q-1, q+1), (q, q), (q, q^2), (q^2, q^3)\},$$

where q is a prime power.

Donald Higman showed in [21; Theorem 3.2] that, if $d = 4$ and $2 \leq n_h$, $n_k \leq n_h^2$. For a proof, see also [43; Theorem 5.3.7(iv)].

Theorem 12.5.3 *Assume that* $d = 6$. *Then the following hold.*

(i) *We have* $n_S = (n_h + 1)(n_k + 1)(n_h^2 n_k^2 + n_h n_k + 1)$.

(ii) *We have* $m_{st} = n_h^3 n_k^3$ *and*

$$m_{\lambda_h} = \frac{n_k^3(n_h^2 n_k^2 + n_h n_k + 1)}{n_h^2 + n_h n_k + n_k^2},$$
$$m_{\lambda_k} = \frac{n_h^3(n_h^2 n_k^2 + n_h n_k + 1)}{n_h^2 + n_h n_k + n_k^2},$$
$$m_{\chi_1} = \frac{n_h n_k (n_h + 1)(n_k + 1)(n_h^2 n_k^2 + n_h n_k + 1)}{2((n_h^2 n_k + n_h - n_h n_k + n_k + n_h n_k^2) + (n_h - 1)(n_k - 1)n)},$$
$$m_{\chi_2} = \frac{n_h n_k (n_h + 1)(n_k + 1)(n_h^2 n_k^2 + n_h n_k + 1)}{2((n_h^2 n_k + n_h - n_h n_k + n_k + n_h n_k^2) - (n_h - 1)(n_k - 1)n)}.$$

PROOF. (i) This equation follows from Theorem 9.1.7(ii) in the same way as the one in Theorem 12.5.2(i).

(ii) All five equations are obtained from (i), Proposition 12.4.3(ii) and Theorem 9.1.7(ii) in the same way as the ones in Theorem 12.5.2(ii).

Assume that $d = 6$ and that $\{1\} = O_\vartheta(S)$. Then, if S is isomorphic to a known example, we have

$$(n_h, n_k) \in \{(q, q), (q, q^3)\},$$

where q is a prime power.

Willem Haemers and Cornelis Roos showed in [16] that, if $d = 6$ and $2 \le n_h$, $n_k \le n_h^3$. For a proof, see also [43; Theorem 5.3.8(iv)].

Theorem 12.5.4 *Assume that* $d = 8$. *Then the following hold.*

(i) *We have* $n_S = (n_h + 1)(n_k + 1)(n_h n_k + 1)(n_h^2 n_k^2 + 1)$.

(ii) *We have* $m_{st} = n_h^4 n_k^4$ *and*

$$m_{\lambda_h} = \frac{n_k^4(n_h n_k + 1)(n_h^2 n_k^2 + 1)}{(n_h + n_k)(n_h^2 + n_k^2)},$$
$$m_{\lambda_k} = \frac{n_h^4(n_h n_k + 1)(n_h^2 n_k^2 + 1)}{(n_h + n_k)(n_h^2 + n_k^2)},$$
$$m_{\chi_1} = \frac{n_h n_k (n_h + 1)(n_k + 1)(n_h n_k + 1)(n_k^2 n_k^2 + 1)}{4((n_h^2 n_k + n_h - 2n_h n_k + n_k + n_h n_k^2) + (n_h - 1)(n_k - 1)\sqrt{2n})},$$
$$m_{\chi_2} = \frac{n_h n_k (n_h + 1)(n_k + 1)(n_k^2 n_k^2 + 1)}{2(n_h + n_k)},$$
$$m_{\chi_3} = \frac{n_h n_k (n_h + 1)(n_k + 1)(n_h n_k + 1)(n_k^2 n_k^2 + 1)}{4((n_h^2 n_k + n_h - 2n_h n_k + n_k + n_h n_k^2) - (n_h - 1)(n_k - 1)\sqrt{2n})}.$$

PROOF. (i) This equation follows from Theorem 9.1.7(ii) in the same way as the one in Theorem 12.5.2(i).

(ii) All six equations are obtained from (i), Proposition 12.4.3(ii) and Theorem 9.1.7(ii) in the same way as the ones in Theorem 12.5.2(ii).

Assume that $d = 8$ and that $\{1\} = O_\vartheta(S)$. Then, if S is isomorphic to a known example, we have $(n_h, n_k) = (q, q^2)$, where $q = 2^e$ for some odd positive integer.

Donald Higman showed in [21; Theorem 3.2] that, if $d = 8$ and $2 \leq n_h$, $n_k \leq n_h^2$. For a proof, see also [43; Theorem 5.3.9(iv)].

Theorem 12.5.5 *Assume that $d = 12$. Then $n_h = 1$, and the following hold.*

(i) *We have $n_S = 2(n_k + 1)^2(n_k^4 + n_k^2 + 1)$.*

(ii) *We have $m_{st} = n_k^6$, $m_{\lambda_h} = n_k^6$, $m_{\lambda_k} = 1$, and*

$$\begin{aligned} m_{\chi_1} &= \frac{n_k}{6}(n_k + 1)^2(n_k^2 + n_k + 1), \\ m_{\chi_2} &= \frac{n_k}{2}(n_k + 1)^2(n_k^2 - n_k + 1), \\ m_{\chi_3} &= \frac{2n_k}{3}(n_k^4 + n_k^2 + 1), \\ m_{\chi_4} &= \frac{n_k}{2}(n_k + 1)^2(n_k^2 - n_k + 1), \\ m_{\chi_5} &= \frac{n_k}{6}(n_k + 1)^2(n_k^2 + n_k + 1). \end{aligned}$$

PROOF. (i) This equation is follows from Theorem 9.1.7(ii) in the same way as the one in Theorem 12.5.2(i).

(ii) All eight equations are obtained from (i), Proposition 12.4.3(ii) and Theorem 9.1.7(ii) in the same way as the ones in Theorem 12.5.2(ii).

12.6 Polarities

Throughout this section, the letter L stands for a set of two involutions. We assume that S is a Coxeter scheme with respect to L and that S has finite valency. Instead of d_L we just write d.

For any two elements l in L and j in $\{0, \dots, d\}$, $r_j(l)$ has the same meaning as in the previous section.

An automorphism a of S is called a *polarity* if $1_X = (a_X)^2$ and $1_S \neq a_S$.

Let a be a polarity of S. Then, for any two elements h and k in L with $h \neq k$, $ha = k$ and $ka = h$.

Lemma 12.6.1 *Let l be an element in L, and let a be a polarity of S. Then we have the following.*

(i) *For each element j in $\{0, \ldots, d\}$, $r_j(l)a = r_j(la)$.*

(ii) *Let j be an element in $\{1, \ldots, d\}$. Assume that X contains an element x with $xa \in xr_j(l)$. Then j is even or $j = d$.*

PROOF. (i) This follows immediately from the definition of $r_j(l)$.

(ii) Let x be an element in X such that $xa \in xr_j(l)$. Then $x \in xar_j(l)^*$ and

$$x = xa^2 \in xar_j(l) = xar_j(l)a = xar_j(la);$$

see (i). Therefore, $r_j(l)^* = r_j(la)$. But, by definition, $la \neq l$. Therefore j is even or $j = d$; cf. Lemma 10.2.1(i).

Let a be a polarity of S. Note that, for each element l in L, $n_{la} = n_l$. Thus, as $la \neq l$, we have that, for each element l in L, $n = n_l$. In the following, we shall write n instead of n_l.

Recall that $\mathrm{Fix}_X(a)$ is our notation for the set of all elements x in X such that $xa = x$.

Lemma 12.6.2 *Let a be a polarity of S, and assume d to be even. Then $|\mathrm{Fix}_X(a)| = n^{d/2} + 1$.*

PROOF. We set $Y := \mathrm{Fix}_X(a)$.

By hypothesis, d is even. Therefore, there exists an integer e such that $2e = d$.

Our first claim is that, for each element x in X, there exist elements y in Y, l in L, and i in $\{0, \ldots, e\}$ such that $x \in yr_i(l)$.

Let x be an element in X. Then there exist elements j in $\{0, \ldots, d\}$ and l in L such that $xa \in xr_j(l)$. Thus, by Lemma 12.6.1(ii), j is even. Let i be an integer such that $2i = j$.

We are assuming that a is as polarity of S. Thus, $\{r_j(l)\} = r_i(l)r_i(la)^*$. Thus, as $xa \in xr_j(l)$, $xa \in xr_i(l)r_i(la)^*$. Thus, there exists an element y in $xr_i(l)$ such that $xa \in yr_i(la)^*$. Thus, by Lemma 12.6.1(i),

$$ya \in xar_i(la) \cap xr_i(l).$$

Let us denote by s the element in S which satisfies $ya \in ys$. Then

$$s \in r_i(l)^* r_i(l) \cap r_i(la)^* r_i(la).$$

Thus, referring to Lemma 10.1.1 and Lemma 10.2.1(i) we obtain $s = 1$. It follows that $y \in Y$. Thus,

$$x \in yr_i(l)^* \subseteq Y\{r_i(l), r_i(la)\}.$$

Since x has been chosen arbitrarily in X, we have proved our first claim.

Let y and z be elements in Y with $y \neq z$. Then there exist elements j in $\{1, \ldots, d\}$ and l in L such that $z \in yr_j(l)$. Thus, by Lemma 10.2.1(i), $j = d$.

So far, we have seen that, for any four elements h and k in L with $h \neq k$ and i and j in $\{1, \ldots, e\}$, we have either $i = e = j$ or $\emptyset = Y\{r_i(h)\} \cap Y\{r_j(k)\}$.

Conversely, let y be an element in Y, and let x be an element in $yr_e(l)$. Then there exists a uniquely determined element z in Y such that $x \in ze_s(la)$.

Recall that, for any two elements l in L and i in $\{1, \ldots, s\}$, $n_{r_i(l)} = n^i$. Therefore,

$$n_S = |Y|(1 + 2n + \ldots + 2n^{e-1} + n^e) = |Y|(n+1)\sum_{i=0}^{e-1} n^i$$

and

$$n_S = (n+1)\sum_{i=0}^{d-1} n^i = (n+1)(n^e+1)\sum_{i=0}^{e-1} n^i.$$

Thus, we conclude that $|Y| = n^e + 1$.

Let us assume that S possesses a polarity a. We set $1\zeta := 1_S$, $a\zeta := a_S$, and $A := \{1, a\}$. From Lemma 5.2.2(iii) we know that ζ is a group homomorphism from A to $\mathrm{Stc}(S)$. Thus, we may consider the semidirect product $S_\zeta A$ of S and A with respect to ζ; cf. Section 7.3.

For any two elements l in L and j in $\{1, \ldots, d\}$, we set

$$r_j^+(l) := r_j(l)_\zeta 1$$

and

$$r_j^-(l) := r_j(l)_\zeta a.$$

Note that

$$\{r_j^-(l) \mid l \in L,\ 0 \leq j \leq d\} \cup \{r_j^+(l) \mid l \in L,\ 0 \leq j \leq d\} = S_\zeta A.$$

Let C be an algebraically closed field of characteristic 0. From Theorem 9.4.8(iii) we know that CX is a $C(S_\zeta A)$-module. In the following proposition, we assume d to be even and completely compute the character of $C(S_\zeta A)$ afforded by this module.

Proposition 12.6.3 *Assume that S possesses a polarity a, and set $A := \{1, a\}$. Let C be an algebraically closed field of characteristic 0, and let us denote by ϕ the character of $C(S_\zeta A)$ afforded by CX. Let l be an element in L, and let j be an element in $\{1, \ldots, d\}$. Then we have the following.*

(i) *If $j \neq 0$, $\phi(\sigma_{r_j^+(l)}) = 0$.*

(ii) *If d is even and j is odd,* $\phi(\sigma_{r_j^-(l)}) = 0$.

(iii) *If both d and j are even,* $\phi(\sigma_{r_j^-(l)}) = (n^{d/2}+1)n^{j/2}$.

PROOF. (i) For each element x in X,

$$x\sigma_{r_j^+(l)} = x\sigma_{r_j(l)};$$

see Theorem 9.4.8(i). Therefore, if $j \neq 0$,

$$\phi(\sigma_{r_j^+(l)}) = \chi_{CX}(\sigma_{r_j(l)}) = 0.$$

(ii) Assume that d is even and that j is odd. Then, by Lemma 12.6.1(ii), X has no element x with $xa \in xr_j(l)$. On the other hand, as j is assumed to be odd, we obtain from Theorem 9.4.8(iv) that

$$\phi(\sigma_{r_j^-(l)}) = |\{x \in X \mid xa \in xr_j(l)\}|.$$

Thus, $\phi(\sigma_{r_j^-(l)}) = 0$.

(iii) Assume that d and j are even. Let i be the integer with $2i = j$, and let x be an element in X. Note that $xr_i(l)$ contains a fixed point of a if and only if $xa \in xr_j(l)$. Thus, referring to Theorem 9.4.8(iv) we obtain from Lemma 12.6.2 that

$$\phi(\sigma_{r_j^-(l)}) = (n^{d/2}+1)n^i.$$

This proves (iii).

The first part of the following theorem was proved by Stanley Payne; cf. [34; Theorem 2]. Its second part is due to Udo Ott [33; Satz 1].

The proof of Theorem 12.6.4 is an application of Theorem 9.1.7(ii) to one of the non-linear irreducible characters of $S_\varsigma A$ and to the character of $C(S_\varsigma A)$ afforded by CX in place of ϕ and ψ. The left hand side of the equation in Theorem 9.1.7(ii) is easy to compute since we know from Proposition 12.6.3(i), (ii) that ϕ vanishes on most of the values of $S_\varsigma A$.

Theorem 12.6.4 *Assume that S possesses a polarity and that $2 \leq n$. Then we have the following.*

(i) *If $d = 4$, $\sqrt{2n}$ is an integer.*

(ii) *If $d = 6$, $\sqrt{3n}$ is an integer.*

PROOF. Let h and k be elements in L such that $\{h, k\} = L$. Let a be a polarity of S, set $A := \{1, a\}$, and set $R := S_\varsigma A$.

From the definition of R we obtain $|R| = 4d$, and from Corollary 7.3.4(ii) we obtain that $r_1^+(h)$ and $r_0^-(h)$ are involutions of R.

From Corollary 7.3.4(ii) we also obtain that R is a Coxeter scheme with respect to $\{r_1^+(h), r_0^-(h)\}$ satisfying

$$\{r_2^-(h)\} = r_1^+(h)r_0^-(h)r_1^+(h)$$

and

$$\{r_2^-(k)\} = (r_0^-(h)r_1^+(h))^2 r_0^-(h).$$

Moreover, if $4 \leq d$, we have

$$\{r_4^-(h)\} = (r_1^+(h)r_0^-(h))^3 r_1^+(h)$$

and

$$\{r_4^-(k)\} = (r_0^-(h)r_1^+(h))^4 r_0^-(h).$$

Finally, if $6 \leq d$, we have

$$\{r_6^-(h)\} = (r_1^+(h)r_0^-(h))^5 r_1^+(h).$$

Let C be an algebraically closed field of characteristic 0, let Q denote the smallest subfield of C, and let Z denote the smallest unitary subring of C. We set $I := I_C(Z)$.

Recall that C_{2d} is our notation for the set of all elements c in $C \setminus \{-1, 1\}$ satisfying $c^{2d} = 1$. Since C is assumed to be algebraically closed, C_{2d} contains an element c such that, for each integer i with $1 \leq i \leq 2d - 1$, $c^i \neq 1$. We fix one of these elements and call it c. We define χ to be the non-linear irreducible character of $C(R)$ which satisfies $c \in C_\chi$; cf. Lemma 12.4.4.

Let us denote by ϕ the character of $C(R)$ afforded by CX.

(i) Assume that $d = 4$. From Proposition 12.4.2(i) (applied to R in the role of S) we compute that

$$\chi(\sigma_{r_2^-(h)}) = (n-1)\sqrt{2n} = \chi(\sigma_{r_2^-(k)})$$

and that $\chi(\sigma_{r_4^-(h)}) = 0$. Moreover, by Proposition 12.6.3(iii),

$$\phi(\sigma_{r_2^-(h)}) = (n^2+1)n = \phi(\sigma_{r_2^-(k)}).$$

Therefore, referring to Proposition 12.6.3(i), (ii) we obtain

$$\begin{aligned}\sum_{r \in R} \frac{1}{n_{r^*}} \chi(\sigma_{r^*})\phi(\sigma_r) &= 2n_S + \frac{1}{n^2}\chi(\sigma_{r_2^-(h)})\phi(\sigma_{r_2^-(h)}) + \frac{1}{n^2}\chi(\sigma_{r_2^-(k)})\phi(\sigma_{r_2^-(k)}) \\ &= 2n_S + \frac{2}{n^2}(n-1)\sqrt{2n}\phi(\sigma_{r_2^-(h)}) \\ &= 2n_S + \frac{2}{n}(n-1)(n^2+1)\sqrt{2n}.\end{aligned}$$

Since we assume that $1 \neq n$, we conclude from Theorem 9.1.7(ii) that $\sqrt{2n} \in Q$. Therefore, by Lemma 8.2.5, $\sqrt{2n} \in Z$.

(ii) Let us now assume that $d = 6$. From Proposition 12.4.2(i) (applied to R in the role of S) we compute that

$$\chi(\sigma_{r_2^-(h)}) = (n-1)\sqrt{3n} = \chi(\sigma_{r_2^-(k)}),$$

that

$$\chi(\sigma_{r_4^-(h)}) = n(n-1)\sqrt{3n} = \chi(\sigma_{r_4^-(k)}),$$

and that $\chi(\sigma_{r_6^-(h)}) = 0$. Moreover, by Proposition 12.6.3(iii),

$$\phi(\sigma_{r_2^-(h)}) = (n^3+1)n = \phi(\sigma_{r_2^-(k)}),$$

that

$$\phi(\sigma_{r_4^-(h)}) = (n^3+1)n^2 = \phi(\sigma_{r_4^-(k)}).$$

Therefore, referring to Proposition 12.6.3(i), (ii) we obtain

$$\begin{aligned}
\sum_{r \in R} \frac{1}{n_{r^*}} \chi(\sigma_{r^*})\phi(\sigma_r) &= 2n_S + \frac{1}{n^2}\chi(\sigma_{r_2^-(h)})\phi(\sigma_{r_2^-(h)}) + \frac{1}{n^2}\chi(\sigma_{r_2^-(k)})\phi(\sigma_{r_2^-(k)}) \\
&\quad + \frac{1}{n^4}\chi(\sigma_{r_4^-(h)})\phi(\sigma_{r_4^-(h)}) + \frac{1}{n^4}\chi(\sigma_{r_4^-(k)})\phi(\sigma_{r_4^-(k)}) \\
&= 2n_S + \frac{2}{n^2}(n-1)\sqrt{3n}\phi(\sigma_{r_2^-(h)}) \\
&\quad + \frac{2}{n^3}(n-1)\sqrt{3n}\phi(\sigma_{r_4^-(h)}) \\
&= 2n_S + \frac{2}{n^2}(n-1)\sqrt{3n}(n^3+1)n \\
&\quad + \frac{2}{n^3}(n-1)\sqrt{3n}(n^3+1)n^2 \\
&= 2n_S + \frac{4}{n}(n-1)(n^3+1)\sqrt{3n}.
\end{aligned}$$

Thus, by Lemma 8.2.5, $\sqrt{3n} \in Z$.

References

[1] Arad, Z., Erez, Y., and Muzychuk, M.: On even generalized table algebras, *J. Algebraic Combin.* **17**, 163–170 (2003)

[2] Artin, E.: Zur Theorie der hyperkomplexen Zahlen, *Abh. Math. Sem. Univ. Hamburg* **5**, 251–260 (1927)

[3] Bang, S., Hirasaka, M., and Song, S.-Y.: Semidirect products of association schemes, *J. Algebraic Combin.* **21**, 23–38 (2005)

[4] Blau, H., Zieschang, P.-H.: Sylow theory for table algebras, fusion rule algebras, and hypergroups, *J. Algebra* **273**, 551–570 (2004)

[5] Brauer, R.: Zur Darstellungstheorie der Gruppen endlicher Ordnung. II, *Math. Z.* **72**, 25–46 (1959)

[6] Damerell, R. M.: On Moore geometries. II, *Math. Proc. Cambridge Philos. Soc.* **90**, 33–40 (1981)

[7] Dedekind, R.: Ueber die Composition der binären quadratischen Formen (Supplement X), pp. 423–462 in: *Vorlesungen über Zahlentheorie* (P. G. Lejeune-Dirichlet) 2nd edition, Vieweg, Braunschweig (1871)

[8] Dedekind, R.: Ueber die von drei Moduln erzeugte Dualgruppe, *Math. Ann.* **53**, 371–403 (1900)

[9] Erdös, P., Rényi, A., and Sós, V. T.: On a problem of graph theory, *Studia Sci. Math. Hungar.* **1**, 215–235 (1966)

[10] Feit, W., Higman, G.: The nonexistence of certain generalized polygons, *J. Algebra* **1**, 114–131 (1964)

[11] Ferguson, P. A., Turull, A.: Algebraic decompositions of commutative association schemes, *J. Algebra* **96**, 211–229 (1985)

[12] Fraenkel, A.: Über die Teiler der Null und die Zerlegung von Ringen, *J. Reine Angew. Math.* **145**, 139–176 (1914)

[13] Fuglister, F. J.: On finite Moore geometries, *J. Combin. Theory Ser. A* **23**, 187–197 (1977)

[14] Fuglister, F. J.: The nonexistence of Moore geometries of diameter 4, *Discrete Math.* **45**, 229–238 (1983)

[15] Glauberman, G.: Central elements in core-free groups, *J. Algebra* **4**, 403–420 (1966)

[16] Haemers, W., Roos, C.: An inequality for generalized hexagons, *Geom. Dedicata* **10**, 219–222 (1981)

[17] Hanaki, A.: Representations of association schemes and their factor schemes, *Graphs Combin.* **19**, 195–201 (2003)

[18] Hanaki, A.: Characters of association schemes and normal closed subsets, *Graphs Combin.* **19**, 363–369 (2003)

[19] Hanaki, A., Uno, K.: Algebraic structure of association schemes of prime order, *J. Algebraic Combin.*

[20] Higman, D. G.: Coherent configurations. I. Part I: Ordinary representation theory, *Geom. Dedicata* **4**, 1–32 (1975)

[21] Higman, D. G.: Invariant relations, coherent configurations, and generalized polygons, pp. 347–363 in: *Combinatorics* (M. Hall, Jr. and J. H. van Lint, eds.) Reidel, Dordrecht (1975)

[22] Hirasaka, M.: On quasi-thin association schemes with odd number of points, *J. Algebra* **240**, 665–679 (2001)

[23] Hirasaka, M., Muzychuk, M.: Association schemes generated by a non-symmetric relation of valency 2, *Discrete Math.* **244**, 109–135 (2002)

[24] Hirasaka, M., Muzychuk, M., and Zieschang, P.-H.: A generalization of Sylow's theorems on finite groups to association schemes, *Math. Z.* **241**, 665–672 (2002)

[25] Hirasaka, M., Zieschang, P.-H.: Sufficient conditions for a scheme to originate from a group, *J. Combin. Theory Ser. A* **104**, 17–27 (2003)

[26] Hölder, O.: Zurückführung einer beliebigen algebraischen Gleichung auf eine Kette von Gleichungen, *Math. Ann.* **34**, 26–56 (1889)

[27] Hoffman, A. J., Singleton, R. R.: On Moore graphs with diameters 2 and 3, *IBM J. Res. Develop.* **4**, 497–504 (1960)

[28] Isaacs, I. M.: *Algebra, A Graduate Course*, Brooks/Cole Publishing Company, Pacific Grove (1993)

[29] Kilmoyer, R., Solomon, L.: On the theorem of Feit-Higman, *J. Combin. Theory Ser. A* **15**, 310–322 (1973)

[30] Maschke, H.: Beweis des Satzes, dass diejenigen endlichen linearen Substitutionsgruppen, in welchen einige durchgehends verschwindende Coefficienten auftreten, intransitiv sind, *Math. Ann.* **52**, 363–368 (1899)

[31] Muzychuk, M., Zieschang, P.-H.: On association schemes all elements of which have valency 1 or 2, *Discrete Math.*

[32] Noether, E.: Hyperkomplexe Größen und Darstellungstheorie, *Math. Z.* **30**, 641–692 (1929)

[33] Ott, U.: Eine Bemerkung über Polaritäten eines verallgemeinerten Hexagons, *Geom. Dedicata* **11**, 341–345 (1981)

[34] Payne, S. E.: Symmetric representations of nondegenerate generalized n-gons, *Proc. Amer. Math. Soc.* **19**, 1321–1326 (1968)

[35] Rassy, M., Zieschang, P.-H.: Basic structure theory of association schemes, *Math. Z.* **227**, 391–402 (1998)

[36] Sylow, L.: Théorèmes sur les groupes de substitutions, *Math. Ann.* **5**, 584–594 (1872)

[37] Tits, J.: *Buildings of Spherical Type and Finite BN-Pairs*, Springer Lecture Notes in Math. **386**, Berlin Heidelberg New York (1974)

[38] Tits, J.: A local approach to buildings, pp. 519–547 in: *The Geometric Vein (The Coxeter Festschrift)* (C. Davis, B. Grünbaum, F. A. Sherk, eds.) Springer, Berlin Heidelberg New York (1981)

[39] Wedderburn, J. H. M.: On hypercomplex numbers, *Proc. London Math. Soc. (2)* **6**, 77–118 (1908)

[40] Weiss, R. M.: *The Structure of Spherical Buildings*, Princeton University Press, Princeton and Oxford (2003)

[41] Zassenhaus, H.: *Lehrbuch der Gruppentheorie*, Teubner, Leipzig und Berlin (1937)

[42] Zieschang, P.-H.: Homogeneous coherent configurations as generalized groups and their relationship to buildings, *J. Algebra* **178**, 677–709 (1995)

[43] Zieschang, P.-H.: *An Algebraic Approach to Association Schemes*, Springer Lecture Notes in Math. **1628**, Berlin Heidelberg New York (1996)

[44] Zieschang, P.-H.: On maximal closed subsets in association schemes, *J. Combin. Theory Ser. A* **80**, 151–157 (1997)

[45] Zieschang, P.-H.: Transversals for association schemes, *J. Math. Soc. Japan* **52**, 465–482 (2000)

[46] Zieschang, P.-H.: The exchange condition for association schemes, *Israel J. Math.*

Index

GPSR Compliance

The European Union's (EU) General Product Safety Regulation (GPSR) is a set of rules that requires consumer products to be safe and our obligations to ensure this.

If you have any concerns about our products, you can contact us on ProductSafety@springernature.com

In case Publisher is established outside the EU, the EU authorized representative is:

Springer Nature Customer Service Center GmbH
Europaplatz 3
69115 Heidelberg, Germany

Batch number: 10584469

Printed by Printforce, the Netherlands